The Environmental Legacy of War on the Hungarian-Ottoman Frontier, c. 1540–1690

The Environmental Legacy of War on the Hungarian-Ottoman Frontier, c. 1540–1690

András Vadas

Routledge
Taylor & Francis Group
LONDON AND NEW YORK

First published in 2023 by Amsterdam University Press Ltd.

Published 2025 by Routledge
4 Park Square, Milton Park, Abingdon, Oxon OX14 4RN
605 Third Avenue, New York, NY 10158

Routledge is an imprint of the Taylor & Francis Group, an informa business

ISBN: 9789463727938 (hbk)
ISBN: 9781041187783 (pbk)
ISBN: 9781003706021 (ebk)

NUR 685

Cover illustration: Hadtörténeti Intézet és Múzeum [Military History Institute and Museum] (Budapest, Hungary) H III a 24 [Karte des ungarischen Kriegsschauplatzes vom Jahre 1556.]

Cover design: Coördesign, Leiden

DOI: 10.5117/9789463727938

In memory of Papa

Table of Contents

List of Figures

Note on Names

If a settlement lies outside the territory of present-day Hungary, I always use its official current place name. To make identification easier in the index, I will also refer to the Hungarian or other relevant forms of the settlement name. If a settlement is lost or integrated into a modern settlement with a different name, I will use its medieval or early modern name. For rivers having sections in present-day Hungary, I will use the Hungarian names unless they have an English version. For rivers outside of present-day Hungary, I will use the form of the name used in the relevant countries. For the historical names of counties, I will use their Hungarian forms. I will use the English forms of the names of kings and queens but will always indicate their title in Hungary.

Acknowledgments

Before turning to a discussion of the early modern environments in the Carpathian Basin, this is the place for me to thank several people without whom the present book would never have been completed, or certainly not in this form. First and foremost, my thanks go to Balázs Nagy, my former supervisor who became a friend to whom I can always turn. I am grateful to have become his colleague and roommate at Eötvös Loránd University. I am equally thankful to the faculty at the Medieval Studies Department, Katalin Szende, Alice M. Choyke, and József Laszlovszky at Central European University for their support.

I would also like to thank the readers of parts of this book in earlier versions, such as Zsuzsanna Újváry (Pázmány Péter Catholic University), András Pálóczi-Horváth (Károli Gáspár University of the Reformed Church in Hungary), and Péter Dominkovits (Sopron Town Archive, Hungarian National Archives). Géza Pálffy (Institute of History of the Research Centre for the Humanities) helped me by drawing my attention to and/or sharing some of the sources I used here. I am indebted to Péter Szabó (Department of Vegetation Ecology of the Institute of Botany of the Czech Academy of Sciences in Brno) and Dénes Saláta (Institute of Nature Conservation and Landscape Management, Szent István University, Gödöllő) for their help with some of the questions on historical forestry. I must also mention the helpfulness and patience of the archivists at the Hungarian National Archives in Budapest. I am most thankful to the librarians of the CEU–ELTE Medieval Library, Ágnes Havasi, Petra Verebics, and Borbála Lovas for overlooking my delay (adding up to months, in some cases) in returning books.

Apart from the Eötvös Loránd Unversity and Central European University, other institutions and scholarships were essential in completing the present work. The "Towards a Europe of Knowledge" research scholarship (*Európai Léptékkel a Tudásért*, TÁMOP-4.2.1.B-09/1/KMR) provided me with the opportunity to spend a month at the Rachel Carson Center for Environment and Society at Munich in 2010 and then to do archival research in Vienna in 2011. The scholarship provided by the European Society for Environmental History for two of their summer schools in 2010 and 2011 in Yvelines and Venice were both important for my engagement with environmental history. I also received a scholarship from the University of Vienna to do archival and library research in the summer of 2015. In the final stages a scholarship by Henan University also was a great help.

This book certainly would never have been completed without the general support of the Hungarian Fulbright Commission and the Department of History at Georgetown University. I am grateful to Gábor Ágoston at Georgetown University for supporting my application, as well as other colleagues at the department – John McNeill and Dagomar Degroot – for their suggestions and for putting me in touch with other colleagues in the U.S. The five months I had the pleasure of working in Washington D.C. were the most peaceful working period one could wish.

Finally, I cannot express how much the support of my family means to me as they helped me to become a historian. Without Nagyi, I do not think I would have engaged in studying history. I am thankful to my parents for always supporting my career decisions. My last words go to Virág, who provides me with the perfect environment to live and work in.

Nagykovácsi, August 2022

1 Introduction

Abstract
The chapter sketches the main questions of the book and provides its main chronological and geographic frameworks. It focuses on the problem of the interrelation between wars and the environment and addresses the question of the environmental transformation caused by the lasting military conflict between the Kingdom of Hungary ruled by the Habsburgs and the Ottoman Empire in the sixteenth and seventeenth centuries.

Keywords: Environmental history, Kingdom of Hungary, Ottoman Empire, Carpathian Basin, military history

The concept behind this book stems from an article I wrote more than ten years ago. I became interested in the environmental history of the Middle Ages and the early modern period and decided to write a piece on the "great famine" of the 1310s in the Carpathian Basin.[1] While browsing through the sources from that period, I encountered some references to the impact of military campaigns on local economies and landscapes. The problems that military campaigns caused are, of course, neither specific to the 1310s nor the Carpathian Basin, and the environmental effects of periods of war have been studied in various contexts. Oddly enough, one of the best studies written thus far on environmental disturbances of medieval warfare looks at the same period, the early fourteenth century, but does not focus on Central Europe but instead on the British Isles and looks at the intertwined history of the Scottish wars, the Great Famine, and the cattle plague that devastated the island.[2] However, ten years ago when I was working on the

1 András Vadas "Documentary Evidence on the Weather Conditions and a Possible Crisis in 1315–1317: Case Study from the Carpathian Basin," *Journal of Environmental Geography* 2, no. 3–4 (2009): 23–29.
2 Philip Slavin, "Warfare and Ecological Destruction in Early Fourteenth-Century British Isles," *Environmental History* 19 (2014): 528–550, idem, "Ecology, Warfare and Famine in Early Fourteenth-Century British Isles: A Small Prolegomenon to a Big Topic," in *Guerra y carestía en*

Vadas, A., *The Environmental Legacy of War on the Hungarian-Ottoman Frontier, c. 1540–1690.* Taylor & Francis Group, 2023
DOI: 10.5117/9789463727938_CH01

fourteenth-century crisis and their military aspects in the Kingdom of Hungary, significantly less literature was available on pre-modern wars and the environment than nowadays. Nevertheless, I tried to access basic literature on the environmental impacts of wars in general. One of the first seemingly relevant articles I found was Joseph Hupy's essay, "The Environmental Footprint of War".[3] Though Hupy's overview referred to pre-modern wars and their environmental implications, it mostly focused on the possible environmental impacts of warfare after the introduction of the systematic use of smokeless gunpowder and, most importantly, the chemical weapons used in the twentieth century. He identified three types of war-related environmental disturbances:

(1) Environmental disturbance and destruction from weaponry;
(2) Direct consumption of resources: timber, water, and food to support armies;
(3) Indirect consumption by military complexes.[4]

He provides the reader with examples of the long-term impacts of the First World War and the Vietnam War. With some of the pre-modern cases that he mentions, he emphasizes the pre-modern warfare use of scorched-earth tactics – in England mostly referred to as 'chevauchee' – which caused abrupt local environmental and landscape transformations. However, he failed to provide examples for the applicability of the above three categories to pre-modern wars in general. Because of the seemingly limited applicability of his theory for my case study at that time, I did not use the article very much when studying the fourteenth-century environmental disturbances and their connections to wars. However, when I engaged in analyzing sixteenth- and seventeenth-century wars in the Carpathian Basin, the three types of environmental disturbances caused by wars that Hupy identified started to seem applicable in a pre-modern context as well.

Hupy applies his theory to modern wars, which due to the nature of modern weaponry tend to be shorter than those fought in pre-modern times. This of course means that modern studies mostly focus on the immediate impacts caused by the armies marching through, or the direct destruction

la Edad Media, ed. Pere Benito i Monclús (Lleida: Milenio, 2015), 85–99, and Philip Slavin, *Experiencing Famine: A Fourteenth-Century Environmental Shock in the British Isles* (Environmental Histories of the North Atlantic World [EHNAW], 4) (Turnhout: Brepols, 2019).

3 Joseph P. Hupy, "The Environmental Footprint of War," *Environment and History* 14 (2008): 405–421.

4 Hupy, "The Environmental Footprint," 406.

such as burning plowlands and pastures, deliberate destruction of infrastructures, or the cutting of supply chains. Slavin's article as well as further case studies point out that the above three elements were also present in pre-modern warfare, however, there are other kinds of disturbances that can also be associated with these conflicts. My goal in this book is to explore a different kind of war than those discussed by such scholars as Hupy or Slavin, one that lasted almost two centuries.

Probably the single most significant political change in the history of Hungary until the twentieth century – or even including what followed the First World War[5] – that triggered transformations in land use and the settlement network occurred in the sixteenth and seventeenth centuries, when the political unity of the medieval Kingdom of Hungary, forming most of the Carpathian Basin, came to an end for a long period and the Ottoman Empire gradually took control over the center of the basin area. In schoolbooks as well as different textbooks, the Ottoman war period is considered the biggest cataclysm in the history of Hungary along with the Mongol invasion of 1241–1242. To some extent, both are remembered in the same terms, with the warring parties plundering much of the Kingdom of Hungary and leaving empty lands behind.[6] However, because of the long-lasting presence – lasting well more than 150 years – of the Ottomans in or in the immediate surroundings of the Carpathian Basin, it has frequently been argued that they had a long-term impact on the environment. This book aims to provide a deeper understanding of the environmental legacy of the Ottomans' presence in the central parts of the basin area, with special regard to the impacts of the recurrent military conflicts during this time.

The main questions I am addressing in this book are, first, how the Ottoman-Hungarian wars affected the landscapes of the frontier zone in the Carpathian Basin, and second, how the environment was used in the military tactics of the opposing realms. The book intends to explore the dynamic interplay between war, environment, and local society in the early modern period. By doing so, I hope to demonstrate that it is just as valid to look at how pre-modern wars impacted the environments as it is to examine the environmental effects of the American Civil War, the two world wars, and the Korean or Vietnam Wars.

5 Géza Pálffy, "Mohács radikálisabb változásokat hozott, mint Trianon" [Mohács brought more radical change than did Trianon]. Online document: Transindex, 2015. http://vilag.transindex.ro/?cikk=25515 (last accessed: 29 June 2020).

6 For more information on the Mongol invasion, see *Tatárjárás* [Mongol invasion] (Nemzet és emlékezet), ed. Balázs Nagy (Budapest: Osiris, 2003).

In recent years, a good number of studies and edited volumes have addressed the problems of war environments, mostly showing interest in the impacts of military campaigns and to some extent the environmental consequences of a frontier area. Although the number of works in the field is rapidly rising, there are only about a few dozen works that directly focus on the environmental consequences of warfare and even fewer that touch upon non-modern warfare and its impacts. This may indirectly lead to the assumption that the environmental effects of war can be best understood through modern warfare.[7] The case studies mostly discuss the environmental impacts of warfare from the American Civil War onwards, through the two world wars, to the Vietnam and the Gulf Wars.[8] Despite the clearly different environmental impacts of the Battle of Gettysburg and the napalm attack in the Vietnam War, most studies – except for the long-term perspectives applied when looking at the wars on the Korean Peninsula[9] – share a common feature: they all discuss relatively short war periods and mostly focus on battlefields and not extensive areas such as hinterlands and the impact of war on these environments.[10] Using a common framework, most studies examine the impacts of modern weaponry on the environment. As environmental history in general grew rapidly in the United States, followed by a time gap by European environmental history, it is rather self-evident why most studies address modern-age problems.

It is, perhaps, not an overstatement to assume that the most popular topic in non-contemporary American history is the Civil War (1861–1865); as such, it is no surprise that the study of the environmental impacts of

7 For the most recent overview of the field, see Lisa M. Brady, "War from the Ground Up: Integrating Military and Environmental Histories," in *A Field on Fire: The Future of Environmental History*, eds. Mark D. Hersey and Ted Steinberg (Tuscaloosa, AL: The University of Alabama Press, 2019), 250–262.

8 For a comprehensive bibliography of the topic, see http://environmentandwar.com/bibliographies/ (last accessed: 17 April 2021) at the Environment and War website edited by Richard Tucker.

9 E.g., Mark Fiege, "Gettysburg and the Organic Nature of the American Civil War," in *Natural Enemy, Natural Ally: Toward an Environmental History of War*, eds. Richard P. Tucker and Edmund P. Russell (Corvallis, OR: Oregon State University Press, 2004), 93–109, and Arthur H. Westing, "The Environmental Aftermath of Warfare in Viet Nam," *Natural Resources Journal* 23 (1983): 365–389; for Korea, see Lisa M. Brady, "Life in the DMZ: Turning a Diplomatic Failure into an Environmental Success," *Diplomatic History* 32 (2008): 585–611, and the special issue of the *Journal of Asian Studies* 77, no. 2 (2018): War and Environment in Korean History.

10 The notable exceptions include the works of Brady as well as the studies to be quoted in this and the following footnotes. *Environmental Histories of the First World War*, eds. Richard Tucker et al. (Cambridge: Cambridge University Press, 2018).

warfare mostly focused on Civil War battlefields,[11] where the first steps to a more complex understanding of warfare environments unfolded. Lisa M. Brady's *War Upon the Land* is pioneering in its consideration of long-term landscape change brought about by three Southern campaigns of the Union army.[12] Brady's work is thus far one of the very few monographs to touch upon the environmental history of war not only by examining the history of the devastation of the land but also by looking primarily at the role of nature in military tacti cs and the role that understanding environmental conditions played in the Union winning the war. 'How did pollution affect landscapes, for instance, turning fertile lands into wastelands? How did depopulation allow reforestation in different areas affected by war or military operations? How do war landscapes become sites of memory?' These questions, including some concerning fundamentally different processes, have only been addressed by a handful of studies so far.[13]

The environmental legacy of warfare in the last three decades, much influenced by the Gulf War experiences, has become an important topic in the environmental history of the modern period. The long-term legacy of warfare and nuclear explosions has provided scholars with excellent laboratories for human–nature interactions. The examples described in different volumes thus far have concerned the nineteenth to the twenty-first centuries. Is there a way to understand pre-modern war environments and their environmental legacies?

In an introductory essay to *Natural Enemy, Natural Ally*, Richard Tucker gave one of the best overviews of the existing research directions in the environmental history of wars.[14] His essay not only touches upon modern

11 *The Blue, the Gray, and the Green. Toward an Environmental History of the Civil War*, ed. Brian Allen Drake (Athens, GA: The University of Georgia Press, 2015).

12 Lisa M. Brady, *War upon the Land: Military Strategy and the Transformation of Southern Landscapes during the American Civil War*, foreword by Paul S. Sutter (Environmental History and the American South) (Athens, GA: University of Georgia Press, 2012).

13 See Lisa M. Brady, "Life in the DMZ: Turning a Diplomatic Failure into an Environmental Success," *Diplomatic History* 32 (2008): 585–611. See also many of the studies in: *Militarized Landscapes: From Gettysburg to Salisbury Plain*, eds. Chris Pearson, Peter A. Coates, and Tim Cole (London: Continuum, 2010), especially those in Part III. Like parts of the areas affected by the Civil War that became national parks or memorial parks, the region of Chernobyl after the 1986 nuclear accident soon became one of the richest areas in wildlife in East-Central Europe. Cf. T.G. Deryabina et al., "Long-Term Census Data Reveal Abundant Wildlife Populations at Chernobyl," *Current Biology* 25, no. 19 (2015): R824–R826. On this issue, see recently: Kate Brown, *Manual for Survival: A Chernobyl Guide to the Future* (New York: W.W. Norton, 2020).

14 See also most recently, with many of the points recapitulated: Richard P. Tucker, "War and the Environment," in *A Companion to Global Environmental History*, eds. John R. McNeill and Erin Stewart Mauldin (Chichester: Wiley Blackwell, 2012), 319–339.

warfare but also considers the problems of pre-modern war environments. As he puts it:

> Throughout the pre-modern world, many conflicts took the form of frontier wars fought between non-state societies, two states, or as wars of conquest pursued by an ambitious power on its periphery. Often protracted and intermittent, these wars were similar in many ways to modern guerilla warfare and counter-insurgency, though they did not produce the devastation that is caused by today's counter-insurgency weapons.[15]

Tucker saw the environmental impact of long-lasting wars at frontiers as one of the issues worth considering. These wars, lasting in many cases for decades, were fundamentally different in nature from modern warfare. The war that took place in the Carpathian Basin, which is discussed in the coming chapters, was certainly one of the long-lasting ones. As I will argue, the constant presence of military troops and a military population in the frontier zone had a lasting impact on local environments. This resonates well with Tucker's quote above regarding the character of long-lasting frontier wars, but I will argue in the following chapters that these wars nonetheless equally have the potential of causing major transformations in land use and as a consequence of transforming the ecological conditions of major areas.

The few studies that have addressed the problem of pre-modern war environments discuss the following three topics: first, the impact of scorched-earth (or chevauchée) tactics on the environment, second, deforestation and wars, and last, the environments of buffer or frontier zones between powers.[16] The environmental impact of scorched-earth tactics has been studied in different contexts from ancient times onwards, but there is a gap in the comprehensive scholarship of its application in early modern warfare.[17]

15 Richard P. Tucker, "The Impact of Warfare on the Natural World: A Historical Survey," in *Natural Enemy, Natural Ally*, 24.

16 I will not go into the discussion of the impact of climate and weather or the impact of different diseases on warfare in the pre-modern period, as they fell out of the scope of the present book. Cf. Dagomar Degroot, "The Frigid Golden Age: Experiencing Climate Change in the Dutch Republic, 1560–1720," (PhD diss., York University, 2014), idem, *The Frigid Golden Age: Climate Change, the Little Ice Age, and the Dutch Republic, 1560–1720* (Cambridge: Cambridge University Press, 2018) and John R. McNeill, *Mosquito Empires: Ecology and War in the Greater Caribbean, 1620–1914* (New York: Cambridge University Press, 2012), all with overviews of existing literature on these topics.

17 See John McNeill, "Woods and Warfare in World History," *Environmental History* 9 (2004): 401, and Tucker, "War and the Environment," 321. See also: Jan Phillip Bothe, "How to 'Ravage' a

This tactic, as is demonstrated in Hungarian research, was seldom used in the Ottoman war period. Many of the sources from the period, however, show that it was very much present in the potential military arsenal of the military leadership, which consisted mostly of non-Hungarian aristocrats.[18]

The second problem mentioned above, the use of forest resources in war periods, is much more important in the context of the present book. Wars consume forest resources in many different ways. One of them was the use of timber and wood to build different war machines. From ancient times, chariots were used extensively in warfare. This may have had an impact on forest resources in Assyria and parts of the Peloponnesus or Egypt[19] but certainly did not affect the whole of the Mediterranean Basin. The use of timber in shipbuilding may have had larger-scale impacts on the forest resources in the same area. The Ottomans in the period of their early expansion were not among the maritime powers, but from the sixteenth century onwards they built up one of the largest navies of the Mediterranean and had ambitious plans to create a major military transportation network using the rivers in East-Central Europe.[20]

Ottomans as well as Hungarians used the dense river network of the frontier area both to protect the frontiers and for supply and logistics.[21] The number of supplies used by the fleets in the Carpathian Basin was negligible

Country. Destruction, Conservation and Assessment of Natural Environments in Early Modern Military Thought," *Hungarian Historical Review* 7 (2018): 510–540.

18 Géza Pálffy, "Scorched-Earth Tactics in Ottoman Hungary: On a Controversy in Military Theory and Practice on the Habsburg-Ottoman Frontier," *Acta Orientalia Academiae Scientiarum Hungaricae* 61 (2008): 181–200.

19 On the lack of forest resources in Ottoman Egypt, see: Alan Mikhail, *Nature and Empire in Ottoman Egypt: An Environmental History* (Cambridge: Cambridge University Press, 2011), 124–169.

20 See the fundamental work of Colin Imber, "The Navy of Süleyman the Magnificent," *Archivum Ottomanicum* 6 (1980): 211–282. See more recently the studies of the volume: *The Ottomans and the Sea* (Oriento Moderno, XX/1), ed. Kate Fleet (Rome: Skilliter Centre for Ottoman Studies, 2001) and Giancarlo Casale, *The Ottoman Age of Exploration* (New York: Oxford University Press, 2010).

21 On the Hungarian fleet at Lake Balaton, see Ferenc Végh, "A balatoni 'hadiflotta' a török korban" [The 'Navy' on Lake Balaton in the Ottoman Period], *Hadtörténelmi Közlemények* 129 (2016): 27–56. Comprehensive studies of both the Hungarian and Ottoman fleet in the Carpathian Basin are yet to be done. See nonetheless on the Ottoman fleet: Klára Hegyi, *A török hódoltság várai és várkatonasága*, 3 vols. [Castles and garrisons of Ottoman Hungary] (História Könyvtár. Kronológiák, Adattárak, 9) (Budapest: História and MTA Történettudományi Intézet, 2007), vol. I, 101–104. See also for another frontier region of the Ottoman Empire from an environmental history perspective: Faisal H. Husain, "Changes in the Euphrates River: Ecology and Politics in a Rural Ottoman Periphery, 1687–1702," *Journal of Interdisciplinary History* 47 (2016): 1–25, idem, *Rivers of the Sultan: The Tigris and Euphrates in the Ottoman Empire* (New York: Oxford University Press, 2021), as well as Gül Şen, *Jordan as an Ottoman Frontier Zone in the Sixteenth–Eighteenth Centuries* (Ulrich Haarmann Memorial Lecture, 15) (Berlin: EB-Verlag Dr. Brandt, 2018).

compared to the timber consumption of the major military complexes, most of all the Arsenale at Venice, which required systematic accounting and protection of forest resources in a major area north of Venice, the so-called Terraferma.[22] To keep track of available forest resources, from the sixteenth century onwards the Venetian administration regularly ordered a count of the oaks in the Terraferma area. Of course, the problem is not limited to the Mediterranean: shipbuilding was of crucial importance to countries ranging from France through the British Empire to Korea in the early modern period.[23]

Scholarship both in Hungary and elsewhere dealing with the same period has found some other forms of wood consumption such as firewood needs for gunpowder production and gun founding, the need for timber for earth and wood fortifications, siege machinery, road construction for military campaigns, etc. Most of these will be discussed below, and earth and wood fortifications seem to bring the most controversial results in the scholarly literature. The varied forms of war-related wood consumption and the difficulties of understanding their importance may explain why relatively few studies have addressed forest resources in the context of pre-modern war environments so far.[24] Finally, some studies have raised the problem of using the environment in frontier protection and organization. Different landscapes and contexts have been studied, such as Qing China, the Southern Russian borderlands, or the Flemish coastal area in the late medieval period and the early modern times.[25] Nonetheless, there is a lack of comparative

22 Karl Appuhn, *A Forest on the Sea: Environmental Expertise in Renaissance Venice* (Baltimore: Johns Hopkins University Press, 2009).

23 See Paul Bamford, *Forests and French Sea Power, 1660–1789* (Toronto: Toronto University Press, 1956), Robert G. Albion, *Forests and Sea Power: The Timber Problem of the Royal Navy, 1652–1862* (Cambridge, MA: Harvard University Press, 1936), and John S. Lee, "Postwar Pines: The Military and the Expansion of State Forests in Post-Imjin Korea, 1598–1684," *The Journal of Asian Studies* 77 (2018): 319–332.

24 For a highly comprehensive overview, see McNeill, "Wood and Warfare." See furthermore: Greg Bankoff, "Wood for War: The Legacy of Human Conflict on the Forests of the Philippines, 1600–1946," in *War and the Environment: Military Destruction in the Modern Age*, ed. Charles Closmann (College Station, TX: Texas A&M University Press, 2009), 32–48.

25 David A. Bello, *Across Forest, Steppe, and Mountain: Environment, Identity, and Empire in Qing China's Borderlands* (Cambridge: Cambridge University Press, 2016). See also: Alfred J. Rieber, *The Struggle for the Eurasian Borderlands: From the Rise of Early Modern Empires to the End of the First World* (New York: Cambridge University Press, 2014), Geoffrey Parker, *Global Crisis: War, Climate Change and Catastrophe in the Seventeenth Century* (New Haven, CT: Yale University Press, 2013), 523–526, John F. Richards, *The Unending Frontier: An Environmental History of the Early Modern World* (Berkeley, CA: University of California Press, 2005), esp. 242–272. See: also: Adriaan M. J. Kraker, "Flood Events in the Southwestern Netherlands and Coastal Belgium, 1400–1953,"

studies in this field, and neither the Habsburg nor the Ottoman Empire's borderlands have been studied from this perspective.[26]

The above overview of environmental histories of pre-modern warfare is anything but exhaustive, mainly because the number of works dedicated to the problem is surprisingly low. The problem has been identified by several scholars in recent decades for different environments, but the fundamental question – that is, what kind of impact a long-lasting medieval or early modern war period may have had on local environments – has only partially been addressed. Also, too few studies have approached the problem from a bottom-up perspective, that is, how local societies were affected by the frontier organization. Because of the nature of the sources used, almost all the above studies have looked at the policies of forming the frontier. This book examines the phenomenon from a different point of view.

1.1 Frontier, Border – Do They Mean Anything?

Today, when thinking of borders, most people probably imagine lines that divide the different polities on maps. Some might even think of the different colors used by mapmakers to indicate the states on political maps. Pre-modern maps were certainly dissimilar, and very few borders were defined by lines.[27] Some basic questions must therefore first be clarified. What makes the study area described above a frontier zone? Was there a well-defined border between the Kingdom of Hungary and the Ottoman Empire in the sixteenth or seventeenth centuries? Were there borders

Hydrological Sciences–Journal des Sciences Hydrologiques 51 (2006): 913–929 and idem, "War, Climatic Stress and Environmental Degradation during the Fifteenth and Sixteenth Centuries. The Case of the North Flemish Coastal Landscape in the Estuary of the Western Scheldt," in *The Dance of Death in Late Medieval and Renaissance Europe: Environmental Stress, Mortality and Social Response*, eds. Andrea Kiss and Kathleen Pribyl (New York: Routledge, 2019), 66–85.

26 For partial treatment of the problem in these contexts, see Husain, "Changes in the Euphrates River," and idem, *Rivers of the Sultan*.

27 For a recent treatment of the formation of linear frontiers in the Early Middle Ages, see: Florin Curta, "Linear Frontiers in the 9th Century: Bulgaria and Wessex," *Quaestiones Medii Aevi Novae* 16 (2011): 15–32. See also Hans-Jürgen Karp's fundamental work on the topic: *Grenzen in Ostmitteleuropa während des Mittelalters. Ein Beitrag zur Entstehungsgeschichte der Grenzlinie aus dem Grenzsaum* (Forschungen und Quellen zur Kirchen- und Kulturgeschichte Ostdeutschlands, 9) (Cologne and Vienna: Böhlau, 1972). On the problem of borders and frontiers in early modern studies, see the introduction to *Grenzen und Grenzüberschreitungen. Bilanz und Perspektiven der Frühneuzeitforschung* (Frühneuzeit-Impulse, 1), eds. Christine Roll, Frank Pohle, and Matthias Myrczek (Vienna, Cologne, and Weimar: Böhlau, 2010) by Christine Roll, "Grenzen und Grenzüberschreitungen in der Frühen Neuzeit – eine Einführung in die Forschung," 13–22.

between polities at all in this period that the different actors were aware of and which necessitated different practices on either side? Before sketching out the chief theme of the book and the question it aims to raise and answer, these basic problems must be discussed, as the following chapters focus on the concept of frontiers.

The terms frontier and border are often used in everyday speech, but as is usual with general terms like these, it is rather difficult to define what they actually mean. Anglo-Saxon historiography has long been obsessed with the problem of frontiers. Following in the footsteps of the highly influential thesis of Frederick Jackson Turner on the role of the frontier in the formation of the American democracy published in 1893, frontier history and frontier studies became core teaching areas in U.S. colleges and universities.[28] Turner's thesis was a milestone in the discussion of frontiers in the social sciences. Historians of medieval and early modern Europe as well as Ottomanists have also been intrigued by the problem of frontiers. While a frontier in Europe had stood between two polities by the late medieval period, the case was different in the eighteenth and nineteenth centuries in the United States, an area of constant expansion and a zone of passage.[29] Although frontiers as understood by Turner had little to do with frontiers in Europe in the pre-modern period, attempts have been made to apply Turner's thesis to the German *Ostsiedlung*.[30] In French and German historiography, for a long time the most influential concepts – apart from Turner's thesis – were rooted in the geographical thinkers of the nineteenth century. Building on many of the ideas of Friedrich Ratzel, the famous historian of the *Annales*, Lucien Febvre made important contributions to the understanding of what frontier and border meant in pre-modern Europe. He points out that historians and geographers like to think of frontiers as borderlines despite the limited applicability of this concept before the origin of modern states

28 Frederick Jackson Turner, "The Significance of the Frontier in American History," in *Proceedings of the State Historical Society of Wisconsin at its Forty-First Annual Meeting, Held December 14, 1893* (Madison, WI: Democrat Print Co., 1894), 79–112. Frederick Jackson Turner, *The Frontier in American History* (New York: Henry Holt and Co., 1920), Ch. 1.

29 Daniel Power, "Introduction," in *Frontiers in Question: Eurasian Borderlands, 700–1700*, eds. Daniel Power and Naomi Standen (Basingstoke: Macmillan, 1999), 2.

30 James Westfall Thompson, "Profitable Fields of Investigation in Medieval History," *American Historical Review* 18 (1913): 490–504. A classic example of the application of this thesis to the European Middle Ages is the study of A. R. Lewis, "The Closing of the Medieval Frontier 1250–1350," *Speculum* 33 (1958): 475–485. Cf. Florin Curta, "Introduction," in *Borders, Barriers, and Ethnogenesis: Frontiers in Late Antiquity and the Middle Ages*, ed. idem (Turnhout: Brepols, 2005), 4, and Andrzej Janecek, "Frontiers and Borderlands in Medieval Europe. Introductory Remarks," *Quaestiones Medii Aevi Novae* 16 (2011): 8.

and administrations in the latter part of the early modern period. He also highlights that the meaning of the French *frontière* has varied from the Middle Ages onwards even within the French context, and the words of the same root – the English *frontier* or the Spanish *frontera* – also have quite different meanings.[31] Studies on frontier histories in the second half of the twentieth century created an abundance of definitions and diversified the understanding of both frontier and border, as a result of which they were used in a wide range of specific contexts. In many cases, the same volumes have published studies on physical frontiers, frontier societies, and frontiers of a certain phenomenon such as a religion or a custom. Apart from an attractive title in most cases, the studies had little to do with each other. Applying different ideas of frontiers and borders, the introductory essays to these volumes try to give some theoretical overview of the concepts and thus provide the most important basis for interpreting frontiers and borders.[32]

The area the present book focuses on can be understood as a frontier from several seemingly different angles. To note but a few, with the advance of the Ottomans, the area of the Kingdom of Hungary came to border on a new empire not only in a political sense but also in a religious one, an idea often thematized in medieval and early modern literary works.[33] This book also interprets the frontier in a very down-to-earth way. The word 'frontier' reflects the military-political position of an area and its impact on the local economies and environments. There are at least two fundamental aspects that characterized frontiers before modern times.[34] First, contemporary actors perceived the area as a frontier. As will be discussed in Chapter 2, this is certainly true for the area in question; not only did the local societies think of the area as a frontier zone but so did the Habsburg administration at Vienna and the Sublime Porte at Istanbul.[35] Second, apart from the sometimes fixed

31 Lucien Febvre, "Frontière," *Bulletin du Centre international de synthèse. Section de synthèse historique* no. 5 (1928): 31–44. See also: idem, *A Geographical Introduction to History* (London and New York: Kegan Paul, Trench, Trubner and Co. and Alfred A. Knopf, 1932), 296–306.

32 E.g., *Medieval Frontiers: Concepts and Practices*, eds. David Abulafia and Nora Berend (Burlington, VT: Routledge, 2002), *Frontiers in Question, Borders, Barriers, and Ethnogenesis, Frontier and Border Regions in Early Modern Europe*, eds. Raingard Esser and Steven G. Ellis (Hannover: Wehrhahn Verlag, 2013), *Menschen und Grenzen in der Frühen Neuzeit*, eds. Wolfgang Schmale and Reinhard Stauber (Innovationen. Bibliothek zur Neueren und Neuesten Geschichte, 2) (Berlin: Verlag Spitz, 2000) and other volumes quoted above.

33 Paul Srodecki, *Antemurale Christianitatis: Zur Genese der Bollwerksrhetorik im östlichen Mitteleuropa an der Schwelle vom Mittelalter zur Frühen Neuzeit* (Husum: Matthiesen Verlag, 2015) has an exhaustive bibliography on the topic.

34 E.g., Nora Berend, "Preface," in *Medieval Frontiers*.

35 See section 2.1.

borders from the medieval period onwards, as is indicated by several case studies, there is an easily definable feature of frontier zones: their militarized nature. Such zones are surrounded by numerous fortifications which are designed to protect the hinterland and control the opposing power. Different forms of frontiers have been identified in recent scholarship; the case of the area discussed here provides an example par excellence of an unstable frontier region with extensive defensive features.[36]

I argue in the chapters to follow that most of the central part of the Carpathian Basin – called Transdanubia (*Dunántúl*) and the Great Hungarian Plain (*Alföld*) – should be categorized as a frontier or contact zone where the military and economic practices differed significantly from those in the core areas of the surrounding polities – the Kingdom of Hungary, the Ottoman Empire, and slightly to the east, the satellite state of the latter, the Principality of Transylvania. As mentioned, scholarship has shown increasing interest in cultural and religious frontiers in recent decades, but the topic of political frontiers in pre-modern frontier studies has elicited limited attention.[37] Although the context of the present book is the political-military frontier, the primary goal is not to follow how the political situation was changing, an approach that has to a large extent already been taken by others in the past decades,[38] but rather to discuss the impacts the political-military organization had on the environment of the examined area. As part of the discussion, this volume will also consider not only the immediate frontier but also their hinterlands, the history of which is intertwined with the war zone.[39]

1.2 The Development of the Ottoman-Hungarian Frontier – The Scene

The study of the environmental history of the early modern period in Hungary is greatly affected by the periodization that structures political

36 Eduardo Manzano Moreno, "The Creation of a Medieval Frontier: Islam and Christianity in the Iberian Peninsula, Eighth to Eleventh Centuries," in *Frontiers in Question*, 35. Most importantly, see Naomi Standen, "Introduction. Nine Case Studies of Pre-Modern Frontiers," in *Frontiers in Question*, 23.

37 See its criticism in medieval and early modern contexts in respectively: Curta, "Introduction," 9, and Raingard Esser and Steven G. Ellis, "Introduction," in *Frontier and Border Regions*, 12.

38 See the author cited in the next section.

39 Cf. *The Resilient City in World War II. Urban Environmental Histories*, eds. Simo Laakkonen et al. (Cham: Springer International Publishing and Palgrave MacMillan, 2019).

history. The archives in Hungary follow various chronological conventions of political history, as a result of which entirely different structures apply to the study of the periods before and after the Battle of Mohács, the decisive Ottoman defeat of the Hungarians on 29 August 1526. This rigid structure allows little room for discussing long-term processes in the late medieval and early modern periods such as changes in land-use patterns, vegetation, and so on. The study of the period of the Ottoman presence in the Carpathian Basin has traditionally been divided among the Principality of Transylvania, the Kingdom of Hungary, and the Ottoman Empire, which in many cases requires different language skills and research methods. Furthermore, different archival systems must be understood and different questions have been raised in the context of these three political entities.

Before turning to the actual changes in the environmental conditions in the Ottoman-Hungarian frontier zone, a brief overview of the political environment in which these changes occurred is necessary. The immediate political context of the present book is the fall of the medieval Kingdom of Hungary and the partial occupation of the country by the Ottomans. The late medieval Kingdom of Hungary, in the period between 1490 and 1526, was under the rule of the Jagiellonian dynasty. Until recently, the period had been retrospectively regarded as a crisis period with weak royal power and a period of complete disrepair.[40] According to the traditional narrative of Hungarian historiography, this crisis led to the loss of more and more fortifications to the Ottomans at the southern ends of the country, which culminated in the major defeat at the above-mentioned battlefield of Mohács on the 29th of August in 1526.[41] In the past few years, a group of young scholars have begun studying the personnel of the leading elite and

40 See the recent studies of Tibor Neumann shedding new light on the rule of mostly King Vladislas II. Most importantly: Tibor Neumann, "Királyi hatalom és országgyűlés a Jagelló-kor elején" [Royal power and diets in the beginning of the Jagiellonian period], in *Rendiség és parlamentarizmus Magyarországon: A kezdetektől 1918-ig* [Estates and parliaments in Hungary: from the beginnings to 1918], eds. Tamás Dobszay et al. (Budapest: Argumentum Kiadó, 2013), 46–54.

41 On the Battle of Mohács, see Gábor Ágoston, "Mohács," in *The Seventy Great Battles of All Time*, ed. Jeremy Black (London: Thames & Hudson, 2005), 100–112, János B. Szabó and Ferenc Tóth, *Mohács 1526. Soliman le Magnifique prend pied en Europe central* (Paris: Économica, 2009). On the period in general: Géza Perjés, *The Fall of the Medieval Kingdom of Hungary: Mohács 1526 – Buda 1541* (War and Society in East Central Europe, 26 = Atlantic Studies on Society in Change, 56 = East European Monographs, 255) (Boulder, CO: East European Monographs, 1989), Géza Pálffy, *The Kingdom of Hungary and the Habsburg Monarchy in the Sixteenth Century* (East European Monographs, 735 = Center for Hungarian Studies and Publications Series, 18), translated by J. Thomas and Helen D. DeKornfeld (Boulder, CO and Wayne, NJ: Social Science Monographs, Center for Hungarian Studies and Publications, Inc., 2009), 35–52.

the royal courts in the Jagiello period, which in the long run will hopefully lead to a complete re-assessment of royal power in that period. Their preliminary results already indicate that the interpretation of this period as a crisis very much stems from the association of the Jagiello kings with the Ottomans' expansion towards this part of Central Europe.[42] This stems from the fact that earlier research drew mostly on narrative sources, but the legal evidence that constitutes the overwhelming majority of the written material surviving from this period was to a large extent omitted. Written after the defeat at Mohács and during the presence of the Ottomans in Hungary, the chroniclers saw the period of King Matthias (1458–1490) as a heyday because the Ottoman advancement came to a halt during his reign, partly due to internal struggles in the Empire. Compared to Matthias's reign, the periods in which Vladislas II (1490–1516) and Louis II (1516–1526) ruled were considered to have paved the road to the defeat at Mohács and the loss of Hungary's independence. As it has been argued more recently, the Ottoman advancement was probably inevitable in the early sixteenth century, the question instead being when – rather than if – it would take over Hungary.[43]

The decades after the Battle of Mohács were one of the most critical periods in the history of the Kingdom of Hungary: apart from the recurrent Ottoman campaigns in the territory of Hungary and Croatia, a serious succession crisis unfolded as well. Both John Szapolyai (1526–1540) and Ferdinand I of Habsburg (1526–1564) were crowned as kings of Hungary, the former in 1526 and the latter in 1527. Both coronations were considered lawful, as they fulfilled the coronation requirements (i.e., crowned at Fehérvár with the Hungarian Holy Crown by the archbishop of Esztergom – or, in its vacancy or absence, the eldest bishop). The next one and a half decades brought civil war to the territory of the Kingdom of Hungary, with recurrent military

42 E.g., see the studies of Tibor Neumann on the Szapolyais.

43 Ferenc Szakály, "Phases of Turco-Hungarian Warfare before the Battle of Mohács (1365–1526)," *Acta Orientalia Academiae Scientiarum Hungaricae* 33 (1979): 65–111; Pálffy, "The Origins and Development of the Border Defence System," idem, "The Habsburg Defense System in Hungary Against the Ottomans in Sixteenth Century: A Catalyst of Military Development in Central Europe," in *Warfare in Eastern Europe, 1500–1800* (History of Warfare, 72), ed. Brian J. Davies (Leiden and Boston: Brill, 2012), 35–61, Ágoston, "Defending and Administering the Frontier," and the studies in *Fight Against the Turk in Central-Europe in the First Half of the 16th Century*, ed. István Zombori (Budapest: METEM, 2004). For the Ottoman-Hungarian military struggles up to 1526, see most recently Tamás Pálosfalvi, *From Nicopolis to Mohács: A History of Ottoman-Hungarian Warfare, 1389–1526* (The Ottoman Empire and its Heritage, 63) (Leiden and Boston: Brill, 2018) (all with much relevant literature not quoted here).

campaigns not only by the Ottomans but also by the two royal armies.[44] By the beginning of the 1530s, after John I had sworn an oath to the Ottoman sultan, Suleyman the Magnificent (1520–1566), it became evident to Ferdinand I that despite his military superiority over John I, he had no other choice than to try to find a way to solidify his power in the areas that were under his military control. These were the western and northern parts of the former Kingdom of Hungary. In the rest of the territories – most importantly over Transylvania – he had to accept the independent rule of John I. The armistice concluded in 1533 between Suleyman and the Habsburg brothers (Ferdinand I and the Holy Roman Emperor, Charles V) led to new Ottoman military tactics towards East-Central Europe and the Habsburg areas. Instead of a rapid occupation by military campaigns, which were attempted in 1529 and 1532 and had caused significant destruction in the Western areas of the Carpathian Basin, a gradual occupation of the Kingdom of Hungary became their dominant military strategy.[45] The Treaty of Oradea (1538) between John I and Ferdinand I again changed the political situation, as it would have allowed the Habsburgs to inherit the areas that were in the hands of the then childless, aging John I. These areas included the capital, Buda, which remained in the hands of John for the coming years, while Ferdinand took possession of many important strongholds in its immediate neighborhood including Visegrád, Esztergom, and Vác. This went against the plans of Suleyman, who instead of direct campaigns against Vienna saw greater potential in a permanent occupation of the central part of the Carpathian Basin, including Buda. The death of John I in 1540 created a new political situation in the Carpathian Basin, which indirectly led to the Ottoman occupation of Buda in 1541. In the central part of the former Kingdom of Hungary, the Ottoman Empire created its northernmost administrative unit, the vilayet of Buda. This did not mean that the eastern areas of the former Kingdom of Hungary were also integrated into the Ottoman Empire. The posthumous son of John I, John II Szapolyai (or John Sigismund) – elected king of Hungary 1540–1571 and prince of Transylvania in 1571 – was also elected king of Hungary, although his kingship was not recognized by the Hungarian, Croatian, and Slavonian estates. His rule was accepted only in Transylvania, which in the Middle Ages had been governed independently.[46]

44 Pálffy, *Kingdom of Hungary*, 41–48.

45 See most recently: Pál Fodor, *The Unbearable Weight of Empire. The Ottomans in Central Europe – A Failed Attempt at Universal Monarchy (1390–1566)* (Budapest: MTA BTK Történettudományi Intézet, 2015), esp. 56–94.

46 Teréz Oborni, "From Province to Principality: Continuity and Change in Transylvania in the First Half of the Sixteenth Century," in *Fight Against the Turk*, 165–180, eadem, "Between

The 1540s was the period when the division of the former realm into three parts was crystallized, and despite a short attempt in the first half of the 1550s to reunite Transylvania with the Habsburg territories, the former polity became a semi-independent Ottoman satellite state from 1556 and remained as such for the entire length of the period this book focuses on.[47] In the western part of the former Kingdom of Hungary, the occupation of Buda and the Danube valley running north-south opened up new perspectives to extend Ottoman authority over large areas of the central part of the Carpathian lowlands. In the following period – from the early 1540s to 1566 – important fortifications fell to the Ottomans, which created a turbulent frontier zone between the Ottoman Empire and the Kingdom of Hungary (ruled by the Habsburgs).[48] In this period, no major campaigns against Vienna were initiated by the sultan, but important fortifications were besieged one by one in Transdanubia as well as on the northern edges of the Great Hungarian Plain. This period – usually referred to as the "period of fortress wars" – put constant pressure on the areas this book concerns itself with. Every few years, parts of the Transdanubian territories were beleaguered by military troops, causing significant damage to the local economies.[49]

In 1566, the last campaign in the life of Suleyman put an end to the expansion of the Ottomans for a relatively long period. The territories occupied by Suleyman were recognized by the Habsburgs in the Treaty of Adrianople signed in 1568. This eight-year peace treaty concluded between Selim II (1566–1574) and the representatives of King Maximilian I (1564–1574) consolidated the situation and won some time for the Habsburgs to work out a long-term defense strategy to protect the remaining parts of the Kingdom

Vienna and Constantinople: Notes on the Legal Status of the Principality of Transylvania," in *The European Tributary States of the Ottoman Empire in the Sixteenth and Seventeenth Centuries* (The Ottoman Empire and Its Heritage: Politics, Society and Economy; 53), eds. Gábor Kármán and Lovro Kunčević (Leiden and Boston: Brill, 2013), 67–89, and most recently, see the studies in the volume: *Isabella Jagiellon, Queen of Hungary (1539–1559). Studies*, eds. Ágnes Máté and Teréz Oborni (Budapest: Bölcsészettudományi Kutatóközpont, 2020).

47 Teréz Oborni, "Le royaume des Szapolyai, du royaume de Hongrie orientale a la principauté de Transylvanie (1541–1571)," *Histoire, Economie et Société Époques Moderne et Contemporaine* 34, no. 3 (2015): 65–77.

48 Pálffy, "The Origins," idem, "The Habsburg Defense System in Hungary," and Fodor, *The Unbearable Weight of Empire*, esp. Ch. 2.

49 Éva Simon, *A hódoltságon kívüli „hódoltság". Oszmán terjeszkedés a Délnyugat-Dunántúlon a 16. század második felében* [Ottomans outside of Ottoman Hungary. Ottoman expansion in the southeast Transdanubia in the second half of the sixteenth century] (Budapest: MTA Bölcsészettudományi Kutatóközpont, 2014), Pálffy, *The Kingdom of Hungary*, Ch. 6.

of Hungary, with Vienna in its hinterland. As scholarship has shown, this new defense strategy was to a large extent planned by a talented military officer from Alsace, Lazarus Freiherr von Schwendi.[50] His role shows that the Habsburgs were thinking of creating a lasting plan to block the further advancement of the Ottomans towards Hungary.

In the second half of the 1560s and in the 1570s, the defense system of the Kingdom of Hungary was solidified by the construction or rebuilding of more than a hundred fortifications from the Adriatic Sea, through Transdanubia, to Upper Hungary. This chain of fortifications was a huge financial burden to the Habsburgs. A similar system was also built on the Ottoman side, although with significantly fewer fortifications.[51] These fortifications were built in an area that had not previously been contested by the two powers. Accordingly, this process resulted in a significant reorganization of the local landscapes. The period between 1568 and 1591 – in which a new war broke out between the Ottoman Empire and the Habsburgs that also involved Transylvania – as well as the period after 1606 is mostly referred to as the period of "wars of the peace years" (*Kleinkrieg*) with an interruption at the turn of the sixteenth century, which brought a rather bloody war involving all three polities in the Carpathian Basin, the Fifteen Years' War, or Long War (1591–1606).

Despite changes in the organization of the military administration, especially following the major war council of 1577 held in Vienna, the most important feature of the period between 1568 and 1691 as well as the period between 1606 and 1663 for the present analysis is the permanent presence of major garrisons on both sides of the frontier. The key problems in the frontier

50 Thomas Niklas, *Um Macht und Einheit des Reiches: Konzeption und Wirklichkeit der Politik bei Lazarus von Schwendi, 1522–1583* (Husum: Matthiesen, 1995) and more recently: Géza Pálffy, "Un penseur militaire alsacien dans la Hongrie au XVIe siècle: Lazare baron von Schwendi (1522–1583)," in *La pensée militaire hongroise à travers les siècles*, eds. Hervé Coutau-Bégarie and Ferenc Tóth (Paris: Economica, 2011), 41–59.

51 On the costs and financing of the Habsburg defense system, see Peter Rauscher, "Kaiser und Reich. Die Reichstürkenhilfen von Ferdinand I. bis zum Beginn des Langen Türkenkriegs (1548–1593)," in *Finanzen und Herrschaft. Materielle Grundlagen fürstlicher Politik in den habsburgischen Ländern und im Heiligen Römischen Reich im 16. Jh.* (Veröffentlichungen des Instituts für Österreichische Geschichtsforschung, 38), eds. Friedrich Edelmayer, Maximilian Lanzinner, and Peter Rauscher (Vienna: Oldenbourg, 2003), 45–83, esp. 62, Pálffy, *The Kingdom of Hungary*, 129–134, István Kenyeres, "A török elleni küzdelem finanszírozása Buda elestétől a drinápolyi békéig" [The financing of the anti-Ottoman struggles from the fall of Buda to the treaty of Adrianople], in *Mozgó frontvonalak. Háború és diplomácia a várháborúk időszakában, 1552–1568* [Moving front lines. War and diplomacy in the period of the fortress wars, 1552–1568] (Studia Agriensia, 35), eds. Györgyi Bujdosné Pap, Ingrid Fejér, and Ágota H. Szilárd (Eger: Dobó István Vármúzeum, 2017), 19–40 and special issue of *Századok* on military-fiscal state in Central Europe 152, no. 5 (edited by István Kenyeres).

zone were the recurrent raids on both sides. These raids, with captives and various goods taken, provided a huge income to the usually underpaid mercenaries.[52] Even though there was no further Ottoman advancement to new parts of the former Kingdom of Hungary in this period, a relatively broad strip of the central part of the Carpathian Basin was regularly exposed to raids by smaller military troops.[53]

The period between 1591 and 1606 again changed the status quo, and apart from the continuing presence of the fortifications' garrisons (which housed significantly more soldiers than they did before or after), major military campaigns took place both in Transdanubia and on the edges of the Great Hungarian Plain.[54] The war brought pressure on a new scale to the people who lived in the plain areas of the Carpathian Basin. The military campaigns of the 1540s to the 1560s brought large armies to the territory of the Kingdom of Hungary, but such campaigns only took place every few years and started only in the spring and ended in the autumn. However, this time large armies overwintered in the Carpathians, putting immense pressure on local societies. The Tatars and Heiducks and other military troops were associated with violent acts against the civilian population.[55] The period brought a demographic crisis in many parts of the Basin.[56] The war concluded

52 Géza Pálffy, "A rabkereskedelem és rabtartás gyakorlata és szokásai a XVI–XVII. századi török–magyar határ mentén. (Az oszmán–magyar végvári szokásjog történetéhez)" [Practices and customs of captive-trade and captive keeping along the 16th–17th-century Ottoman-Hungarian frontier (To the customary law of the Ottoman-Hungarian frontier region)], *Fons. Forráskutatás és Történeti Segédtudományok* 4 (1997): 5–78 and Ilona Tarkó, "Rabkereskedelem és anyagi kultúra a XVI–XVII. században a Batthyány család levéltára alapján" [The Ottoman Slave Trade and Material Culture in the Sixteenth to Seventeenth Century According to the Family Archive of the Batthyánys] (PhD diss., Pázmány Péter Catholic University, 2012).

53 See the works of Péter Illik on the topic. See also the special issue of the *Hungarian Historical Review* 4, no. 2 (2015) [Cultures of Christian-Islamic Wars in Europe (1450–1800)], edited by Gabriella Erdélyi.

54 Caroline Finkel, *The Administration of Warfare: The Ottoman Military Campaigns in Hungary, 1593–1606* (Vienna: VWGÖ, 1988), Jan Paul Niederkorn, *Die europäischen Mächte und der 'Lange Türkenkrieg' Kaiser Rudolfs II. 1593–1606* (Archiv für Österreichische Geschichte, 135) (Vienna: Verlag der Österreichischen Akademie der Wissenschaften, 1993).

55 For the Tatar presence: Mária Ivanics, *A Krími Kánság a tizenöt éves háborúban* [The Crimean Tatar Khanate in the Fifteen Years' War] (Kőrösi Csoma kiskönyvtár, 22) (Budapest: Akadémiai, 1994). On French soldiers in the Carpathian Basin: Péter Sahin-Tóth, "La France et les français face à la 'longue guerre' de Hongrie (1591–1606)," 2 vols. (PhD diss., Université François-Rabelais, 1997), and Brian Sandberg, "Going Off to the War in Hungary: French Nobles and Crusading Culture in the Sixteenth Century," *Hungarian Historical Review* 4 (2015): 346–383.

56 Géza Dávid, "Magyarország népessége a XVI–XVII. században" [The population of Hungary in the 16th–17th centuries], in *Magyarország történeti demográfiája (896–1996)* [Historical demography of Hungary], ed. József Kovacsics (Budapest: KSH, 1997), 141–171, and István H. Németh,

with slightly more territorial gain on the side of the Ottomans than on that of the Habsburgs. For some years, several fortifications changed hands both in the Great Hungarian Plain and in Transdanubia (e.g., Győr, Pápa, Tata, Fehérvár), but when signing the peace treaty, the Ottomans took possession of a very important fortification, Kanizsa. As will be argued in Chapters 2 and 3, the loss of this fortress to the Ottomans had a significant role in the transformation of the landscape in the central part of western Transdanubia. The Habsburgs, however, managed to take back some important castles in the Northern Hungarian Mountain areas and Upper Hungary such as Fiľakovo, Szécsény, or Nógrád, but more importantly, the war showed the equalization of the military potential of the two major realms. The period following the peace treaty of Zsitvatorok in 1606 brought a period of peace much like in the last third of the sixteenth century. In the seventeenth century, with the Habsburgs involved in the Thirty Years' War (1618–1648) in Central Europe and the longer periods without a threat of a major Ottoman campaign, there was a slow but steady decrease in the garrisons of the frontier fortifications. Instead of the 22,000 troops during the sixteenth century, only about 17,000 soldiers were present in the Habsburg-Hungarian borderline fortifications, and the number of fortifications also decreased from c. 120 to less than 90.[57] The short war period of 1663 to 1664 and the capture of one of the most important strongholds of the Habsburgs in defending Vienna, Nové Zámky in present-day Slovakia, led to a rapid reorganization of the frontier north of the Danube. Nonetheless, it did not seriously impact Transdanubia, where the fortification system remained basically the same from 1606 to the 1680s. The last Ottoman effort to capture Vienna in 1683 and its lasting yet unsuccessful siege set the stage for the recapture of the Kingdom of Hungary. The 1680s and the 1690s brought a series of major Habsburg military campaigns during which most of the territories that were under the authority of the late medieval kings of Hungary were regained by the Habsburgs. The first phase of the re-conquering war was concluded in 1699 with the Treaty of Karlowitz, which except for the almost entirely uninhabited Banate (Temesköz) restored

"Háború és népesség a kora újkori Magyarországon (16–17. század)" [War and population in early modern Hungary, Sixteenth to Seventeenth centuries], in *Történeti demográfiai évkönyv 2001* [Historical demography yearbook, 2001], eds. Tamás Faragó and Péter Őri (Budapest: KSH Népességtudományi Kutatóintézet, 2001), 129–141.

57 Géza Pálffy, "The Origins and Development of the Border Defence System Against the Ottoman Empire in Hungary (Up to the Early Eighteenth Century)," in *Ottomans, Hungarians, and Habsburgs in Central Europe: The Military Confines in the Era of the Ottoman Conquest* (The Ottoman Empire and its Heritage, Politics, Society and Economy, 20), eds. Géza Dávid and Pál Fodor (Leiden, Boston, and Cologne: Brill, 2000), 59.

the borders of the medieval Kingdom of Hungary in the south. By this time, the military campaigns in the central part of the Carpathian had drawn to a close and the political-administrative structure of the area was under a full re-organization, which only took place in the 1710s, however, following a Hungarian war of independence between 1703 and 1711.

The nature of the war discussed in the previous pages is somewhat different from that of modern-time wars, which seldom lasted more than a few years. Although its intensity differed from that of modern wars, recurrent military operations took place for more than 150 years in the central part of the Carpathian Basin, significantly impacting the environment. The book addresses the ways this war changed the landscape, with special emphasis on the transformation along the frontier. To do so, we need to first deepen our understanding of the type of landscape that was prevalent in the area, which is discussed with the use of case studies.

1.3 Transdanubia and the Great Hungarian Plain – The Setting of the War

The Kingdom of Hungary at the end of the Middle Ages covered most of the Carpathian Basin. The area was a well-defined geographical unit that was also relatively well-protected from many directions. From the northwestern to the southern edges, most of the basin is bordered by the Carpathian Mountains. The mountains are well above 2000 meters in many areas, which limited access to a few passes from the north, east, and southeast. To the west, the Basin is bordered by the foothills of the Alps and on the south by the Sava and Danube rivers, which provided better access to the central parts of the Basin.

The basin is divided up by its main rivers, the Danube and the Tisza. The westernmost area, west of the Danube which flows from north to south, is called Transdanubia. The area between the Danube and the Tisza is usually referred to as Danube–Tisza Interfluve (*Duna–Tisza-köze*), while the areas east of the Tisza (bordering the highlands in Transylvania) is called Transtisza (*Tiszántúl*) (see Fig. 1.1). The term Transdanubia was used from the Ottoman period onwards. Its name derives from the fact that, viewed from Bratislava, the new capital of the Kingdom of Hungary was located on the other side (the left bank) of the Danube.[58] The Ottoman presence in the early

58 Pál Engel, *The Realm of St. Stephen. A History of Medieval Hungary, 895–1526*, trans. Tamás Pálosfalvi (London and New York: I.B. Tauris, 2001), xxiii. For Bratislava as capital, see Géza Pálffy,

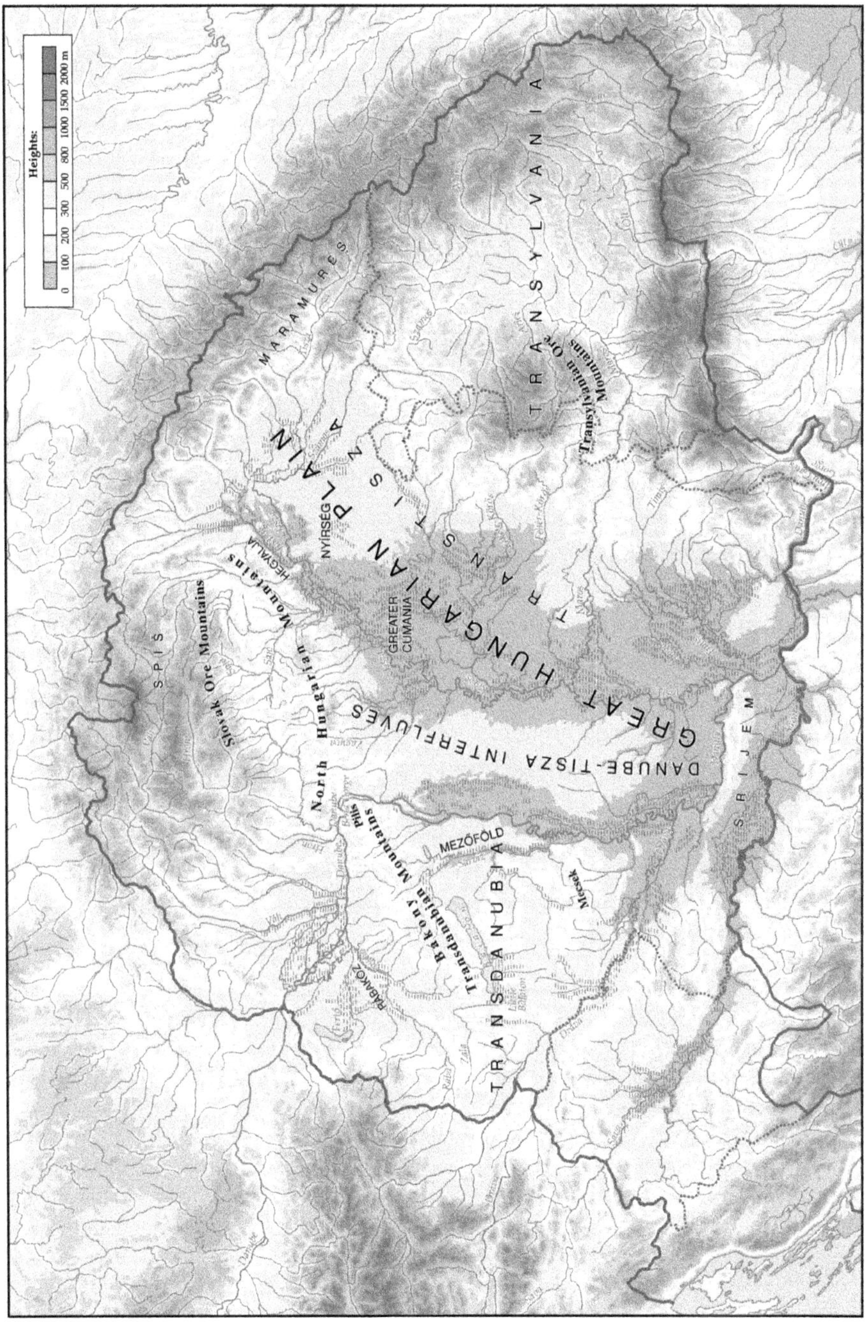

Figure 1.1 Major geographical units of the Carpathian Basin (drawn by Béla Nagy)

modern period covered the whole of the Danube–Tisza Interfluve, and major parts of both the Transdanubia and the Transtisza regions were also under their authority. The easternmost area of the Carpathian Basin is a highland scattered by mountains that in the early modern period belonged to the Principality of Transylvania. The northern part of the basin, again a highland with numerous mountain ranges, was called Upper Hungary (covering mostly present-day Slovakia) in the early modern period. This territory was particularly important both in the medieval and the early modern period because apart from having been rich in forests, it had significant precious metal deposits such as copper and other ferrous metals.

The central part of the basin area – covering a small part of Transdanubia, the whole of the Danube–Tisza Interfluve, and the Transtisza region – was an almost completely flat lowland called the Great Hungarian Plain. Considerable parts of this lowland were not suited to crop production, partly because they either belonged to the lower floodplain areas and were recurrently inundated or because they were prone to salination. There were areas such as the southern part of the Danube–Tisza Interfluve, however, which had one of the highest yielding crop fields in the Carpathian Basin in the past millennium.

Difficult to defend, the lowland of the Great Hungarian Plain was by the second half of the sixteenth century almost fully controlled by the Ottomans, and a new military frontier came into existence along the edges of the plain area (see Fig. 1.2). This new frontier ran from the Adriatic Sea through Croatia and divided Transdanubia into two parts. It crossed the lowland areas of the Danube–Tisza Interfluve and ran south to the line of the Danube on the eastern margin of the plains.[59] This more than 1,000-kilometer-long

"A Magyar Királyság új fővárosa: Pozsony a XVI. században" [The new capital of the Kingdom of Hungary: Bratislava in the 16th century], *Fons. Forráskutatás és Történeti Segédtudományok* 20 (2013): 3–76.

59 On the organization of the frontier on the two sides, see Pálffy, "The Origins," William O'Reilly, "Border, Buffer and Bulwark. The Historiography of the Military Frontier, 1521–1881," in *Frontiers and the Writing of History, 1500–1850*, eds. Steven G. Ellis and Raingard Esser (Hanover: Wehrhahn, 2006), 229–244 (both with rich reference to the existing literature). See for the Ottoman standpoint: Hegyi, *A török hódoltság várai*, Gábor Ágoston, "Ottoman Conquest and the Ottoman Military Frontier in Hungary," in *A Millennium of Hungarian Military History*, 85–110, Gábor Ágoston, "The Ottoman Empire and Europe," in *The Oxford Handbook of Early Modern European History: 1350–1750. Volume II. Cultures and Power*, ed. Scott M. Hamish (Oxford: Oxford University Press, 2015), 612–637, and Gábor Ágoston, "Defending and Administering the Frontier: The Case of Ottoman Hungary," in *The Ottoman World*, ed. Christine Woodhead (Milton Park, Abingdon, and Oxon: Routledge, 2012), 220–236. See also: Mark L. Stein, *Guarding the Frontier Ottoman Border Forts and Garrisons in Europe* (London and New York: I.B. Tauris, 2007), esp. 13–28.

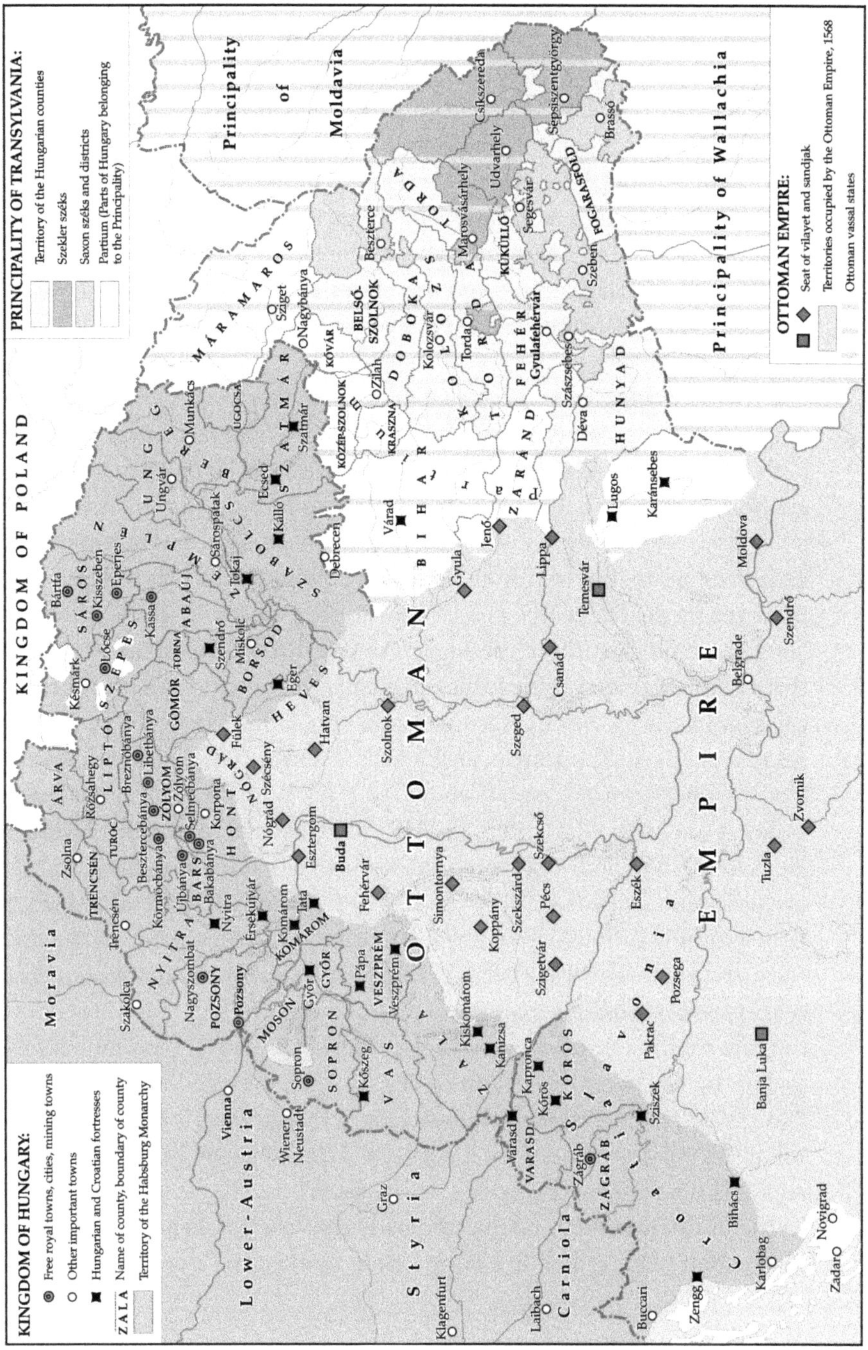

Figure 1.2 The Ottoman Empire and the Kingdom of Hungary in the second half of the sixteenth century (drawn by Béla Nagy)

frontier and its environmental transformations are the main focus of the present book. Being rather long, this frontier called for a complex defense system on the sides of both the Hungarians and the Ottomans. A detailed study of the environmental history of the entire frontier lies outside the limits of a work like this one, so I have chosen western Transdanubia as a study area because it is the best documented in sources as well as equipped with the densest network of fortifications from the mid-sixteenth century, potentially putting more pressure on the environment than anywhere else in the Basin. However, as mentioned earlier, I will have a brief look at both the hinterlands on the Ottoman side of the border – that is, the plains of the Great Hungarian Plain – and the hinterlands in Transdanubia, which belonged to the authority of the Hungarian kings during the whole of the period studied here.

Geographically, Transdanubia can be characterized as a mosaic landscape, unlike the Danube–Tisza Interfluve or the Transtisza region, which both belong to the lowland of the Great Hungarian Plain. Changes in the environment during the presence of the Ottomans in the Great Hungarian Plain have already been addressed, but the processes in this area may have been quite different from what can be observed in Transdanubia.[60] Bordered by the Dráva River on the south, the Danube on the east and the north, and the foothills of the Alps to the west, Transdanubia has at least three quite distinctive landscapes. The eastern and southern areas belong geographically to the Great Hungarian Plain called Mezőföld and Dráva Plain respectively. These areas, just like the Danube–Tisza Interfluve, seldom reach the height of 200 meters above sea level. While the soil of the Dráva valley is not particularly rich, the Mezőföld, thanks to its thick layers of loess, has rich agricultural potential. Both areas were under the authority of the Ottomans from as early as the 1540s onwards. The middle third of Transdanubia is made up of low hills, usually ranging between 200 and 700 meters in height. They became strategically important in the sixteenth century, as the frontier between the two powers ran along this hilly region, roughly from the southwest to the northeast, cutting Transdanubia into two parts. The northwestern third of Transdanubia, the Little Hungarian Plain (*Kisalföld*, meaning little lowlands in Hungarian), is again flat, with small hills scattered over the landscape. Almost 40,000 km^2 in area, this region is the focus of the present analysis. While Chapters 2 and 3 focus on the Little Hungarian Plain and mostly the valley of its most significant river called Rába, which to a large extent marked the frontier of the two

60 See section 4.1.

polities in the seventeenth century, the emphasis of Chapter 4 is on a more extensive geographical unit: the whole of Transdanubia, with a look at the Great Hungarian Plain as well.

The Rába River originates in the Eastern Alps (Friesbacher Alps) and runs into the Mosoni Danube by the town of Győr. With its catchment area of roughly 10,000 km^2 and a length of 300 kilometers, the Rába is the third largest river in present-day Hungary and the most important right-bank tributary of the Danube between the Enns and Dráva rivers. Its size cannot be compared to the major tributaries of the Danube in the Carpathian Basin such as the Sava or the Tisza rivers, as its average discharge is only around 80 m^3/s at Győr.[61] Most of its water comes from the Austrian part of its catchment, and therefore the flood regime of the river is closely connected with the snowmelt and the precipitation maxima in the Alpine region.[62] The section of the river most thoroughly examined in this book is important in terms of its flood discharge because the Rába flows through a more extensive plain area after reaching the foothills of the Alps there. Here, on the Little Hungarian Plain, the river has enough space to meander, while upstream (in present-day Austria) it flows along a rather narrow riverbed. This area was chosen as the focus of this study because the nature of the location and the hydromorphological conditions of the river made it an important strategic point in the frontier zone. From the Fifteen Years' War onwards, defense was often built on rivers, particularly the River Rába.[63]

The almost two centuries of war and recurrent military activities from the early sixteenth to the early eighteenth centuries accompanied fundamental

61 Gergely Szalay and Endre Szilágyi, *Magyarország vizeinek műszaki-hidrológiai jellemzése: Mosoni-Dunaág, Rába* [The technological – hydrological analysis of the waters of Hungary. Mosoni Danube, Rába] (Magyarország vizeinek műszaki–hidrológiai jellemzése) (Budapest: Vízgazdálkodási Intézet, 1989), Pál Ambrózy, "A Felső-Rába vízgyűjtőjének éghajlati jellegzetességei" [Climatic specificities of the catchment area of the Upper Rába], *Vízügyi Közlemények* 79 (1997): 498–517, László Goda and Vilmos Vasvári, "A Felső-Rába vízjárásának statisztikai jellemzése" [Statistical analysis of the water-regime of the Upper Rába], *Vízügyi Közlemények* 79 (1997): 518–538, and Zoltán Károlyi and Sándor Somogyi, "Felszíni vízfolyások" [Above ground waters], in *A Kisalföld és a nyugat-magyarországi peremvidék* [Little Hungarian Plain and the Western Hungarian Highlands] (Magyarország tájföldrajza, 3) (Budapest: Akadémiai, 1975), 104, and 107–111.

62 Heinz Bergmann et al., *Hydrologische Monographie des Einzugsgebietes der Oberen Raab – A Felső-Rába vízgyűjtőjének hidrológiai monográfiája* (Schriftenreihe zur Wasserwirtschaft, 23) (Graz and Budapest: Technisches Universität Graz, 1996).

63 *Zrínyi-Újvár. A Seventeenth-Century Frontier Defensive System on the Edge of the Ottoman Empire*, eds. Gábor Hausner and András Németh (Budapest: Dialóg Campus, 2020), and József Kelenik, "A kanizsai övezet és természetföldrajzi adottságai a XVI. század 70-es éveinek végén" [The Kanizsa region and its geographical conditions in the 1570s], in *Végvár és környezet*, 163–174.

transformations to the economy, the demography, the ethnic composition, as well as the religious life of the Carpathian Basin. While these aspects have all been at least partially examined by scholars, the subject of how the environmental conditions changed in this period has not been the focus of many scholarly works.

1.4 The Book's Concept

The transformation of the environmental conditions of the Carpathian Basin and the frontier zone in the sixteenth and seventeenth centuries has usually been considered in the context of economic and political crises. As discussed above, the Late Middle Ages were seen as a period of a political crisis. Similarly, scholars usually perceived the period after the expulsion of the Ottomans from the central basin area as a general crisis in which the Habsburg authorities had to make huge efforts to "rebuild" the country, which also included interventions to the environment. This entailed channeling rivers, draining marshlands, turning fallows to plowlands, etc. These efforts were interpreted as answers to the crisis directly or indirectly associated with the Ottoman presence. Without going into much detail on the notion of *crisis*, which has been widely discussed in recent scholarship, it is certainly worth considering the above-mentioned processes from a different perspective and looking at the phenomenon of adaptation (or resilience), which has also attracted attention in recent environmental history studies.[64] The change in the political and economic structures and the environmental conditions in the Ottoman period can be understood as a crisis but also as a new challenge requiring different responses from both local societies and larger polities to which different groups adapted in dissimilar ways.[65]

64 See, e.g., Daniel Curtis, *Coping with Crisis. The Resilience and Vulnerability of Pre-Industrial Settlements* (Burlington, VT: Ashgate, 2014).

65 Cf. *Natural Disasters, Cultural Responses. Case Studies Toward a Global Environmental History*, eds. Christof Mauch and Christian Pfister (Plymouth: Lexington Books, 2009). In regional context, see Zsolt Pinke, "Alkalmazkodás és felemelkedés – modernizáció és leszakadás: Kis jégkorszaki kihívások és társadalmi válaszok a Tiszántúlon" [Adaptation and Rise – Modernization and Decline: Little Ice Age Challenges and Social Responses on the Trans-Tisza Region (Hungary)] (PhD diss., Pécsi Tudományegyetem, 2014). On the notion of crisis, see furthermore: Ansgar Nünning, "Krise als Erzählung und Metapher: Literaturwissenschaftliche Bausteine für eine Metaphorologie und Narratologie von Krise," in *Krisengeschichte(n): 'Krise' als Leitbegriff und Erzählmuster in kulturwissenschaftlicher Perspektive* (Beihefte der Vierteljahrschrift für

In the few rather basic works on the landscape and environmental history of early modern Hungary, three relatively distinct topics are discussed, usually all in the context of crisis: weather and climate, forests, and water management.[66] Until recently, most of the works revolving around the role the wars played in the Ottoman-period environmental changes started with a discussion of the long-term changes in the climatic conditions.[67] In their narratives, the period of the Ottoman presence in the Carpathian Basin coincided with the most frequently examined climatic shift in the Holocene apart from recent global warming, the Little Ice Age. Despite major differences in its regional periodization and characteristics, the Little Ice Age is considered to have had an impact throughout the Northern Hemisphere and to have contributed to different crises including the so-called global crisis of the seventeenth century.[68] Partly along these lines, research in East-Central Europe, more specifically in Hungary, also considers the entire sixteenth and seventeenth centuries as a period of significant climatic stress and even as a subsistence crisis.[69] Even though promising new research has been conducted in recent years, most of the studies that discuss the Little

Sozial- und Wirtschaftsgeschichte, 210), eds. Carla Meyer, Katja Patzel-Mattern, and Gerrit Jasper Schenk (Stuttgart: Steiner, 2013), 117–144, esp. 126.

66 The few but notable exceptions include: Béla Iványi, *Képek Körmend multjából / Ex praeteritis oppidi Körmend* (Körmendi füzetek, 4) (Körmend: „Rábavidék" nyomda és lapkiadóvállalat, 1943), idem, *Részletek a magyarországi fertőző betegségek történetéből. Adatok a körmendi levéltárból, a pestis XVI–XVII. századi történetéhez /1510–1692/* [On the history of contagious diseases in Hungary. Data on the sixteenth to seventeenth-century history of pestilence in the archive of Körmend (1510–1692)] (Communicationes ex Bibliotheca Historiae Medicae Hungarica. Supplementum, 3) (Budapest: Országos Orvostörténeti Könyvtár, 1965). See also the frontier defense history conferences held at Noszvaj and their proceeding volumes, such as Petercsák and Szabó, *A végvárak és régiók*, and Petercsák and Pető, *Végvár és környezet*, and most recently: Hausner and Németh, *Zrínyi-Újvár*.

67 E.g., Lajos Rácz, "The Price of Survival. Transformations in Environmental Conditions and Subsistence Systems in Hungary in the Age of Ottoman Occupation," *Hungarian Studies* 24, no. 1 (2010): 21–39, Zoltán Péter Bagi, "The Life of Soldiers during the Long Turkish War (1593–1606)," *The Hungarian Historical Review* 4 (2015): 384–417.

68 See most of all Parker, *Global Crisis*. See also Dagomar Degroot's work on the impact of the Little Ice Age on wars in the Dutch Republic: idem, "The Frigid Golden Age."

69 E.g., R. Várkonyi, *Pelikán a fiaival*, eadem, "Környezet és végvár," eadem, "A párbeszéd esélyei," Ágnes R. Várkonyi, "'A természet majd az értelemmel …' Történeti ökológia és a XVIII. századi Magyarország környezeti válsága" ['Nature with conscience…' Historical ecology and the environmental crisis of Hungary in the eighteenth century], in *Környezettörténet: Az utóbbi 500 év környezeti eseményei történeti és természettudományi források tükrében* [Environmental history. The environmental events of the last 500 years in the light of historical and scientific data] (Környezettörténet, 1), ed. Miklós Kázmér (Budapest: Hantken, 2009), 21–54, Rácz, "The Price of Survival," and idem, *The Steppe to Europe. An Environmental History of Hungary in the Traditional Age* (Cambridge: White Horse Press, 2013), 125–177, esp. 137–140 and 174–177.

Ice Age still use old weather compilations such as the one by Antal Réthly,[70] a meteorologist in the mid-twentieth century, and his data as interpreted by Lajos Rácz and others in the 1990s and early 2000s.[71] Even if the main trends sketched out following the footsteps of Réthly were accurate, such as the coincidence in time between the coldest periods of the Little Ice Age in the Carpathian Basin and the coldest periods in Western Europe, at least two points are still problematic in these works. First, they mostly disregard that the Little Ice Age had a wide variety of impacts on many aspects of society, from political life through material culture to settlement networks and economic opportunities. These works argue, usually without an actual understanding of the characteristics of the Little Ice Age in the area they address, that it hurt local economies. Second, scholarship in many cases uncritically attributes individual weather events – like a cold spell, an extremely cold summer or month, etc. – to the Little Ice Age.[72]

70 Antal Réthly, *Időjárási események és elemi csapások Magyarországon 1700-ig* [Weather events and natural disasters in Hungary until 1700] (Budapest: Akadémiai, 1962), idem, *Időjárási események és elemi csapások Magyarországon 1701–1800-ig* [Weather events and natural disasters in Hungary from 1701 to 1800] (Budapest: Akadémiai, 1970), and idem, *Időjárási események és elemi csapások Magyarországon 1801–1900-ig*, 2 vols. [Weather events and natural disasters in Hungary from 1801 to 1900] (Budapest: OMSZ, 1998). For the methodological problems of the data collection, see Andrea Kiss, "Historical Climatology in Hungary: Role of Documentary Evidence in the Study of Past Climates and Hydrometeorological Extremes," *Időjárás* 113 (2009): 317–320.

71 Lajos Rácz, *Climate History of Hungary since 16th Century: Past Present and Future* (Discussion Papers, 28) (Pécs: Centre for Regional Studies of the Hungarian Academy of Sciences, 1999), idem, *Magyarország éghajlattörténete az újkor idején* [Climate history of Hungary in the Modern times] (Szeged: Juhász Gyula Felsőoktatási Kiadó, 2001). See also: Judit Bartholy, Rita Pongrácz, and Zsófia Molnár, "Extremes and Millennial Trends in the Carpathian Basin Using the Rethly Documentary Collection," *Bulletin of the American Meteorological Society* 12 (2004): 3791–3802 and Judit Bartholy, Rita Pongrácz, and Zsófia Molnár, "Classification and Analysis of Past Climate Information based on Historical Documentary Sources for the Carpathian Basin," *International Journal of Climatology* 24 (2004): 1759–1776.

72 See, e.g., Ágnes R. Várkonyi, "Természet és társadalom. A történeti ökológia regionális lehetőségei" [Nature and society. The regional perspectives of environmental history], *A Nógrád Megyei Múzeumok Évkönyve* 26 (2002): 345–373 and eadem, "Az öltözködés filozófiájáról" [On the philosophy of clothing], *Történelmi Szemle* 53 (2011): 503–536. See also: Zoltán Bagi, "Egy kudarc okai: Kanizsa 1601. évi ostroma" [Reasons of a failure: the siege of Kanizsa in 1601], *Aetas* 28, no. 1 (2013): 5–30, Zoltán Péter Bagi, "A folyóvíz, a csapadék és az áradások mint a hadakozást befolyásoló tényezők a tizenöt éves háború időszakában" [Rivers, precipitation, and floods as agents of military campaigns in the period of the Fifteen Years' War], in *Víz és társadalom Magyarországon a középkortól a XX. század végéig* [Water and society in Hungary from the Middle Ages to the end of the 20th century], ed. Gergely Krisztián Horváth (Budapest: Balassi, 2014), 189–206, Bagi, "The Life of Soldiers," and Gábor Ágoston, "Ottoman Conquest and the Ottoman Military Frontier in Hungary," in *A Millennium of Hungarian Military History*, eds. Béla Király and László Veszprémy (Boulder, CO: Atlantic Research and Publications, 2002), 103–107.

The possible impacts of climatic fluctuations on the economic and social changes and political events occurring in the territories discussed here have been raised not only in Hungarian scholarship but recently also in the Ottoman context. In a pioneering work, Sam White attempted to provide a new interpretation to the outbreak of the Celali rebellions in the Ottoman Empire at the turn of the sixteenth century, which took place at the same time as the Fifteen Years' War.[73] A deeper knowledge of the climatic processes would certainly contribute greatly to the evaluation of the environmental influence of the Ottoman-Hungarian wars, as certainly the two were not independent of each other. Early twentieth-century historians attributed major landscape changes to the war, blaming the Ottomans. But then the situation changed, and from the 1980s the supposed impact of the Little Ice Age became an important cornerstone in the narrative of the changes in the landscape. Perhaps now, views are slowly moving from mono-causal reasoning to a more complex but still biased view of the changes in the main environmental conditions including climate. There are as yet no studies that consider the climate fluctuation at the time as a potential trigger of economic transformation, as a challenge to which some communities answered successfully.[74]

Changes in the forest cover and transformation in the waterscapes due to changing management – or a lack of management – in the early modern Carpathian Basin were brought to the attention of scholars by the prominent historian Ágnes R. Várkonyi (1928–2014). After some pioneering works by ethnographers,[75] she was the first to emphasize the potential environmental stress caused by the Ottoman war period. However, while hypothesizing some changes, she never actually studied the relevant source materials to test the validity of these assumptions.[76] According to her ideas, the changes

73 Sam White, *The Climate of Rebellion in the Early Modern Ottoman Empire* (Cambridge: Cambridge University Press, 2011).

74 Edit Sárosi, *Deserting Villages – Emerging Market Towns Settlement dynamics and land management in the Great Hungarian Plain 1300–1700* (Series Minor, 39) (Budapest: Archaeolingua, 2016).

75 Bertalan Andrásfalvy, *A Duna mente népének ártéri gazdálkodása Tolna és Baranya megyében az ármentesítés befejezéséig* [The flood plain economy of the peoples of the Danube valley in Tolna and Baranya counties before the completion of the regulation works] (Tanulmányok Tolna megye történetéből, 7) (Szekszárd: Tolna Megyei Levéltár, 1975), and for its more recent edition: idem, *A Duna mente népének ártéri gazdálkodása: ártéri gazdálkodás Tolna és Baranya megyében az ármentesítési munkák befejezése előtt* [The flood plain economy of the peoples of the Danube valley in Tolna and Baranya Counties before the completion of the regulation works]. ([Budakeszi]: Ekvilibrium, 2007).

76 On the genealogy of research in Hungary: Ágnes R. Várkonyi, "Történeti ökológia," [Historical ecology] in *A történelem segédtudományai* [Auxiliary sciences of history] (A történettudomány

in the political structures, and most importantly the demography of some regions, created a lasting environmental crisis that necessitated the Habsburg administration's major interventions in the landscape of the Carpathian Basin. Although at some points she re-considered the sometimes strikingly anti-Ottoman tendencies in the previous literature, she did not entirely give up the idea that the formation of non-agricultural wastelands in the Great Hungarian Plain (the loss of woodlands or the extension of the water-covered areas) could largely be attributed to the activities of the Ottomans in the Carpathian Basin.[77] She did note, however, that some of the changes that had lasting visible impacts, like the deliberate flooding of areas around border fortifications, were equally the result of Habsburg's military strategy as in that of the Ottomans.

While the origin of the extensive wastelands (the so-called *puszta*) of the Great Hungarian Plain has been the focus of some recent studies, neither the frontier environments nor the hinterlands elsewhere in the lowlands have attracted the attention of scholars.[78] The environmental transformations of the Little Hungarian Plain and the Transdanubian lowlands in general, for instance, have been completely ignored in the scholarly literature even though they could have been seen as a laboratory for the impacts of the war on the environment, as the transformations occurred during the busiest phase of the wars.

In the coming chapters, I will revisit the problem partly raised by R. Várkonyi and analyze what roles the environment played in the Ottoman-Hungarian wars and what changes this lasting military conflict wrought on the natural resources. Chapter 2 looks at how the environment was considered as part of military strategies in the period when the Ottomans gradually

kézikönyve, 1), ed. Iván Bertényi (Budapest: Osiris Kiadó, 2001), 44–65. For the works of Ágnes R. Várkonyi: eadem, *Pelikán a fiaival* [Pelican with his sons] (Budapest: Liget Műhely Alapítvány, 1992), eadem, "Környezet és végvár. Végvárrendszer és a történeti ökológia kérdései a 16–17. századi Magyarországon" [Environment and border defense castles. Border defense system and questions of historical ecology in sixteenth and seventeenth century Hungary], in *A végvárak és régiók a XVI–XVII. században. Tudományos tanácskozás előadásai – Noszvaj, 1991. okt. 17–18.* [Borderline castles and regions in the sixteenth and seventeenth centuries. Proceedings of the conference held at Noszvaj, 17–18 October 1991] (Studia Agriensia, 14), eds. Tivadar Petercsák, and Jolán Szabó (Eger: Heves Megyei Múzeumok Igazgatósága, 1993), 7–27, Ágnes R. Várkonyi, "A párbeszéd esélyei, (A végvárrendszer-kutatások humánökológiai megközelítéséről)" [The chances of dialogue (on the human ecological approach to the borderline fortification research], in *Végvár és környezet*, 7–32 and eadem, "Természet és társadalom – a történeti ökológia regionális lehetőségei" [Nature and society – the regional possibilities of historical ecology], *A Nógrád Megyei Múzeumok Évkönyve* 26 (2002): 345–373.

77 R. Várkonyi, *Pelikán a fiaival*, 46–50, and eadem, "Természet és társadalom," 357–359.

78 Molnár, "Az Alföld erdei." See also Sárosi, *Deserting Villages*.

took possession of parts of Transdanubia and most of the Great Hungarian Plain in the mid-sixteenth century. Using previously unknown sources, it is argued that right from the beginning of the wars in the Carpathian Basin the local Hungarian military leaders as well as the highest-ranking officials of the Habsburgs considered the environment as an organic part of their defense strategies. Chapter 3 looks at the same region almost a century later, in the period when after the fall of Kanizsa and the end of the Fifteen Years' War the previously planned defense strategy, centered around the line of the River Rába, was put into effect. The question tackled in this central chapter is how the implementation of the military strategy influenced the water system of the Rába River.

While Chapters 2 and 3 look at riverine landscapes and the transformation of waterscapes in the frontier zone, Chapter 4 looks at the changes in the usage of another natural resource traditionally associated with the war – wood – and looks at whether or not linking the Ottoman war period to forest loss is grounded. In this chapter, the geographical focus is broader, as it not only looks at the entire region of Transdanubia but also provides some insights into the changes in the forest cover in the hinterlands, both on the Ottoman and the Hungarian sides of the frontier. The chapter's primary focus, however, is to understand the different spheres of wood consumption related to the military activities in the basin area. If ever there was an extensive area where war affected the forest cover, it was Transdanubia in the Carpathian Basin, as it not only experienced most military campaigns but also had the largest number of newly built fortifications and permanent garrisons in the one and a half century between c. 1540 and 1690, when Ottoman authority prevailed at least in parts of the region. Despite the limited geographical and temporal scope of the analyses carried out in the three main chapters of the book, the case studies aim to demonstrate the long-term environmental effects of the Ottoman presence and the recurrent military activities in the central basin area encircled by the Carpathian Mountains. In this way, we can gain a deeper understanding of what a militarized landscape in pre-modern times might entail.

2 From the Center to the Frontier

The Environment of Central Transdanubia in the Early Modern Period

Abstract
The chapter looks at the transformation of the waterscapes along the frontier between the Kingdom of Hungary and the Ottoman Empire in Transdanubia in the sixteenth century in order to understand the role that the waterscape of the Carpathian Basin played in the formation and maintenance of the frontier of the Kingdom of Hungary in the pre-modern period. It argues that from the early sixteenth century onwards, rivers such as the Sava – and from the mid-sixteenth century the Dráva, the Mura, and most importantly the Rába in the central part of Transdanubia – became organic elements in Hungary's military defense.

Keywords: Environmental history, Kingdom of Hungary, Ottoman Empire, frontier, military history, water history

Against the plundering of the Turks, the Rába is a safeguard
but bigger protection is watching.[1]

One of the most frequent truisms in relevant Hungarian studies is the presence of endless swamps in the pre-modern Carpathian Basin. The extent of swamps and marshes along the rivers was, however, reduced in the modern period, mostly in the one and a half centuries from the 1780s onwards, due to large water control projects that have been interpreted in the scholarly literature ever since as masterpieces of engineering. These

1 *"A török rablása ellen oltalom ugyan a Rába, de amellett nagyobb oltalom a vigyázás"* – Lajos Gecsényi, "Elképzelések a Rábaköz önvédelmének megszervezésére a XVII. század közepén" [Ideas about the defense of the Rábaköz region in the mid-seventeenth century], *Soproni Szemle* 45 (1991): 345. The document is a proposal of György Horváth and Miklós Falusy on the protection of the Rábaköz region against Ottoman raids from the mid-seventeenth century.

Vadas, A., *The Environmental Legacy of War on the Hungarian-Ottoman Frontier, c. 1540–1690*. Taylor & Francis Group, 2023
DOI: 10.5117/9789463727938_CH02

projects became symbols of the modernization of the Kingdom of Hungary in the first half of the nineteenth century, the so-called Reform period.[2] Researchers assumed that because the Ottomans made no effort to manage water, areas that before their presence had been suited to growing grain or other commercial crops became temporarily covered by water at the beginning of the eighteenth century, with large areas becoming wastelands with no potential income connected to them. This rather simplistic view was largely based on a map made in the 1930s by Woldemár Lászlóffy, one of the most respected hydrologists of twentieth-century Hungary. His map, published in 1938, entitled "Water-Covered Areas and Wetlands in the Carpathian Basin Before the Beginning of Flood Protection and Drainage Works" (*A Kárpát medence vízborította és árvízjárta területei az ármentesítő és lecsapoló munkálatok megkezdése előtt*; see Fig. 2.1)[3] heavily overestimated the areas that were recurrently covered in water before the beginning of the above-mentioned drainage works. Lászlóffy used the maps of the First and Second Military Surveys of Hungary (carried out in the 1780s and the first half of the nineteenth century respectively) but relied most importantly on a series of maps produced in the 1890s ordered by regional flood-control associations. The latter maps purposefully exaggerated the extent of areas that were prone to flooding, as by doing so more taxpayers were included in the areas of the different associations, which provided the associations with stronger financial background for the projected water regulation works.[4] The bias, of course, hardly means that water was not one of the most important factors in settlement processes and land-use patterns in the central basin areas in either the Great or the Little Hungarian Plain[5] and in many lower flood plain areas of smaller

2 For a critical re-evaluation of this view, see Zsolt Pinke, "Modernization and Decline: An Eco-Historical Perspective on Regulation of the Tisza Valley, Hungary," *Journal of Historical Geography* 45 (2014): 92–105.

3 *A Kárpát-medence vízborította és vízjárta területei az ármentesítő és lecsapoló munkálatok megkezdése előtt.* [Map] [Water-Covered Areas and Wetlands in the Carpathian Basin Before the Beginning of Flood Protection and Drainage Works], ed. Woldemár Lászlóffy (Budapest: Vízrajzi Intézet, 1938).

4 Zsolt Pinke and Beatrix Szabó, "Analysis of the Map of the Ministry of Agriculture: Water-Covered Areas and Wetlands in the Carpathian Basin Before the Commencement of Flood Protection and Draining," in *2. Nemzetközi és 8. Országos Interdiszciplináris Grastyán konferencia előadásai* [Papers presented at the 2nd International and 8th National Grastyán Conference], eds. Virág Rab and Melinda Szappanyos (Pécs: PTE Grastyán Endre Szakkollégium, 2010), 207–217.

5 András Vadas, "Late Medieval Environmental Changes of the Southern Great Hungarian Plain – A Case Study." *Annual of the Medieval Studies at CEU* 17 (2011): 41–60, and more recently: Zsolt Pinke, László Ferenczi, Gyula Gábris, and Balázs Nagy, "Settlement Patterns as Indicators

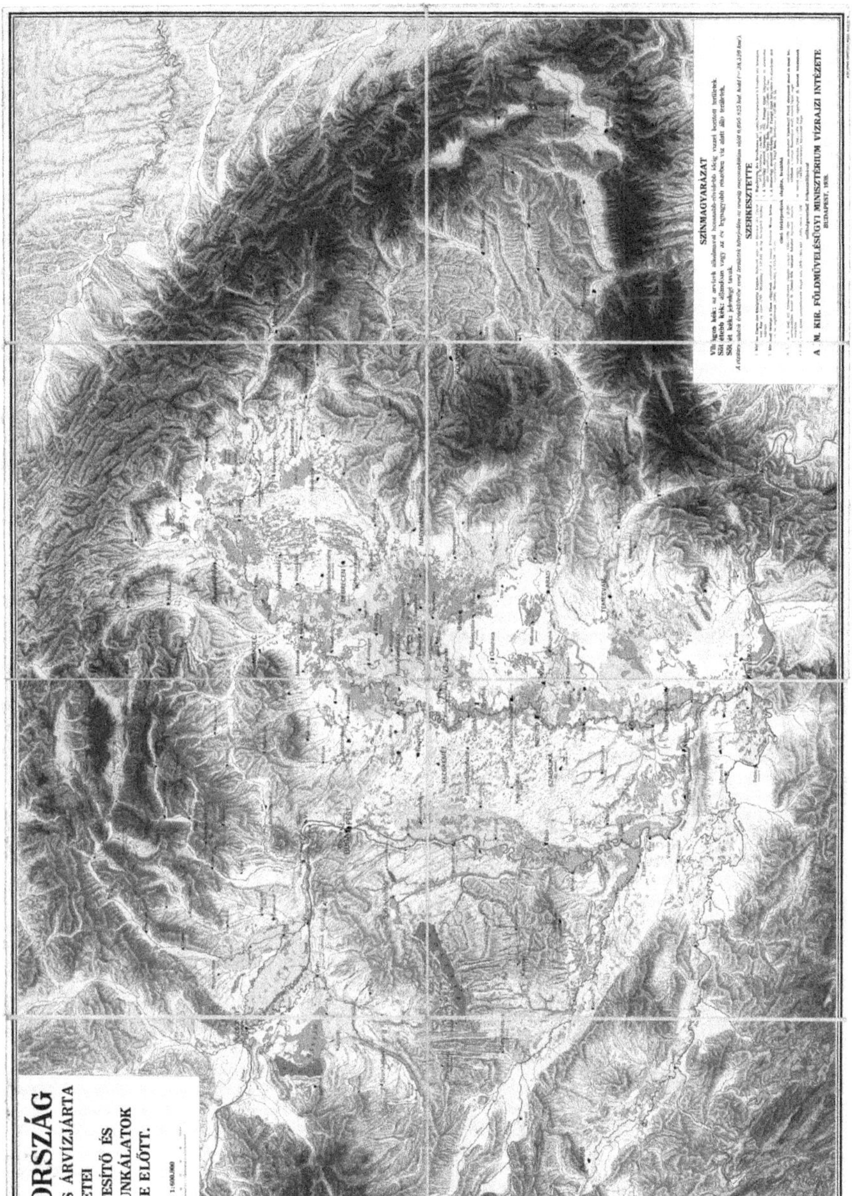

Figure 2.1 The map of Woldemár Lászlóffy, "Water-Covered Areas and Wetlands in the Carpathian Basin Before the Beginning of Flood Protection and Drainage Works."

streams and rivers that regularly left their usually not very deep and broad beds. Because of this rather dense waterscape in parts of the central basin, and the absence of mountains that are difficult to cross, water proved to be the most important resource in military defense. This chapter aims to deepen the understanding of the role played by the waterscape of the Carpathian Basin in the formation and the maintenance of the frontier of the Kingdom of Hungary in the pre-modern period and argues that from the early sixteenth century onwards, rivers such as the Sava, and then from the mid-sixteenth century onwards the Dráva, then the Mura and most importantly the Rába in the central part of Transdanubia were organic elements in Hungary's military defense.

2.1 Why Hydrography Matters: Military Defense and Waterscape in the Kingdom of Hungary

Although studies on the military organization and defense strategies in the Middle Ages and the early modern period lack a detailed review, in many cases they refer to the role that water played in the historical border defense systems of the Kingdom of Hungary. On the western edges of the Kingdom of Hungary, neighboring the strong Holy Roman Empire, there was a clear intention as far back as the period of the state's foundation to block communication if it was necessary. With no major mountains towards the west, the most important mechanism in the frontier defense system was built on different bodies of water. This frontier defense system was called *gyepű* and was based on the use of different obstacles and dams to flood certain areas. Despite the obvious role that landscape plays in this system, studies have seldom addressed the impact of this frontier defense system on local economies and water management systems.[6] Scholars have instead looked at the military aspects and discussed the role of different ethnic groups, most of all the Székelys (an ethnic group

of Water Level Rising? Case Study on the Wetlands of the Great Hungarian Plain," *Quaternary International* 415 (2016): 204–215.

6 On the interrelation of *gyepű* and natural environment, see: Károly Tagányi, "Gyepü és gyepüelve, I–IV" [*Gyepű* and *gyepűelve*], *Magyar Nyelv* 9 (1913): 97–104, 145–152, 201–206, and 254–266, and also István Győrffy, *A feketekőrös-völgyi magyarság települése: az erdélyi magyarság eredete* [The Hungarians of the valley of the Fekete-Körös: The origin of the Hungarians in Transylvania] (Budapest: Fritz Ny., 1914). On the *gyepű*s in general, see István Herényi, *A nyugati gyepü erődítményei: erődítmények, várak Nyugat-Magyarországon* [Constructions of the Western *gyepű*: Fortifications and castles in Western Hungary] (Budapest: Heraldika, 2007).

settled mostly in Transylvania), in the formation and maintenance of this system.[7]

In the lowland areas, hydrological conditions were crucial in maintaining this defense system. The marshlands and the communication channels controlled by frontier fortifications made up a complex landscape-based frontier already from early statehood. However, as the basin area became less prone to attacks from the west, this system was gradually dissolved from the middle of the Árpádian period (1000–1301) onwards.[8] The areas such as the westernmost counties in medieval Hungary (Vas and Zala) had been very sparsely inhabited until the mid-twelfth century, when they started to attract a growing number of people and even chartered settlements came into existence in the area.[9] This defense system from the medieval period – which was meant to prevent enemies crossing from the west (i.e., from the Holy Roman Empire) rather than from the south – was restored in a somewhat different form in the early modern period. In the following pages, I will explore the role that environmental constraints played in the formation of the sixteenth- and seventeenth-century military frontier between the Ottoman Empire and the Kingdom of Hungary. This process was firmly

7 Győrffy, *A feketekőrös-völgyi*, Elek Benkő, *A középkori Székelyföld*, 2 vols. [Székely Lands in the Middle Ages] (Budapest: MTA Bölcsészettudományi Kutatóközpont Régészeti Intézet, 2012), vol. I, 171–197 (with an exhaustive bibliography on the topic), *Székelyföld története* [History of the Székely Lands], 3 vols., eds. Elek Benkő and Teréz Oborni (Székelyudvarhely: MTA Bölcsészettudományi Kutatóintézet; Erdélyi Múzeum-Egyesület; Haáz Rezső Múzeum, 2016), vol. I, esp. 320–347 (the part in question is the work of Elek Benkő and András Sófalvi). In German, see Hansgerd Göckenjan, *Hilfsvölker und Grenzwächter im mittelalterlichen Ungarn* (Quellen und Studien zur Geschichte des östlichen Europa) (Wiesbaden: Franz Steiner, 1972).

8 Esp. Tagányi, "Gyepü és gyepüelve," Gábor Kiss, and Endre Tóth, "A vasvári 'Római sánc' és a 'Katonák útja' időrendje és értelmezése. Adatok a korai magyar gyepűrendszer topográfiájához. I. rész" [The chronology and interpretation of the 'Roman rampart" and the 'Soldiers' Road' at Vasvár. Data on the topography of the gyepű system], *Communicationes Archaeologicae Hungariae 1987*, 101–137, Károly Takács, "Néhány észrevétel Györffy György Árpád-kori történeti földrajzának legújabb kötetéhez" [Some notes on the new volume of György Györffy's Árpádian-period Historical Geography], *Aetas* 14, no. 3 (1999): 101–107. On *gyepűs* is Vas County, see Elek Kalász, *A szentgotthárdi apátság birtokviszonyai és a ciszterci gazdálkodás a középkorban* [Estate structure of the Szentgotthárd abbey and Cistercian economy in the Middle Ages] (Tanulmányok a magyar mezőgazdaság történetéhez, 5) (Budapest: Sárkány Nyomda, 1932), 15–31 and Attila Zsoldos, "A vasi várispánság felbomlása" [Dissolution of the royal county district of Vas], *Vasi Szemle* 54, no. 1 (2000): 27–46.

9 Katalin Szende, "Towns Along the Way. Changing Patterns of Long-Distance Trade and the Urban Network of Medieval Hungary," in *Towns and Communication. 2. Communication between Towns. Proceedings of the Meeting of the International Commission for the History of Towns (ICHT) London 2007 – Leeds 2008*, eds. Hubert Houben and Kristjan Toomaspoeg (Galatina: Mario Congedo Editore, 2011), 166–167, and 197–199.

linked to environmental conditions, and the Habsburg–Hungarian military organization applied techniques very similar to those the Hungarians had used against the Holy Roman Empire centuries earlier.

Here I investigate how contemporaries on the Hungarian side identified the environmental elements that could contribute to the defense of Hungary against the Ottomans. Cartographic and most importantly different documentary sources allow a nuanced understanding of the environmental background of the Kingdom of Hungary, which from the late sixteenth century onwards partly coincided with the area along and north of the Rába River. Some recent studies have drawn attention to the fact that it was not only the Kingdom of Hungary and the Habsburgs who tried to organize the frontier by using natural protective elements extensively but also the Ottomans.[10] This makes it of critical importance to understand how the Ottoman Empire thought of its frontiers, and how much the Ottomans were aware of the geographical conditions when they organized their frontiers.[11] Several studies, most recently by Gábor Ágoston and other Ottomanists, have highlighted the fact that the Habsburgs were not alone in taking measures to learn about the geographical conditions of the frontier zones, especially those around the potential conflict zones of their empire. The Ottomans also did so, not only here but in the Balkans and in Asia Minor as well.[12] They had to be aware of the geographical conditions to identify the

10 Most importantly, Hegyi, *A török hódoltság várai*. On the geographic knowledge of the Ottomans, see: Lajos Fekete, "A hódoltság-kori törökség Magyarországra vonatkozó földrajzi ismeretei" [Geographical knowledge of Ottomans on Hungary], *Hadtörténelmi Közlemények* 31 (1930): 1–17 and 134–154. See also, with an emphasis on the problem of frontier and environment on both sides: Gábor Ágoston, "Where Environmental and Frontier Studies Meet: Rivers, Forests, Marshes and Forts along the Ottoman–Hapsburg Frontier in Hungary," in *The Frontiers of the Ottoman World* (Proceedings of the British Academy, 156), ed. Andrew C.S. Peacock (Oxford: Oxford University Press, 2009), 57–79.

11 For an overview of the frontiers of the empire, see: Peacock, *The Frontiers of the Ottoman World*, and Dariusz Kołodziejczyk, "Between Universalistic Claims and Reality: Ottoman Frontiers in the Early Modern Period," in *The Ottoman World*, ed. Christine Woodhead (London and New York: Routledge, 2012), 205–219, and 471–529.

12 Gábor Ágoston, "Birodalom és információ: Konstantinápoly, mint a koraújkori Európa információs központja" [Empire and Information: Constantinople as the information center of early modern Europe], in *Az értelem bátorsága. Tanulmányok Perjés Géza emlékére* [The courage of intellect. Essays offered in memory of Géza Perjés], ed. Gábor Hausner (Budapest: Argumentum, 2005), 31–60, Gábor Ágoston, "Information, Ideology, and Limits of Imperial Policy: Ottoman Grand Strategy in the Context of Ottoman-Habsburg Rivalry," in *The Early Modern Ottomans: Remapping the Empire*, eds. Virginia H. Aksan and Daniel Goffman (Cambridge: Cambridge University Press, 2007), 75–103, and Ágoston, "Where Environmental and Frontier Studies Meet." See also: Husain, "Changes in the Euphrates River," and idem, "Flows of Power."

placement of their garrisons. The sources consulted for this book provide deeper insights into the Christian side of the frontier, but where possible I also consulted studies written on the Ottoman organization of the frontier, as it had a great impact on the long-term changes of the hydrogeography, vegetation and other natural features of the Carpathian Basin.

The Ottoman conquest of the Kingdom of Hungary is outlined in section 1.3, with the focus on the use of environmental constraints in the defense of the frontier between the two realms. The Ottoman Empire and the Kingdom of Hungary faced each other many times before the early modern period or even the Battle of Mohács (1526). From the rule of King Louis I the Great (1342–1382) onwards, the Ottoman advance into the Balkans affected the foreign policy of Hungary because its southern frontiers were repeatedly threatened by Ottoman troops. This does not mean, however, that Louis I consciously started to build a defense system. This was not the case until the reign of King Sigismund of Luxemburg (1387–1437), when a network of castles was built or extended along the southern border. Besides the internal political struggles in the empire at the time, the Ottoman threat was mitigated thanks to Wallachia, Serbia, and Bosnia – the remaining tributary states in the Balkans.[13] This defense system was already adjusted to the geographical and hydrological constraints. For a long time, the flow of the Danube and Sava rivers was identified as the most important defense line to protect the Kingdom of Hungary from the Ottoman advance. Both rivers were major obstacles because of their significant flows and the lack of fords. The bridgehead fortifications along this line, most of all in Belgrade at the confluence of the Danube and Sava rivers, were seen as linchpins in the defense for roughly 100 years starting in the early fifteenth century. Even though King Matthias (1458–1490) changed the system significantly, these reforms were institutional rather than geographic. The key fortresses remained the same in the period of his reign. A major change occurred, however, when, after two unsuccessful attempts in 1440 and 1456, the

13 Ferenc Szakály, "A török–magyar küzdelem szakaszai a mohácsi csata előtt, 1365–1526" [Stages of the Ottoman–Hungarian struggles before the battle of Mohács, 1365–1526], in *Mohács. Tanulmányok a mohácsi csata 450. évfordulója alkalmából* [Mohács. Studies in the memory of the 450th anniversary of the battle of Mohács], eds. Lajos Ruzsás and Ferenc Szakály (Budapest: Akadémiai, 1986), 11–57, Pál Engel, "Magyarország és a török veszély Zsigmond korában, 1387–1437" [Hungary and the Ottoman threat in the age of King Sigismund, 1387–1437], *Századok* 128 (1994): 273–287, Gyula Rázsó, "Military Reforms in the Fifteenth Century," in *A Millennium of Hungarian Military*, 54–58. See recently: Emir O. Filipovic, "The Key to the Gate of Christendom? The Strategic Importance of Bosnia in the Struggle against the Ottomans," in *The Crusade in the Fifteenth Century: Converging and Competing Cultures* (Crusades – Subsidia, 8), ed. Norman Housley (London and New York: Routledge, 2016), 151–168.

Ottomans finally captured the fortification of Belgrade in 1521. The repeated attempts by the Ottomans to capture this stronghold illustrate their clear strategic thinking and geographic knowledge. The area at the hinterland of Belgrade, basically the whole of the Great Hungarian Plain, was virtually impossible to protect because geographical conditions did not allow the construction of a defense line across this area due to the shortage of major rivers. After the fall of Belgrade, the Ottomans rapidly captured a series of other key fortifications along this southern defense line. From that time on, the road leading to the decisive Hungarian defeat at Mohács in 1526 was, so to speak, paved.[14]

After the Battle of Mohács, with the appearance of the Habsburgs as one of the claimants to the Hungarian throne in the defense of the kingdom, attempts were made to organize the defense of the remainder of the country along the smaller rivers in the Dráva-Sava Interfluve and along Dráva River itself, but this defense strategy proved to be short-lived.[15] The Dráva later became more important in the defense system of the Ottomans themselves, as it flowed from areas (e.g., Carinthia) that were a potential threat to them.[16] Between the Dráva and Lake Balaton in the center of Transdanubia on to the Transdanubian Mountains, there were no major natural obstacles that could withstand the Ottomans. Still, as one of the prominent Hungarian military historians noted: "the running of the defense line was not determined by the soldiers or the members of the Council of War [*Hofkriegsrat*], but rather by the valleys of rivers and streams, the marshes providing a natural defense and the roads, bridges, and fords crossing them."[17] These were anything but

14 Pálffy, "The Origins," 13–15.

15 Géza Pálffy, "A török elleni védelmi rendszer szervezetének története a kezdetektől a 18. század elejéig. (Vázlat egy készülő nagyobb összefoglaláshoz)" [The history of the Anti-Ottoman defense from the beginnings to the early eighteenth century. (Sketch to an overview in progress)], *Történelmi Szemle* 38 (1996): 184. The Habsburgs indeed were financially involved from 1521. See: *Magyarország melléktartományainak oklevéltára = Codex diplomaticus partium Regno Hungariae adnexarum* (Monumenta Hungariae historica. Első osztály: okmánytárak, 31), eds. Lajos Thallóczy and Antal Hodinka (Budapest: Magyar Tudományos Akadémia, 1903), 34–38 (no. XXXV), and 38–39 (no. XXXVI). See also: Pálffy, "The Origins," 15, esp. note 23.

16 Hegyi, *A török hódoltság várai*, vol. I, 98. See also the following volume: *"per sylvam et per lacus nimios" The Medieval and Ottoman Period in Southern Transdanubia, Southwest Hungary: the Contribution of the Natural Sciences*, eds. Gyöngyi Kovács and Csilla Zatykó (Budapest: Institute of Archaeology, Research Centre for the Humanities, Hungarian Academy of Sciences, 2016).

17 József Kelenik, "Tata helye és szerepe a végvári rendszerben a 16. század utolsó harmadában" [The place of Tata in the borderline defense in the last third of the sixteenth century], in *Tata a tizenötéves háborúban. Tatán 1997. május 23-án megtartott tudományos ülésszakon elhangzott előadások anyaga* [Tata in the Fifteen Years' War. Papers read at the conference held at Tata, 23 May 1997] (Annales Tataienses, 1), eds. János Fatuska, Éva Mária Fülöp, and László ifj. Gyuszi

significant streams, as most of them could be crossed dry-shod in parts of the year. The first river that could potentially pose as an obstacle both to plunderers and major military campaigns in Transdanubia was the Rába. The first time that the Habsburg military elite attributed some strategic importance to the Rába River was in 1537, when the first building work started on the previously modest fortifications of Győr. The importance of the Rába in the Habsburg-Hungarian strategic thinking on the Habsburg-Hungarian site can be deduced by King Ferdinand I's command for Vas, Sopron, and Zala Counties to block the fords by backing up the river in order to hinder the Ottomans from crossing.[18] In 1543, following the fall of Fehérvár, in the middle of a major marshland that not even the plundering Mongols had captured fully in 1242, the routes towards the western and northern parts of Hungary were open to the Ottomans. Despite its relatively small discharge, the Rába River, along with other small rivers in Transdanubia such as the Marcal, were the only natural features that could serve as a western continuation of the Transdanubian Mountains with Lake Balaton at its foothills.

In the period following the defeat at Mohács, it soon became clear to the Habsburg administration that better knowledge of local geographic conditions was essential. Having just established their rule in Hungary, they had not been aware of the defense potential of geographic background information. However, to gain at least partial control over their newly gained but immediately threatened realm, the Habsburgs took immediate action. Some defense concepts were developed on their orders, which helped the Austrian military leadership to become familiar with the situation along the southern borders of the Kingdom of Hungary. These concepts are by far

(Tata: Mecénás Közalapítvány, 1998), 46, József Kelenik, "A kanizsai övezet és természetföldrajzi adottságai a XVI. század 70-es éveinek végén" [The Kanizsa region and its geographical conditions in the 1570s], in *Végvár és környezet*, 163–174, Géza Pálffy, "A magyarországi török és királyi végvárrendszer fenntartásának kérdéséhez" [To the question of the management of the Ottoman and Hungarian border defense systems in Hungary], *Keletkutatás* (Spring 1995): 61–86, Pálffy, *A császárváros*, and idem, *Európa védelmében: haditérképészet a Habsburg Birodalom magyarországi határvidékén a 16–17. században* [In defense of Europe. Military mapping at the edges of the Habsburg Empire in Hungary in the 16th–17th centuries] (2nd ed. Pápa: Jókai Városi Könyvtár, 2000).

18 MNL OL E 142 Fasc. 34 no. 19, Mandate of King Ferdinand I to Vas, Sopron and Győr Counties, 25 January 1537. For the edition of the source, see Vadas, "A Rába-mente környezeti viszonyai," 33–34. On the beginnings of the fortification of Győr in the sixteenth century, see: Lajos Gecsényi, "Győr erődváros kiépítése a 16. század második felében" [The formation of Győr as a garrison town in the second half of the sixteenth century], in idem, *Gazdaság, társadalom, igazgatás. Tanulmányok a kora újkor történetéből* [Economy, society, administration. Studies in the history of the early modern period] (Győr: Győr-Moson-Sopron Megye Győri Levéltára and Győr Megyei Jogú Város Levéltára, 2008), 445–452 and Pálffy, *A császárváros*, 41–44.

the best sources to understand the military role that experts in the early modern period attributed to rivers. Below, the ideas of riverine frontier defense systems will be discussed using some of these concepts as well as other, mostly written sources.

2.1.1 The Historical Environment of a Frontier River – What to Look At?

Understanding the environmental conditions of a river in pre-modern times is a complex task. Historical geography and environmental history studies have different perspectives for studying similar problems. The sources used may also differ fundamentally; scientific, archaeological, historical, and linguistic sources have all been utilized recently to understand the relationships of rivers and societies in different areas and periods.[19] Systematic studies of riverine environments and local societies have mostly been discussed in the Anglo-Saxon context, and most of the case studies concern England, the United States, Canada, and other English-speaking countries.[20] In recent years, a growing interest in similar questions has arisen in other areas, for instance, Central Asia and Central Europe,[21] but there has been very limited research on the role of rivers in early modern frontier studies.[22]

Access to the geographical knowledge of one's lands – or a country's territories in pre-modern times – was fundamentally different before the

19 *Rivers in History: Perspectives on Waterways in Europe and North America*, eds. Christof Mauch and Thomas Zeller (Pittsburgh: University of Pittsburgh Press, 2008), Robert H. Webb, *Requiem for the Santa Cruz: An Environmental History of an Arizona River* (Tucson, AZ: University of Arizona Press, 2014), Robert E. Henshaw, and Frances F. Dunwell, *Environmental History of the Hudson River: Human Uses That Changed the Ecology, Ecology That Changed Human Uses* (Albany, NY: State University of New York Press, 2011). For early modern case studies: *Water History* 5 (2013) [special issue: Dealing with Fluvial Dynamics: A Long-Term, Interdisciplinary Study of Vienna and the Danube] and Leona Skelton, *Tyne after Tyne: An Environmental History of a River's Battle for Survival, 1530–2015* (Cambridge: White Horse Press, 2017).

20 For a recent and comprehensive overview of related, mostly English-language publications on the environmental history of rivers, see Paula Schönach, "River Histories: A Thematic Review," *Water History* 9 (2017): 233–257. For German scholarship, see Martin Schmid, "The Environmental History of Rivers in the Early Modern Period," in *An Environmental History of the Early Modern Period: Experiments and Perspectives*, eds. Martin Knoll and Reinhold Reith (Berlin: LIT Verlag, 2014), 19–26. Most recently: Lukas Werther et al., "On the Way to the Fluvial Anthroposphere—Current Limitations and Perspectives of Multidisciplinary Research," *Water* 13, no. 2188 (2021). doi: 10.3390/w13162188.

21 E.g., *Water History* 5 (2013) as quoted above.

22 See esp. Roll, Pohle, and Myrczek, *Grenzen und Grenzüberschreitungen*.

introduction of systematic cartography in East-Central Europe sometime in the eighteenth century. As discussed above, in the Late Middle Ages and the early modern period, a piece of land was usually described by providing reference points and describing the landscape in words; it was rarely presented visually. This was the case on both the Hungarian and the Ottoman sides of the border. Terriers (*urbaria*) in the Kingdom of Hungary and *defterler* on the Ottoman side of the border both described many features of the land that have usually been done by cartographic tools from the modern times onwards.[23] Although until the mid- or late-sixteenth century, cartography played a relatively limited role[24] in acquiring geographic information, this does not mean that there was no significant knowledge of the landscapes that armies needed for planning their campaigns. This is obvious from the campaigns themselves, which on both sides were usually led in carefully chosen directions and against key fortifications. Without systematic information gathering, this would certainly have been different.

On the Ottoman side, the fifteenth- and sixteenth-century campaigns towards the Balkans and then Hungary show clear strategic thinking and a high level of understanding of the role that different landscape features such as mountains, water bodies, and fortifications could play in the long-term occupation of the region. Similarly, from the mid-sixteenth century on, the Habsburgs started to acquire as much information as possible on areas that might potentially be involved in military campaigns. The information-gathering created new types of documentary evidence that are valuable for modern research. Narrative sources, lists of castles, surveys, maps, and sketches all contributed to elucidating early modern efforts to understand the geographical and environmental conditions of certain regions.[25] Rivers were seen by both sides as natural features that were es-

23 Szabó, "Sources for the Historian," 271–272 and Ágoston, "Where Environmental and Frontier Studies Meet," 63.

24 See, e.g., the Ottoman maps published as appendices to Pálffy, *Európa védelmében*, III–IV.

25 For the Hungarian-Habsburg efforts, see, e.g., Géza Pálffy, "Egy rendkívüli forrás a magyar politikai elit 16. századi földrajzi ismereteiről. Az 1526 és 1556 között török kézbe került magyarországi városok, várak és kastélyok összeírása a Német-római Birodalom rendjei számára" [A Unique Document on the Geographical Knowledge of the Political Elite of the Kingdom of Hungary in the Middle of the Sixteenth Century: the Register of the Hungarian and Slavonian Towns, Fortresses and Castles occupied by the Ottomans from 1526 until 1556 for the Estates of the Holy Roman Empire], in *Várak nyomában*, 177–194, esp. 180–184, Géza Pálffy, *Die Anfänge der Militärkartographie in der Habsburgermonarchie. Die regelmäßige kartographische Tätigkeit der Burgbaumeisterfamilie Angielini an den kroatisch-slawonischen und den ungarischen Grenzen in den Jahren 1560–1570* (Budapest: Ungarisches Nationalarchiv, 2011), Zsolt Török, "Renaissance Cartography in East-Central Europe c. 1450–1650," in *Cartography in the European Renaissance:*

sential in military strategies, both in defense and in providing a hinterland for military campaigns.[26] One of the first major infrastructural works built by the Ottomans in the Carpathian Basin, the pontoon bridge at Osijek over the Dráva River, reflects the role of waters both as a means of transportation and as a natural obstacle.

This thinking was not specific to the Ottomans. As mentioned, the Hungarian-Habsburg military strategies also focused on rivers. In the following, I demonstrate, using the example of the River Rába, the basic ideas of military experts in using rivers and water bodies in general to defend the Habsburg realm. The most important sources utilized in this chapter are river surveys, a less typical kind of source that is of primary importance from the point of view of understanding the role of a river – in this case, the Rába – in frontier protection. Even though the existence of some of the sources discussed here have been known to scholars for a long time, previous studies that referred to these sources focused almost exclusively on their military historical aspects. Géza Pálffy, one of the leading scholars of the political history of early modern Hungary, was the first to draw attention to these sources when he discussed the organization of the military frontier. He used two surviving sixteenth-century river surveys that describe the environment of the Rába; the first is especially valuable, as it presents the main landscape features of the river and its surroundings before the appearance of the Ottomans in the immediate neighborhood in the 1540s. The later survey, dating to 1594, is not as detailed as the first but is important in identifying how the river was manipulated to protect the hinterland.[27] A third, to some extent similar source from the end of

The History of Cartography, vol. 3, part 2, ed. David Woodward (Chicago: University of Chicago Press, 2007), 1806–1851. Most recently, see Robert Born, "The Ottoman Expansion and the Development of Cartography in East-Central Europe (15th–18th Centuries)," *Revue des Études Sud-Est Européennes* 54 (2017): 121–152. In a broader geographical context, see Pinar Emiralioğlu, *Geographical Knowledge and Imperial Culture in the Early Modern Ottoman Empire* (Burlington, VT: Ashgate, 2014), Palmira Brummett, *Mapping the Ottomans: Sovereignty, Territory, and Identity in the Early Modern Mediterranean* (Cambridge: Cambridge University Press, 2015), Ian Manners, *European Cartographers and the Ottoman World* (Oriental Institute Museum Publications, 27) (Chicago: Oriental Inst. Museum of the Universityf of Chicago, 2007). For the Ottoman attempts to acquire geographical knowledge of the area, see, e.g., Fekete, "A hódoltság-kori törökség," Ágoston, "Where Environmental and Frontier Studies Meet," 63–65 and Brummett, *Mapping the Ottomans*, esp. 72. (with reference to a number of relevant works).

26 Palmira Brummett, "The River Crossing: Breaking Points (Metaphorical and 'Real') in Ottoman Mutiny," in *Rebellion, Repression, Reinvention: Mutiny in Comparative Perspective*, ed. Jane Hathaway (Westport, CT and London: Praeger, 2001), 215–231.

27 Géza Pálffy, *Pápa a hosszú török háborúban. A végvár története az 1594–1597. esztendőkben, különös tekintettel a töröktől való visszafoglalására* [Pápa in the Long Turkish War. The history

the seventeenth century will be discussed in the concluding chapter of the book. After the Ottomans were expelled from the central part of the Kingdom of Hungary, surveys were made along other major rivers. The best known and most frequently discussed is the survey and mapping activity of Luigi Ferdinando Marsili (1658–1730), which is best reflected in his major work entitled *Danubius Pannonico-Mysicus, observationibus geographicis, astronomicis, hydrographicis, historicis, physicis perlustratus et in sex tomo digestus*, printed in The Hague in 1726. Mapping along the Danube had military purposes and was carried out following a different agenda. The Rába was surveyed somewhat earlier, around the time of the Treaty of Karlowitz (1699), to assess the problems of water management and to find possible local solutions, because by this time the military role of the river had become negligible.

Further documentary evidence from a diversity of sources is available on the valley of the Rába. The estate holders in the area – particularly the prominent Batthyány family – systematically archived hundreds of private letters referring to the forests, the hydrography along the Rába, different oxbow lakes, streams, mills, bridges, roads, and other features.[28] From the sixteenth century onwards, starting as early as the second half of the century, dozens of terriers (*urbaria*, conscriptions of holdings and duties of their dwellers) have survived that describe the environment of the villages and towns along the river.[29] In the archives of the landowners in the Rába valley – the Batthyánys as well as other members of the high nobility such as the Nádasdys, the Sennyeys, and the Széchényis – terriers or different kinds of inventories can be found from every few years. These documents are also considered here as a control source to the river surveys.

The earliest of the three surveys mentioned above was conducted in 1543 and 1544 by Tamás Nádasdy (1498–1562), chief justice (later palatine)

of the border fortress from 1594 to 19] (Pápa: Jókai Mór Városi Könyvtár, 1997), 20–22. For the edition of the document: Zsolt Baráth, "Oszmán terjeszkedés a Dunántúlon és az 1594. évi Rába felmérés" [Ottoman expansion at the Transdanubia and the 1594 Rába River survey], *Győri Tanulmányok* 39 (2018): 53–76.

28 On the section of the river, see András Vadas, *Körmend és a vizek. Egy település és környezete a kora újkorban* [Körmend and the waters. A settlement and its environment in the early modern period] (ELTE BTK Történelemtudományok Doktori Iskola. Tanulmányok – konferenciák, 5) (Budapest: Történelemtudományok Doktori Iskola, 2013) and András Vadas, "Vízgazdálkodás és háborús védekezés. Csákány és a Vas megyei Rába-mente a kora újkorban (1600–1658)" [Water management and border protection at Csákány and along the Rába in Vas County between 1600 and 1658], in *Víz és társdalom*, 207–245.

29 On terriers, see Péter Szabó, "Sources for the Historian of Medieval Woodland," in *People and Nature*, 271–272.

of Hungary, at the order of King Ferdinand I of Habsburg. This exceptionally detailed source gives a step-by-step survey of both banks of the Rába River. Hungarian research used the source as early as the beginning of the twentieth century because of its Hungarian place names and numerous vernacular terms.[30] More importantly, Károly Tagányi also used this source when he described the above-discussed *gyepű* system.[31] After Tagányi, perhaps the first to use the source was Géza Pálffy, who drew attention to the importance of this survey in the context of the formation of the captaincy of Győr and the organization of the frontier on the Habsburg-Hungarian side against the Ottomans.[32] Instead of its content, he found the pure existence of the document relevant, as it was one of the first sources to demonstrate how the Habsburgs planned to defend Vienna in the long run. At that time, the Ottomans had yet to occupy the southern part of Transdanubia, which makes the document on the Habsburgs' plan particularly interesting, as it shows that the Habsburgs from an early period on were thinking in terms of a broad buffer zone between the two realms. The geographical-environmental information included in the work was not of primary importance from a military-historical point of view, as the development of the protective structures advised by Nádasdy was halted. In addition, there is limited information about when and how much these plans were ever acted on. This is probably why this document has received relatively little scholarly attention. Nonetheless, using this source allows us to outline quite accurately the possible forms of border protection. In the early seventeenth century, after the fall of Kanizsa in the Fifteen Years' War, the organization of the frontier followed the principles laid down by this document, which as we shall see in Chapter 3 significantly transformed the river's environment in

30 István Szamota, *Magyar oklevél-szótár. Régi oklevelekben és egyéb iratokban előforduló magyar szók gyűjteménye* [Hungarian charter-dictionary. Collection of Hungarian language words in old charters and other documents] (Budapest: Hornyánszky Viktor Könyvkereskedése, 1902–1906), passim.

31 Tagányi, "Gyepü és gyepüelve, II," 150–151.

32 Pálffy, *Európa védelmében*, 20–28, esp. 20 (note 29), idem, *A császárváros védelmében: a győri főkapitányság története, 1526–1598* [In defense of the imperial city: the captainship of Győr, 1526–1598] (Győr: Győr-Moson-Sopron Megye Győri Levéltára, 1999), 53–54. The survey is also referred to in: Gyula Benczik, "Fentő és góré. A hódoltság kori védelmi munkálatok emlékei a Felső-Rábavidék dűlőneveiben" [*Fentő* and *góré*. The memory of the Ottoman-period defensive works in the field names of the Upper Rába region], *Vasi Honismereti és Helytörténeti Közlemények* no. 1 (1995): 42–48. On the source, see furthermore: András Vadas, "A Rába-mente környezeti viszonyai a 16. század közepén egy 1543–44-es folyófelmérés tükrében" [The environment of the Rába valley in the mid-16th century as reflected in the river survey of 1543–44], *Soproni Szemle* 69 (2015): 21–22.

the decades to follow. Apart from the final survey itself, few other sources tell of its creation. Tamás Nádasdy's mandate from 1543, in which he called on the landholders in the Rába valley to help the surveyors in their work, has also survived. The documents also note the people in charge of the surveying. The importance of the work is reflected in the choice to delegate this task to the castellans of the most significant fortifications of the region (Lőrinc Farkas, castellan of Sárvár, and Fereny Sennyey and Márk Paladin, both castellans of Kapuvár) and Ferenc Polányi, deputy-*ispán* (*vicecomes*, most probably of Vas County).[33] The survey was interrupted at the end of 1543 and was only completed by the end of 1544, but in the second phase, the survey was carried out under the leadership of Sixtus Budor, a well-known nobleman in the entourage of the Nádasdy family.[34]

Even though the survey was completed by the end of 1544 and included important pieces of advice on how to protect the northwestern part of Transdanubia and Vienna in its hinterland, this advice was not always put into practice. Several letters exchanged between Pál Várday, the archbishop of Esztergom, Niklas Graf zu Salm, the royal chief commandant, and Tamás Nádasdy from 1546 demonstrate that at least part of the works proposed in the survey were yet to be completed.[35] The same can be presumed from the decisions of a local noble gathering held in January 1547, as one of the resolutions at the gathering called for the nobility along the River Rába to construct the defensive works suggested by the survey.[36]

33 MNL OL E 185 No. 2. (44. d.) fol. 24. See the edition of the text: Vadas, "A Rába-mente."

34 On Sixtus Budor: Bendefy, "Középkori magyar hossz- és területmértékek," 63–64, Irén Őriné Bilkei, "A zalavári és a kapornaki konventek hiteleshelyi tevékenysége és ügyfelei, a megyei nemesség a Mohács utáni évtizedekben" [The activity of the chapters of Zalavár and Kapornak as places of authenticity, and its clients, the nobility of the county in the decades after the battle of Mohács] (PhD diss., Eötvös Loránd Tudományegyetem, 2007), 56. For his signature, ibid., 117. On the Budor family: Géza Pálffy, "Egy szlavóniai köznemesi família két ország szolgálatában: a budróci Budor család a XV–XVIII. században" [A Slavonian lesser noble family in the service of two countries: The Budor family of Budróc in the 15th–18th centuries], *Hadtörténelmi Közlemények* 115, no. 4 (2002): 923–1007. For Sixtus: ibid., 935. See further: Béla Iványi, *A körmendi levéltár memorabiliái 1352–1698*. (*Acta Memorabilia in tabulario gentis principum de Batthyány reperibilia*). (Körmendi füzetek, 2) (Körmend: Rábavidék, 1942), 77–78.

35 On the role of Várday, Niklas zu Salm, and Nádasdy in the defense of the country: György Laczlavik, "Várday Pál helytartói működése, 1542–1549" [Pál Várday as procurator, 1542–1549], *Levéltári Közlemények* 83, no. 1–2 (2012): 41–46.

36 *...et sub poena birsagii medii fioreni super eos, qui huiusmodi dimensionem facere negligerent, tociens quociens id facere recusaverint, tamdiu, donec loca ipsis decisa et ostensa reformaverint extorquenda, fieri conclusimus* – MNL OL E 142 Fasc. 23. no. 47. For its edition: *Magyar országgyűlési emlékek történeti bevezetésekkel*, III. (1546–1556.) [Hungarian dietary records with introductions] (Magyar történelmi emlékek 3. osztály. Országgyűlési emlékek, 3), ed. Vilmos Fraknói (Budapest:

The most important value of the 1543–1544 survey lies in the fact that it gives more details about the settlements, buildings, and vegetation along the Rába than any other source up to the late eighteenth century, after which the sheets of the First Military Survey provide a similarly detailed image.[37] In the sixteenth century, the principles of map-making were different from those known to modern understanding. In the mid-sixteenth century, the Hungarian noble elite (and probably the Habsburg administration) as well as the Habsburgs, who had ordered the survey, were still thinking in written descriptions of areas. Perhaps at that time, depictions functioned better in strategic planning than maps. In 1543 and 1544, the two banks of the Rába River were surveyed separately, with the surveyors identifying different landmarks as reference points. They sometimes used similar marks utilized in perambulations, such as a cross carved in a tree.[38] Most of the reference points are settlements, bridges, water mills, and fords, which are mostly easy to identify, but what some other points signify should still be clarified.

The second survey mentioned above was carried out half a century later than the first. This also had a direct military purpose, as the presence of the Ottomans was an everyday experience along the Rába by the mid-1590s, which marks the beginning of the Fifteen Years' War. The son of Tamás Nádasdy, Ferenc (1555–1604), was ordered to conduct a thorough survey of the fords, mills, and fortifications along the river; this survey is shorter than the previous one, and there is little reference to environmental features.[39] The document refers almost exclusively to buildings that were to be used in the defense of the left bank of the river in case of military pressure. It was carried out in the critical period of the Fifteen Years' War, immediately after the fall of one of the linchpins in defending Vienna, the garrison town of Győr. After the fall of this fortified town in 1594, the route was yet again paved for the Ottomans to siege Vienna, the "imperial city" that had been their goal for almost three-quarters of a century by then.[40] Just like

Ráth and Magyar Tudományos Akadémia, 1876), 73–77. On the diet itself, see ibid., 65–66 and Géza Pálffy, "Pápa szerepe a XVI. századi végvárrendszerben" [Role of Pápa in the 16th-century defense system], in *Tanulmányok Pápa város történetéből II.* [Studies in the history of the town of Pápa, II], ed. István Hermann (Pápa: Pápa Város Önkormányzata, 1996), 82.

37 *Az Első Katonai Felmérés 1763–1785* [First Military survey, 1763–1785] [DVD-Rom] (Budapest: Arcanum Adatbázis Kft., 2006). Available online: http://mapire.eu/hu/map/firstsurvey/ (last accessed: 28 April 2022).

38 On this, see Lajos Takács, *Határjelek, határjárás a feudális kor végén Magyarországon* [Boundary marks and perambulation at the end of the feudal age in Hungary] (Budapest: Akadémiai Kiadó, 1987), 51–53.

39 Pálffy, *Pápa a hosszú*, 20–22, and Baráth, "Oszmán."

40 Pálffy, *Pápa a hosszú* and moreover Pálffy, *A császárváros.*

the 1543–1544 survey, the second survey has seldom been referred to by historians and has only fairly recently been published in its full length. The greatest value of this document is that it provides at least some hints about the extent to which the protective measures ordered in 1543 and 1544 became effective by the end of the century. The survey, which is indeed a proposal for defensive measures, consists of two parts. The first is a relatively short, mostly Hungarian-language summary of why the Rába is of key importance in defending the remaining parts of the Kingdom of Hungary. The second half of the document is a Latin-language list of the fords and bridges from Kapuvár to Szentgotthárd, followed by a list of the fortifications near the Rába (*Denominationes Vadorum et Pontium ab Arce Kapu, quae Germanicae Felbach appellatur usque ad S. Gothardum penes decursum fluvii Raba ex interiori parte fluvii*). As indicated by the title, defensive structures of the time were only listed from the left bank, which indicated the role attributed to the right bank settlements by the river.

This defense proposal was not the only one carried out around the end of the sixteenth century. Because of the rapidly changing political circumstances of the Fifteen Years' War, many similar proposals were written by members of the aristocracy in Transdanubia, especially following the above-discussed fall of Kanizsa, one of the most important strongholds in the southern part of the region, in 1600. After the fall of this fortification, the Rába valley's strategic importance became even more evident, and therefore most of these proposals treated the areas on the right bank of the Rába as potentially belonging to the Ottoman Empire in the long run. This is clear from another proposal by the same Ferenc Nádasdy, written in 1600:

> The other presidiaries [towns with garrisons], like those at Körmend and Sárvár, should remain where they are to defend the inner part [left bank] of the River Rába from the plunderers and Turkish soldiers *because if the other bank* [right bank] *of the Rába is under Ottoman rule, which is hard to avoid* as Kanizsa, Kiskomár and Somogy are not protected from the inner garrisons, they [the Ottomans] will plunder the inner side of the Rába as well. Also, for those at the border forts, the outer banks give protection, and the inner presidiaries can prove to be useful against great plundering, and the Turks may feel more threat from the enemy facing them.[41]

41 *Az teóbj az praesidiariusoknak, mind Keörmendinek, mind Saruarjnak ugian azon helekben köllene lennik, hogj az belsö feltis az Raban innet öriznek, mind rablotul, mind Chiata Theöreöktwll,* merth ha az Raba tulso fell holdult lezen, kit nehez megh otalmazni attul, *Kanisaj, Kiskomarj es Somogj Theöreök ellen, köuetkeznek, az raba innenczeö feltis el puztiltana az Theörök, ha nem*

The importance of the 1594 proposal should not be exaggerated, but under the new political circumstances this was one of the first proposals that faced the fact that the Ottomans were a constant threat to the settlements on the left bank of the Rába. As the document shows, the concept intended for presentation to Emperor Rudolf II (1576–1612) was worked out together by György Zrínyi (1549–1603) and the chief captain of Transdanubia, Ferenc Nádasdy, who followed Zrínyi in this position. Unlike the plan written in 1543–1544, as Géza Pálffy has demonstrated, this defense proposal served as the general guideline when the Habsburgs had to reorganize the defense of Transdanubia after the fall of Győr. To replace the captainship of Győr, the Magyaróvár-Sárvár district defense unit was established based on this document.[42] The two documents together are essential for identifying how the Hungarians, and indirectly the Habsburgs, thought of the use of a river – in this case, the Rába – and its environment for defending the remaining part of the Kingdom of Hungary. These sources became even more valuable considering that this new system of defense, which was heavily built on the river as an obstacle (as I shall argue in Chapter 3) had a lasting impact on the environment around the river.

2.1.2 Riverine Frontiers Against the Ottomans – The Forms of Defense

Until modern times, one of the easiest ways to hamper military campaigns was to flood extant areas. As discussed regarding the *gyepű*, this was known to Hungarians from their settling in the Carpathian Basin (in the early tenth century) onwards. Although references are rare in the existing scholarly literature, the artificial flooding of communication routes mentioned above was practiced in the early modern period.[43] To give but one example, János

örzenek az belseö praesidiomokbul, azonkiueöl az kwlseö felnekis az Veghbelieknek oltalomra ualo segetsegre, derek rablasok ellen haznossan erkezhetnenek az belseö praesidiariusok, az Theöreökis inkab tartana az elötte ualo ellensegtöl. – ÖSta KA HKR, Registratur, 1601. März. No. 179. fol. 13. The plan, along with two other ones, were published by Éva Simon, "Magyar nagybirtokosok tervezetei a Kanizsával szembeni végvidék kiépítéséről" [The plans of the Hungarian aristocracy on the construction of the border defense line against Kanizsa], in *Zalai történeti tanulmányok* [Studies in the history of Zala County] (Zalai gyűjtemény, 42), ed. Csaba Káli (Zalaegerszeg: Zala Megyei Levéltár, 1997), 61–86, the quoted part ibid., 69 (the source here is quoted in the transcription of Éva Simon; emphasis added).

42 Pálffy, *A császárváros.*

43 The artificial flooding of the valleys in Zala County appeared in the defense plans in the noble gatherings of the county already in the 1570s. On this, see: Simon, "Magyar nagybirtokosok," Kelenik, "A kanizsai övezet," 74–75, and László Vándor, "Kanizsa története a honfoglalástól a város

Keczer, captain of the fortification of Csákány (discussed in detail in Chapter 3) along the Rába, suggested that the landlord of the settlement should reconstruct one of his abandoned fish ponds because on several occasions the Ottomans had attacked Csákány and Körmend (another town to be discussed below) through the basin of this dried out pond.[44] Backing up rivers and flooding different areas, especially those which could potentially serve as Ottoman military routes, was not exceptional. Most frequently, small streams were diverted and dams were built, which usually served more than one purpose. The stream diversions and dams provided a more consistent water level and more effective power for mills built beside these dams. The use of this form of riverine frontier defense along the Rába and smaller streams in Transdanubia was self-evident at the time, as many water mills had been in operation in this area from the late twelfth and early thirteenth centuries onwards. In the Rába valley, maintaining a stable, relatively high water level allowed easy crossing at only a limited number of fords and bridges. These places were usually associated with minor earth and wood fortifications or at least some kind of watchtower (referred to as *góré* in the Ottoman period).[45]

Because of its discharge, the Rába River was instrumental in defending the hinterland only if its water was backed up to high levels. The Habsburg defense forces recognized this, as is reflected in the above-presented sources from the 1540s onwards. It is even more evident that the Rábaköz region (between the former Répce and Rába rivers; see Fig. 2.2) was difficult to defend in any other way, as is clear from the exchange of the above letters

török alóli jelszabadulásáig" [The history of Kanizsa from the Hungarian conquest to its liberation from the Ottoman occupation], in *Nagykanizsa. Városi monográfia*, vol. I. [Nagykanizsa. Town monograph, I], ed. József Béli (Nagykanizsa: Nagykanizsa Megyei Jogú Város Önkormányzata, 1994), 288–290.

44 The letter of János Keczer to Ádám Batthyány, MNL OL P 1314 no. 24 516 (24 July 1651).

45 On the different forms of riverine frontier protection by the Rába, see Zsolt Gellén, "A Rába védelmében" [In defense of the Rába], in *Táj és történelem. Tanulmányok a történeti ökológia világából* [Landscape and history. Studies in historical ecology], ed. Ágnes R. Várkonyi (Budapest: Osiris, 2000), 232–254, Benczik, "Fentő és góré," and Zsolt Baráth, "A Rába mint védelmi vonal a 17. században. – Védelmi munkálatok és létesítmények a folyó Vas megyei szakaszán" [Rába as a defense line – protection measures and constructions in the Vas County-section of the river], *Vasi Honismereti és Helytörténeti Közlemények* no. 1 (2014): 31–53, and idem, "A Rába védelmi vonal és Szentgotthárd térsége a 17. század derekán" [The Rába defense line and the surroundings of Szentgotthárd in the mid-17th century], in *1664 – A szentgotthárdi csata* [1664 – the battle of Szentgotthárd] (Szombathely: Szülőföld Kiadó, 2015), 257–295. On the social and administrative organization behind the defense strategy, see Péter Dominkovits, "Folyóvizek és a XVII. századi vármegyei közigazgatás, bíráskodás" [Rivers and 17th century county administration and jurisdiction], in *Víz és társadalom Magyarországon*, 155–188.

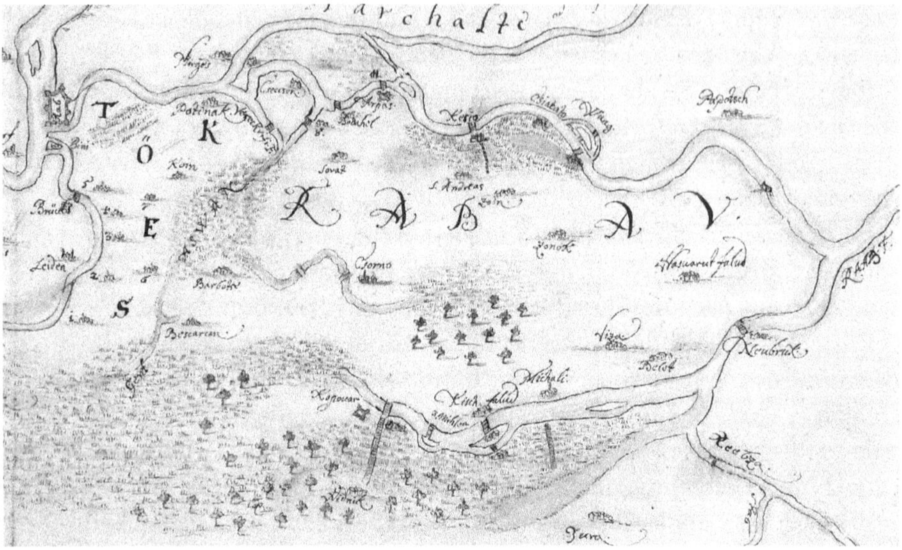

Figure 2.2 The Rábaköz on the detail of a manuscript map from 1673[46]

between Pál Várday and Niklas Graf zu Salm from 1546. Similar strategic thinking crystallized in some of the letters by Tamás Nádasdy. According to one from 1558, Nádasdy and the other aristocrats in the region planned to back up the Rába on a 30-kilometer stretch between Bodonhely and Győr to defend the left bank, namely, the Rábaköz region, against growing Ottoman pressure.[47] The desire to build the defense on the river as a natural obstacle came up from time to time in either aristocratic circles or at the noble gatherings in the counties by the river. The nobles of Sopron County, for instance, decreed in 1579 that the fords of the river in the Rábaköz region had to be put under the supervision of appointed nobles, who were to be in charge of backing up the waters if necessary.[48] In 1643, the problem of

46 *Plan über die Schüt und Raaben* (1673). HIM B IX a 674/6. For the area and the transformation of the environmental conditions, see Takács, "Árpádkori csatornarendszerek," idem, "Néhány észrevétel." See also Imre Göcsei, *Kapuvári-Rábaközföldrajza* [Geography of the Rábaköz area around Kapuvár] (Értekezések a Magyar Királyi Horthy Miklós Tudományegyetem Földrajzi Intézetéből, 5 [35]) (Szeged: [N.p.], 1943).

47 *Comiserat Sacra Caesarea Maiestas nobis clementer, ut cum architecto consilium inirem, quo in loco fluvius Raba vocata* [sic!] *possit commodius intumesci, per quam intumefactionem aqua versus Bodonhel et ultra distaret, ne quispiam transire possit* – MNL OL E 185 Missiles, letter of Adam Gall to Tamás Nádasdy, 31 October 1558. The source is quoted in Pálffy, *A császárváros*, 123.

48 András Komáromy, "Sopron vármegye végzései a malmokról és a Rábaköz védelmezéséről" [The decrees of Sopron County on the mills and the defense of the Rábaköz], *Történelmi Tár* [NS] 9 (1908): 61–63 and Péter Tóth, *Sopron vármegye közgyűlési jegyzőkönyveinek regesztái*, I.

the damming of Rába arose again in the context of defending the Rábaköz. This time it was the palatine himself, Miklós Esterházy (1582–1645), acting also as the *comes* of Sopron County, who ordered a review of the different fords and dams along the Rába in the Rábaköz.

It was not only the river that served as an important element in the defense system but also the nearby forests. Various ways of protection relied on the presence of trees along the watercourse.[49] The first two types of defensive measures discussed here were known as *bevágás* (literally meaning 'cutting') and *fentőzés* (meaning some sort of pole) in Hungarian, followed by another form of defense, the construction of watchtowers (*góré*s).

Bevágás is a method of using felled trees at places along a river where the water current and the riverbed otherwise would allow an easy crossing. The 1543–1544 Rába survey, the minutes of the noble gatherings of Vas and Sopron Counties as well as several letters mention this form of border protection along the Rába.[50] In 1619, the land steward of the village of Csákány, Benedek Károl, informed his landlord Ádám Batthyány (to be discussed in detail below in Chapter 3) that it was impossible to transport promised wares because the "river had been cut" earlier. The Rába was not only unsuitable for crossing, the bridge by the settlement had also been destroyed as a result of this work.[51] A few decades later, in 1655, Bernát Csány, the captain of the same fortress Csákány, referred in a letter to felling trees that bordered the Rába into the water to hinder easy crossing by Ottoman troops. He also noted the need for the river to be looked after more regularly.[52] The problem of insufficient care given to guarding the frontier keeps reoccurring in the sources, most importantly in private letters sent by local administrators and military personnel. Despite the constant presence of military forces in the region, it was the local tenant peasant (*iobagio* or *jobbágy*[53]) that were driven to guard the river, and it was they who had to fell the trees to

1579–1589 [Regestas of the noble gathering of Sopron County, I, 1579–1589] (Sopron: Sopron Város Levéltára, 1994), 24 (no. 58).

49 On the defense against the Ottomans and its connections to the forests, see: Benczik, "Fentő és góré," 42–48. On the problem of the different forms of protection with the example of the Rábaköz region, see: ÖSta HHSta Csáky Fasc. 103. no. 11 and Fasc. 104. No. 86 and 98. For their edition, see Gecsényi, "Elképzelések a Rábaköz."

50 Such as the letter of Benedek Károl to Ferenc Batthyány, MNL OL P 1314 no. 24 059 (18 May 1619) and the letters of Bernát Csány to Ádám Batthyány, MNL OL P 1314 no. 8799 (9 July 1655) and No 8802 (2 August 1655). See also: Benczik, "Fentő és góré," 43.

51 The letter of Benedek Károl to Ferenc Batthyány, MNL OL P 1314 no. 24 059 (18 May 1619).

52 The letter of Bernát Csány to Ádám Batthyány, MNL OL P 1314 no. 8799 (9 June 1655).

53 For the term in a Hungarian context, see János M. Bak, "Servitude in the Medieval Kingdom of Hungary," in *Forms of Servitude in Northern and Central Europe Decline, Resistance, and Expansion*

protect the hinterland of the fortress.[54] Despite all the efforts, however, when water levels were particularly low, it was impossible to protect the banks only by guarding them and destroying some fords. At times, the Rába had dozens of fords, as is reflected in some of the letters as well as in the survey of 1543–1544.[55]

It may have been crucial in terms of border protection to keep the river mostly impassable, but it was certainly damaging to the local economy. Many of the settlements on the left bank of the Rába had holdings on the right bank – meadows, plowlands, and forests (including some of the richest gallery forests along the riverbank) –that they wanted to use.[56] In some cases, the local tenant peasants, having had enough of the loss of these resources, began to take the felled trees out of the rivers so that they could reach the other bank. In other instances, the same bondsman who had been forced to fell trees into the river was the same person who later pulled them out.[57] This happened in 1639 on the domain of another prominent Hungarian noble family, the Csákys: "People living by the Rába had cut the river ... Some, like the peasants of Csáky, went into the river and pulled out the trees, and they are now crossing the river."[58] Similar problems occurred in the neighboring Sopron County:

> At the Rába from Bodonhely to the castle of Kesző many fords were created by vagrants and some disobedient people, by activities which had already been banned early on and which represent great threats to the Rábaköz. Therefore, the noble county must decree ... that nobles, as well as peasants, should have the freedom to catch the people trying to cross at banned fords.[59]

This quotation comes from a letter sent by Péter Káldy, a confidant of Ádám Batthyány and later chief commander of Batthyány's personal army, supposedly to inform his landlord of the growing threat. He wrote this letter

(Medieval Texts and Cultures of Northern Europe, 9), eds. Paul Freedman and Monique Bourin (Turnhout: Brepols, 2005), 387–400 and Engel, *Realm of St. Stephen*, passim.

54 On the protective measures taken to keep the Ottoman plundering away from the Rábaköz, see: Gecsényi, "Elképzelések a Rábaköz."

55 The letter of István Potyondi to Ádám Batthyány, MNL OL P 1314 no. 38 296 (n.d.). For the forms of border protection, see Gellén, "A Rába védelmében," 237–239.

56 The letter of Bernát Csány to Ádám Batthyány, MNL OL P 1314 no. 8802 (2 August 1655).

57 The letter of Bernát Csány to Ádám Batthyány, MNL OL P 1314 no. 8803 (4 August 1655).

58 The letter of Péter Káldy to Ádám Batthyány, MNL OL P1314 no. 23 211 (6 December 1639).

59 Komáromy, "Sopron vármegye," 66.

to him because by this time Batthyány was one of the leading figures in the military hierarchy of the Kingdom of Hungary, and Káldy, a prominent member of the administration of Vas County, must have hoped that Batthyány would react by ordering the necessary works to be carried out again. The problem of pulling previously cut trees out of the river at Csörötnek (a village upstream from Csákány) was also a reoccurring subject of debates in the noble gatherings of Vas County.[60] River cuttings had their annual rhythm. As with much water-related work, they were usually carried out just after the end of the spring floods, as before that, the usually high water levels and the icing of the river made Ottoman raids unlikely.

The other widespread form of protection against raids was the so-called *fentő*. There is evidence for the presence of this form of a defensive structure at several settlements along the river Rába, including Körmend[61] and Csákány,[62] which will be discussed in more depth in the next chapter. *Fentő* was similar to *bevágás* to the extent that one of its elements also comprised felling trees into the riverbed. Nevertheless, a significant difference between the two defensive methods was that trees were felled on the side of the enemy, not on the side meant to be protected. This method in the case of the Rába meant that piles of trees called *fentő* were deposited on the left bank. Branches were then placed among the piles, and the gaps were filled with mud. When finished, these constructions looked somewhat similar to walls of earth and wood fortifications (discussed in Chapter 4 in detail) and created considerable obstacles to crossing the river, but their construction required significantly more effort than *bevágás* and was therefore never used on long sections of the riverbank. In the minutes of the noble gatherings of Vas County, *fentő*s are mentioned on several occasions, and based on one of the entries it is likely that this method of border protection originated from well before the period of the Ottoman wars.[63]

*Góré*s were to some extent less direct means of border protection. They were erected along the relatively close frontiers – in some cases in line of sight of each another – to ease communication and control. Some were

60 The letter of László Csáky to Ádám Batthyány, MNL OL P 1314 no. 8448 (17 August 1643).

61 The letter of Gáspár Francsics to Ádám Batthyány, MNL OL P 1314 no. 15 159 (28 March 1647). See, furthermore, other letters of Francsics: MNL OL 1314 no. 15 301. (10 March 1651) and 15 303 (20 March 1651).

62 The letter of Benedek Károl to Ádám Batthyány, MNL OL P 1314 no. 24 142. (21 November 1624) and Bernát Csány to Ádám Batthyány, MNL OL P 1314 no. 8803 (4 August 1655). See also: Benczik, "Fentő és góré," 43.

63 See: Tóth, *Vas vármegye*, vol. II, 136 (no. 1362), and 171 (no. 1508).

built on low elevations near the Rába, and some were hidden in wooded areas.[64] No systematic analysis has been carried out on the number of *górés*, but to indicate how many such watchtowers may have existed along the Ottoman border, even one settlement can serve as an example. At least four *górés* stood along the Rába in a few kilometer-long section in the boundaries of the above-mentioned small settlement of Csákány in the seventeenth century, which indicates that they may have been built in high numbers during the Ottomans' presence. The maintenance and construction of these wooden buildings were the duties of the local tenant peasants and smallholders. Because of the lack of their central administration, the maintenance of the *górés* was frequently halted, and many of them fell into disrepair in the periods and regions with less frequent plunders. The *góré* in itself was not, of course, enough to defend the hinterland but was important in providing warnings of Ottoman plunders.

In wintertime, the forms of border protection at the rivers were fundamentally different from the types discussed above. When the water froze, artificial flooding and the felling of trees into the running water was impossible. The problem of ice bridges in wartime defense appears in sources from the Late Middle Ages onwards when, as noted, Ottomans had already led raids into the southern areas of the Kingdom of Hungary. In 1431, the captain of Belgrade, Frank Thallóczy, warned the *comes* of Keve County to watch out for the freezing of the Danube because of the Ottoman threat. Similarly, on 20 December 1486, Mátyás Várdai, bishop of Bosnia, wrote a letter to his brother, Aladár, in which he invited him to Đakovo (in Croatia) for Christmas. In the kind brotherly invitation, he also noted that he would not be able to go and visit his brother, as the only thing that protected them from the Ottomans was the Sava River, and if it froze over it could threaten the whole town.[65] The Rába, a less significant river than either the Danube or the Sava, had lower water level, and when it froze, it provided hundreds of crossing points.

64 On the *górés* around Győr and the question of supplying them, see Pálffy, *A császárváros védelmében*, 170–172.

65 *Constat enim vobis nos in confinibus Turcorum manere, nuncque nos preter Zawam nihil tutatur, que si congelabitur nos hinc discedere non possumus. Sollicitudines igitur nostras et sollicitudines vos videre cupimus et nihilominus vos videre et vobiscum aliqua peragere. Secus igitur non faciatis. Rogamus ... antequam veniatis, nuncietis* – MNL OL DL 81 962, edited in: A zichy és vásonkeői gróf Zichy-család idősb ágának okmánytára [Cartulary of the ancient branch of the Zichy family of Zich and Vásonkeő], 12 vols., eds. Imre Nagy, Iván Nagy, and Dezső Véghely (Pest: Magyar Történelmi Társulat, 1871–1931), vol. XII, 322–323 (no. 258).

There is limited information on the winter defense strategies along the Rába, but somewhat more documentary evidence testifies to the problem at other rivers in the Carpathian Basin. The sources suggest, for instance, that in 1565 during the siege of Tokaj, which was under the control of the prince of Transylvania (John Sigismund) at the time, holes were cut in the ice covering the Bodrog and Tisza rivers – at the confluence of which the fortification stood – so that soldiers of the attacking Habsburg army would fall through the ice.[66] In November 1572, György Zrínyi mentions in a letter that his tenant peasants were trying to break the ice of the Dráva River at Legrad. Their efforts proved unsuccessful, as the water froze over again within a few days.[67] It seems that ice-breaking was not an unusual practice. A statute from Győr County obliged the tenant peasants and the landlords of the Szigetköz region to take part in the defense by guarding the river during the night and by breaking the ice during daylight hours when the river was frozen.[68] Accordingly, at the beginning of 1594, the people of the Tóköz region (a part of the above-mentioned Rábaköz area) were ordered to break the ice and cut holes through the ice. As noted in the case of the siege of Tokaj, when rivers were about to freeze over, it was usually the local peasantry that was called to cut holes in the ice. A similar case is mentioned in a letter by Ferenc Káldy. Káldy informs Ádám Batthyány that Vas County had ordered the local tenant peasantry of Oszkó to cut holes in the ice of the Rába. Oddly, the village of Oszkó lies on the right bank of the Rába, that is, their work did not serve any apparent purpose for defending the safety of the village.[69]

The landlords in the Rába region took all the measures they could to obstruct the tenant peasants from crossing the river since this could help Ottoman plunderers identify where the fords lay.[70] When the river froze in

66 Miklós Istvánffy, *Magyarok dolgairól írt históriája Tállyai Pál XVII. századi fordításában*, vol. I/2 [The history of Miklós Istvánffy in the 17th-century translation of Pál Tállyai], ed. Péter Benits (Budapest: Balassi Kiadó, 2009), 368–371. See also: Elemér Soós, "A tokaji vár története" [The history of the castle of Tokaj], *Hadtörténelmi Közlemények* 14 (1913): 76.

67 Sándor Takáts, *Emlékezzünk eleinkről*, 2 vols. [Let's remember the ancestors of the Hungarian] ([Budapest]: Genius, [1929]), vol. I, 257.

68 Károly Ráth, "A Győr vármegyei hódoltságról" [On the Ottoman-period in Győr], *Magyar Történelmi Tár* 5, no. 1 (1860): 8, 57, and 66.

69 The letter of Ferenc Káldy to Ádám Batthyány, MNL OL P 1314 no. 23 563 (25 October 1658).

70 The letter of István Keserű to Ádám Batthyány, MNL OL P 1314 no. 26 424 (24 December 1614). See also: Péter Tóth, *Vas vármegye közgyűlési jegyzőkönyveinek regesztái*, vol. II. 1601–1620, 1631–1641 [Regestas of the noble gathering of Vas County, II, 1601–1620 and 1631–1641] (Vas megyei levéltári füzetek, 5) (Szombathely: Vas Megyei Levéltár, 1992), 16 (no. 792), 20 (no. 816), and 182 (no. 1554). For other cases, see: Benczik, "Fentő és góré," 44.

winters – which was frequently the case in the early modern period – the same problem applied.[71] By crossing the river on its ice, peasants demonstrated to the Ottomans that the ice was thick enough to walk on. Landlords who had properties in the region – amongst others the Nádasdy family – tried to prevent local people from crossing the river on ice bridges. The prohibition decreed by the Nádasdys was a subject of debate at several noble gatherings both in Sopron and Vas Counties, as the local tenant peasants rebelled against the will of their landlord. After a decade-long lawsuit, a joint gathering of the nobility of the two counties gave the tenant peasantry the right to use ice bridges on the rivers and stipulated that the goods confiscated by the Nádasdy family as a penalty should be given back.[72]

In the previous paragraphs, I have tried to demonstrate that the effective protection of the Kingdom of Hungary was to a large extent organized along the use of natural features, the most important of which were rivers in the central part of the Carpathian Basin. From the Middle Ages onwards, in different forms, the Hungarians systematically manipulated water bodies to protect the country. To establish the most effective form of border protection along a certain river, it was necessary to gather information on the environment. In the following section, I intend to show how landscape elements were tackled by contemporaries when they were surveying the Rába.

2.1.3 Environmental Conditions and Frontier Protection in the Rába Valley in the Early Modern Period

Instead of a detailed presentation of the river surveys one by one listing the settlements and the built environment referred to in the documents,[73] I will give a summary of the environmental features that were considered by the surveyors and the possible roles that were assigned to the built environment and the landscape. There are four important characteristics that were quite consistently noted by the surveyors in 1543–44: the vegetation; the different

71 Cf. András Vadas, "A Dunára én bizon nem megyek, mert még nem akarok meghalnom" – a Duna jégjelenségei a kora újkorban (1530–1650)" [Ice regime of the Danube in the Early modern period (1530–1650], in *Micae mediaevales III. Fiatal történészek dolgozatai a középkori Magyarországról és Európáról.* (Studies in medieval history from Hungary and Europe, III), eds. Judit Gál et al. (Budapest: ELTE BTK Történettudományok Doktori Iskola, 2013), 219–235.

72 On this, see Éva Turbuly, *Sopron vármegye közgyűlési jegyzőkönyveinek regesztái*, II. 1595–1608 [Regestas of the noble gathering of Sopron County, II, 1595–1608] (Sopron: Győr-Moson-Sopron Megyei Levéltár Soproni Levéltára, 2002), 133 (no. 573) and Tóth, *Vas megye*, vol. II, 65 (no. 1056).

73 This is done in: Vadas, "A Rába-mente környezeti viszonyai."

tributaries and bifurcating branches of the river, the dams, sluices, and other man-made interventions to control the river's flow; and finally the crossing points, either the fords or the bridges that were important from the point of view of military defense.

Trees were vital for many reasons. As discussed above, most of the means of frontier defense by rivers in the Kingdom of Hungary used the surrounding vegetation, either directly or indirectly. This explains why in the 1543–44 survey, systematic references were made to the type of vegetation on the riverbanks. The first way that trees could help a military defense along rivers was by felling them directly into the river along the bank. The survey mentions this kind of intervention having been made at the village of Vica, for instance.[74] The water level of the river was raised by felling trees into the Rába – a clear reference to *bevágás* – to aggravate difficulties in crossing. Similarly, the survey kept referring to the distance of the trees from the banks.

The distance from the river to the wooded areas was viewed differently on the two banks of the river: while on the right bank, the forests had to be as close to the river as possible, it was probably somewhat different in the case of the right bank. While the Ottomans never had permanent control over the right bank of the river and they never used the trees by the Rába River as the Hungarians did, the trees were important in the case of their raids as well. It was crucial for them to find a way to remain hidden for the longest time possible, so the presence of forests along the river close to fords played a significant role. This is probably why at one point in the survey done in 1543–44, the trees on the right bank were referred to as being at a 'satisfactory distance' from the banks (*in bona distancia*), probably meaning far enough that the Ottomans could neither hide before raiding nor monitor the crossing and the presence of guards on the left bank of the river. The survey clearly noted that, because of the necessity of trees in the direct defense strategies, it was important to hide small watchtowers wherever the vegetation was treeless.

The different tributaries, artificial channels, and bifurcating branches of the Rába were also used in the defense strategies in the sixteenth and seventeenth centuries. The example below may clarify the role of the

74 There is a very early reference, dating to 1162, of the mills at Beled and upstream at Vica: *et in villa Vidta* [Vica] *similiter, in villa etiam Velene* [Beled] *unum pratum cum duobus molendinis* – MNL OL DF 201 635. See Károly Mollay, "Pájer Imre: Rábaköz népének védekezése az áradások ellen 1870–1889. Csorna, 1990," [Book review] *Soproni Szemle* 46 (1992): 383. On the mill at Vica, see also: MNL OL DL 90 957.

hydrography of the Rába in the defense of the hinterland. Upstream of the water mill of Beled, the next reference point marked by the 1543–44 survey was important for military defense. This point is a bifurcation of the Rába called Ásvány-Rába (*Rabam Asvan appellatam*, literally meaning 'dug Rába'). This branch of the river existed until the eighteenth century and flowed in the direction of Vitnyéd, hence on maps in the eighteenth century it appears mostly as the Vitnyéd branch of the Rába (*Ramus Arrabonis Wittnydiensis*).[75] There is no clear evidence why this branch of the river was dug, but by artificially or semi-artificially diverting the river to this branch, making it flow towards the extensive marshes of the Hanság, a major area could have been protected from Ottoman raids.[76] The Hanság, the water system that was connected by this channel to the Rába water system, not only formed a huge obstacle to the Ottomans, but in periods of extensive rainfall, it made communication between Sopron and Vienna with Győr, for instance, challenging. It is very likely that this artificial branch was sluiced in some cases to control water levels in large areas between the Ottomans and the hinterland of the Kingdom of Hungary.

This would not have been an exception, as other sluices and dams were also built to control the water levels around Kapuvár and Babót, the first settlement upstream from the mouth of the smaller branch of the Rába in the Hanság marshes. In this area, there are altogether three entities called *sarampos* in the earlier 1543–44 survey. The meaning of this Hungarian vernacular term is not evident, but it may be connected to some defensive structure. In some sources, it refers to bridges that could be drawn ('sorompó' in Hungarian), but in some sources, it appears as something like a sluice gate. It is unlikely that there were three bridges within such a short distance; neither the settlements nor the road network makes it likely. There was a bridge by the castle of Kapu going back to the Middle Ages, and this bridge was in use at the time of the 1594 survey as well. What the other two *sarampos* may have been for is not clear, but they were likely to have been involved in the regulation of the water level.[77] Not only sluices but

75 On this, see Takács, "Néhány észrevétel."

76 Similar geographic names can be attested in the Tóköz region as well: Károly Takács, "Medieval Hydraulic Systems in Hungary: Written Sources, Archaeology and Interpretation," in *People and Nature in Historical Perspective* (CEU Medievalia, 5), eds. József Laszlovszky and Péter Szabó (Budapest: CEU Press and Archaeolingua, 2003), 289–312.

77 The bridge here referred to as *Vntatónak* in a document from 1350: *ponte Vntato dicto supra fluvium Raba nuncupatum existente* – MNL OL DL 4146 (8 September 1350). See: Imre Nagy, *Sopron vármegye története. Oklevéltár.* I.: 1156–1411 [History of Sopron County. Cartulary, I, 1156–1411] (Sopron: Litfass Károly Könyvnyomdája, 1889), 212–213 (no. 167). A seventeenth-century terrier

also dams – probably not that different from the *bevágás* – were built most probably for water control purposes. The yet to be identified place called *Újgát* (meaning 'new dam') in the earlier river survey was probably not a village or a farmstead but rather a dam built to control the water level at a section of the Rába. It was a construction that probably was not connected to a mill, serving solely military defense purposes and not used by the local economies.

While fords and bridges made it easier for the Ottomans to raid, they were necessary for by the local economies and for communication. Situated all along the Rába, these were the most frequently mentioned as points that had to be defended, so both surveys discussed above systematically refer to their locations and how they were or should have been protected. The surveys not only note the more dangerous fords but also make it clear which had been used by the Ottomans.[78] For the defense of the hinterland, many of the bridges that had been used in the Middle Ages were not in use anymore. The one at Beled must have functioned as a bridge from the medieval period onwards, as a toll was mentioned there in the fourteenth century.[79] The bridge was not open here, neither in 1543 nor in 1594. There is no indication about the reason, but it is very likely to be related to the defense of the left bank. From the early seventeenth century, however, as attested by the minutes of the noble gathering of Sopron County, the bridge was in use again, which suggests that it had not been fully destroyed.[80] The bridge survived the seventeenth-century Ottoman wars and was also represented on one of the early detailed maps of the region, drawn by Andreas Ericus Frics in 1761.[81] There are bridges marked as functioning but were marked for demolition by the surveyors for defensive purposes in both sixteenth-century surveys.

The river surveys of 1543–44 and 1594 highlight the main features of the environment of the Rába valley in the sixteenth century (see Fig. 2.3a-b). The first, much more detailed survey describes the environment similar to a

also refers to an opening bridge: *Az Raba hidgian valo emelchos Kapu* ("bridge with a lever over Raba"). MNL OL E 156 Fasc. 12. no. 42. 4. füz. 363. p. (terrier of Kapuvár from 1660). Several similar geographical names are known from the period: *Győr-Moson-Sopron megye földrajzi nevei 1. A kapuvári járás* [Place names in Győr-Moson-Sopron County, I. The surroundings of Kapuvár], eds. Lajos Balogh and Ferenc Ördög (Győr: Apáczai Csere János Tanítóképző Főiskola, 1998), 37.

78 A charter of donation already mentioned a mill place in Rum in the thirteenth century: MNL OL DL 49 548. Edited in: Gusztáv Wenzel, *Árpádkori új okmánytár / Codex Diplomaticus Arpadianus continuatus*, 12 vols. (Pest: Eggenberger, 1860–1874), vol. IX, 60–61 (no. 36).

79 János Belitzky, *Sopron vármegye története*, vol. I [The history of Sopron County, I] (Budapest: Stephaneum, 1938), 801 and 974.

80 Turbuly, *Sopron vármegye közgyűlési*, vol. II, 75 (no. 262) (17 June 1601).

81 MNL OL S 12 Div XI No 30:3-6. See above note: xxxiii.

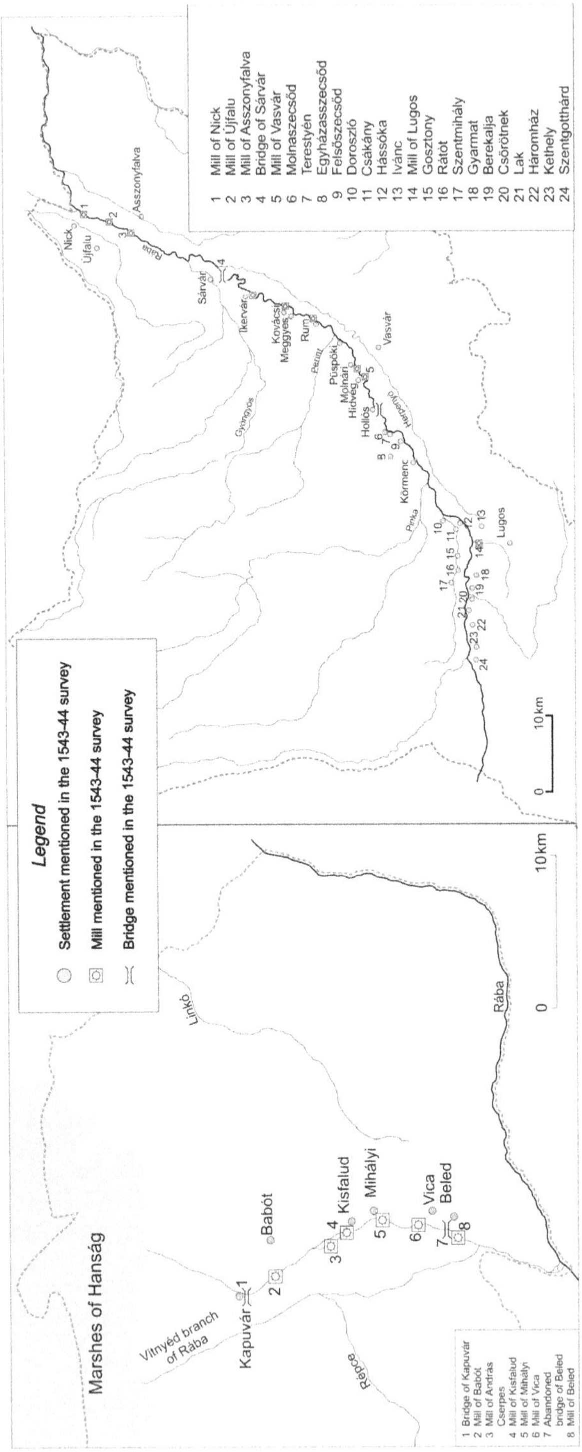

Figure 2.3a-b The settlements, dams, fords, and buildings referred to in the 1543–1544 survey of the Rába River

detailed travel account or map. This is exceptional, as most of the early travel accounts from Hungary describe a smaller area, and the descriptions of the vegetation and water-related constructions are less systematic. Thus, scholars using such sources are forced to reconstruct historical vegetation based on a few, sometimes vague references.[82] The river surveys provide important data for military historians, but that aspect has partly been covered already by Géza Pálffy. These sources are the best Ottoman-period descriptions of the natural vegetation and the built environment in the frontier zone. The system sketched out by the survey of 1543–1544 was certainly key to organizing the protection of the left bank of the Rába. It makes sense why more and more gallery forests were marked as 'forbidden' (*tilalmas* or *prohibita*) and their use restricted to all but the landlords.[83] The clearance of these forests raised the threat of Ottoman plundering considerably.

2.2 Summary

It is clear from the sources discussed above that rivers in pre-modern times played major roles in military defense strategies. No less importantly, the defense systems of the Kingdom of Hungary had long-term impacts on local economies and environments. The different interventions in the flow of the river may have been instrumental in protecting the remaining part of the Kingdom of Hungary, but they also resulted in considerable environmental change in the valleys. The next chapter discusses what kind of long-term impacts these forms of border protection might have had. I discuss the problems of local economies in the period when the Ottoman presence in the region was an everyday experience, that is, the period from the turn of the sixteenth century. The case study area selected is the town of Körmend and its surroundings, including the villages belonging to the same manorial complex. When discussing the local economies, the primary focus will be on understanding the impact of the manipulated riverine environments on farming and how exactly these impacts can be related to the regional defense strategy.

82 Márta Tóber, "Mennyiben tükrözi Bertrandon de la Brocquière útleírása a középkori Homokhátság természeti viszonyait?" [How much the travel account of Bertrandon de la Brocquière reflects the natural conditions of the medieval Homokhátság?], in *A táj változásai a Kárpát-medencében: történelmi emlékek a tájban* [Landscape changes in the Carpathian Basin: historical monuments in the landscape], ed. György Füleky (Gödöllő: Környezetkímélő Agrokémiáért Alapítvány, 2013), 309–314.

83 On forbidden forests, see Szabó, *Woodland and Forests*, 60–62.

3 A Century of Water?

The Rába Valley in the Seventeenth Century

Abstract
Using the example of the town of Körmend and its manorial complex as well as another settlement (Csákány) nearby, this chapter argues that the defense strategy of the Kingdom of Hungary in the first half of the seventeenth century led to a significant transformation of the local environment. Heavily relying on natural obstacles such as the Rába River, the Hungarian defense tactics altered the flood regime of the river, causing serious economic difficulties in the entire valley. These economic difficulties can with a high probability be attributed to the defensive role the river had in the half a century following the Fifteen Years' War.

Keywords: Environmental history, Kingdom of Hungary, Ottoman Empire, frontier, floods, water history

As discussed in Chapter 2, from the middle of the sixteenth century, the Habsburgs as well as members of the Hungarian aristocracy thought of the Rába River and other minor waterways in the central part of Transdanubia as important chains in the frontier between the Ottoman and the Habsburg empires (including the Kingdom of Hungary). Because of this rather special role of the Rába in military defense, other factors besides economic benefits were also taken into consideration when making decisions about farming in the sixteenth and seventeenth centuries. In this chapter, using examples of individual settlements, I will examine the conflicts that were caused by the transformation of the roles of the river Rába and its region. Using the example of the town of Körmend and its manorial complex plus another settlement close by, Csákány, in the first half of the seventeenth century (up to c. 1660), I argue that the defense strategy sketched out in the previous chapter led to a complete transformation of the local environment, most importantly the flood regime of the river, causing serious economic difficulties in the Rába

Vadas, A., *The Environmental Legacy of War on the Hungarian-Ottoman Frontier, c. 1540–1690.* Taylor & Francis Group, 2023
DOI: 10.5117/9789463727938_CH03

River's entire valley. These economic difficulties can with a high probability be attributed to the defensive role the river had in the period discussed.

Two immediate questions arise: why focus on Körmend, and why the first half of the seventeenth century? The reasons are relatively apparent. Körmend was one of the most significant settlements in the frontier zone. This does not mean that it was a major town comparable to free royal towns in Hungary like Bratislava or Sopron, but it was the center of a domain, a small but prosperous estate complex owned by the prominent Batthyány family. In this case, the landlords were of key importance, because in the early modern period, right from the beginning of the seventeenth century and especially after the 1620s, they created one of the most complex manorial administrations in the Kingdom of Hungary, with a highly decentralized accounting system. This generated an incredible amount of documentation, a good part of which has survived. The number of documents preserved from the sixteenth and seventeenth centuries in the archive of the family is comparable to what has been preserved of the archival material for the whole of the medieval Kingdom of Hungary (1000 to 1526).

Körmend thus provides an ideal case study for water-related environmental problems and for the conflict of interests between the local economy and defense needs, as it was a prosperous agricultural and commercial settlement lying on an important road with numerous infrastructural elements that were connected to the river. The town itself was partly defended by an earth and wood fortification and a moat, the water of which came from the Rába. In addition, there was a bridge and a major water mill operating on the river by the town, all of which were affected by changes in the flow of the river. The river itself, coming from the mountains, had a meandering, less expressed valley in this section than upstream. Even after the first major water regulation works in the eighteenth century, the Rába was considered an unregulated and rhapsodic river. The following is a nineteenth-century description of the environment of the river at Körmend:

> The meadows of Körmend have a great location since the floods and the inundations in the narrow valleys from the unregulated flow of the Rába and the Pinka Rivers at the beginning of spring fertilize its lands. However, sometimes around the harvest, the floods from the fast current of the river and its exudations also cause losses. In some years, despite the good location of the meadows, there is a lack in forage.[1]

1 [Körmend] *Rétjei igen kedvező fekvésűek, mert a Rába és a Pinka szabályozatlan futása és ezeknek terjedésükhöz képest szűk medreikbe tavasz elején a Stájer országi hegyekből bőven olvadó*

A somewhat later description from around 1820 depicts the valley of the Rába (and one of its more significant tributaries, the Pinka) in fairly similar words:

> The River Pinka, arriving from Styria, after leaving its nice valley crosses the lowlands and flows at Horvátnádalja, some thirty minutes from Körmend, into the River Rába, a pier of which borders the big, beautiful princely garden of Körmend. These waters leave their beds frequently and flood some parts of the settlement, causing major damages to the mills.[2]

These two descriptions from the period of the major nineteenth-century water management projects on the Rába leave little doubt that Körmend and the surrounding manorial complex with its villages is at a position where changes in the watershed, the gallery forests along the Rába, and, most importantly, the manipulation of the downstream flow of the river could seriously affect local farming and economy. Below, I discuss what sources are most instrumental for understanding the changes in the environmental conditions of the river valley in the early modern period.

The time frame of the analysis is partly arbitrary. The beginning of this period is the early seventeenth century when, after half a century of being a buffer zone, the discussed area finally became an immediate frontier of Hungary and the Ottoman Empire with the Fifteen Years' War. Despite the 1606 peace treaty of Zsitvatorok, which put an end to the long military campaign in the central part of the basin, the threat of conflict became more apparent in the region than ever before. The closing date for the focus of this chapter is not closely connected to any major political development

hó és rohanó eső víz által történni szokott áradások és kiöntések azokat szerfelett termékenyittik; de ellenben később s' némellykor takarulás idejében rohanó víz áradásak és kiöntések kártékony következést is hagynak magok után, … olyannyira, hogy némely esztendőkben a réteknek ily kedvezőfekvésük jóminémüsége mellett is szűkölködni kelletik a takarmányban." – description of Mihály Bendekovics, provisor, and János Harangozó, engineer of Körmend: MNL OL P 1322 Urbáriumok [Terriers] IV. Összeírások [Conscriptions] 97. cs. no. 122 (translation by the author). Edited in: *Szemelvénygyűjtemény Körmend történetének tanulmányozásához* [Sources to the history of Körmend], 2 vols, ed. Éva Kondicsné Kovács (Körmend: Jubileumi Bizottság, 1993–2002), vol. I, 6–7. See also István György Tóth, *Jobbágyok, hajdúk, deákok: a pposite uradalom társadalma a 17. században* [Tenant peasant, hajduk, literates: the society of Körmend in the 17th century] (Értekezések a Történeti Tudományok Köréből [New Series], 115) (Budapest: Akadémiai, 1992), 123.

2 Michael von Kunits, *Topographische Beschreibungen des Königreiches Ungarn und seiner einverleibten Provinzen* (Pesth: Ludwig Landerer Edlen v. Fűskút, 1824), 136–137 (translation by the author). See also: György Tilcsik, "Eine unbekannte topographische Beschreibung der Fürst Philipp Batthyányschen Herrschaft in Körmend aus der ersten Hälfte der 1820er Jahre," *Zeitschrift des Historischen Vereins für Steiermark* 101 (2010): 179–194.

in the area but is rather defined by the available sources – private letters of a noble family with major landholdings in the area – which were abundant until 1659 but became scarcer afterward (for reasons to be discussed below). The 1650s did not bring major changes, although the Ottoman military campaign of 1663–1664, which ended in the peace treaty of Vasvár (1664), can be understood as the beginning of a roughly 20-year period of status quo in the region, which was followed by the last Ottoman siege of Vienna in 1683 and the long period of struggle by the Habsburgs to regain the territory of the Kingdom of Hungary.

3.1 Primary Sources from a Town by the River

After its founding in the mid-thirteenth century, Körmend was always in the possession of prominent members of the Hungarian nobility from the late medieval period onwards.[3] At the beginning of the early modern period, the larger part of the town lay in the hands of the Erdődy family,[4] but at the end of the sixteenth century, the landlords changed fairly frequently. After a raid and plundering by the Ottomans, the campaigns of the Fifteen Years' War, and the Bocskai uprising (from 1604 to 1606),[5] the domain was in a very poor state.[6] From the end of 1604 until the twentieth century, the

3 Zsuzsanna Bándi, *Körmend a középkorban* [Körmend in the Middle Ages] (Körmend: Körmend Város Tanácsa Végrehajtó Bizottsága, 1987), Iványi, *Képek Körmend*, idem, *Körmend története a római kortól a mohácsi vészig* [History of Körmend from the Roman period to the battle of Mohács] (Körmend: Csaba József Honismereti Egyesület, 1999), József Dénes, "Körmend a középkorban" [Körmend in the Middle Ages], in *Körmend története* [The history of Körmend], ed. László Szabó (Körmend: Önkormányzat, [1994]), 32–98.

4 Most of the family's archive was brought to the Hungarian National Archives (MNL OL P 107), but many of the documents related to the history of Körmend was brought to the Östa and has been poorly studied (HHSta Erdődy Lad. 83. Fasc. 1. 94, 95, 96. kt.). Most of the documents here are of a legal rather than economic nature. I am thankful to Bence Péterfi for his help with the archival documents kept in Vienna.

5 The domain was more than the town of Körmend and its borders; six villages (three inhabited by Hungarians, three by Croatians) belonged to it: Molnaszecsőd, Hidashollós, Egyházashollós, and Berkifalu, Harasztifalu, and Horvátnádalja respectively. In the seventeenth century (most of all in the 1630s), a further three villages on the right bank of the Rába also belonged to the domain for a while: Győrvár, Boldogasszonyfa, and Szentiván. Cf. Tóth, *Jobbágyok, hajdúk, deákok*, 11–17, András Koltai, *Batthyány Ádám. Egy pposi főúr és udvara a XVII. század közepén* [Ádám Batthyány. A Hungarian aristocrat and his court in the mid-17th century] (A Győri Egyházmegyei Levéltár Kiadványai. Források – Feldolgozások, 14) (Győr: Győri Egyházmegyei Levéltár, 2012), 538. See the terrier of the domain from 1635 for its biggest extent: MNL OL P 1322 IV. no. 52. (June 1635).

6 On the political situation, see: Péter Dominkovits, *"Egy nemzetek lévén..." A Nyugat-Dunántúl Bocskai István 1605. évi hadjárata idején* ['Being one nation...' The Western Transdanubia in the

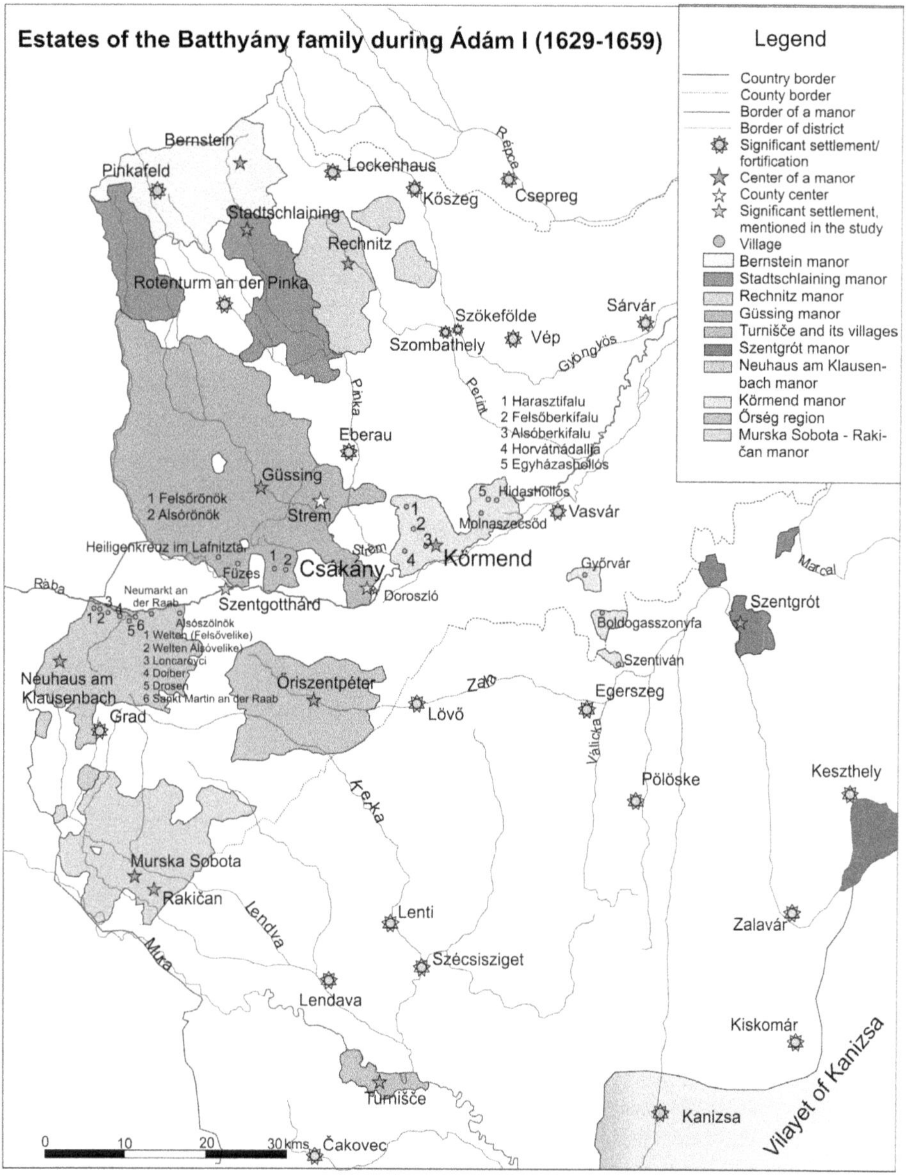

Figure 3.1 The estate complex of the Batthyány family in the middle of the seventeenth century

period of István Bocskai's military campaign] (Budapest: Martin Opitz Kiadó, 2006). According to a document kept in the Hungarian Chamber, because of the Ottoman devastations, the domain of Körmend was worth well under 20,000 Hungarian florins. MNL OL E 15 6 April 1604, no. 17.

town was owned by the Batthyány family.[7] Two members of the Batthyánys – Ferenc (1573–1625) and his son Ádám (1610–1659) – radically reformed the farming methods and the economy of the Batthyány estate complex (for their holdings, see Fig. 3.1).[8] This included the idea of building up a highly professional and decentralized administrative system,[9] which required the heads of the local administration – the *provisors* of the different domains – to keep the landlords informed of all farming, trade, and administrative issues in the settlements under their authority. The documents derived from this administrative system provide a rich resource for research into the early modern history of the environment of the Rába River.

Despite the growing interest among scholars of environmental history in new sources, private letters have so far seldom been used.[10] It is difficult to argue against using them, because their usually exact dating and localization offer precise information, and they occur in high frequency in some cases, like at Körmend. The sixteenth- to seventeenth-century letter collection of the Batthyány family numbers far more than 60,000 letters, and the almost 50,000 extant letters that were sent to Ferenc and Ádám provide a systematic accounting of the manorial complexes. After the death of Ádám in 1659, however, the number of letters sent to the landlords decreased rapidly, and it is no longer possible to reconstruct the environmental circumstances of individual domains based on letters or other extant sources.[11]

7 Tóth, *Jobbágyok, hajdúk*, 17–19 and István György Tóth, "Körmend a kora újkorban (1526–1809)" [Körmend in the early modern period (1526–1809)], in *Körmend története*, 115–117.

8 Koltai, *Batthyány Ádám*, 155–164.

9 See: Ágnes Póka, "A Batthyány-birtokkomplexum igazgatása Batthyány Ádám alatt (1632–1659)" [The administration of the Batthyány estate complex in the period of Ádám Batthyány], in *Tanulmányok Badacsonyból. A Fiatal Levéltárosok Egyesületének konferenciája, Badacsony, 2010. július 9–10.* [Studies from Badacsony. Conference of the Association of Young Archivists, 9–10 June 2010] (Fiatal Levéltárosok Egyesületének kiadványai, 1), eds. Béla Vilmos Mihalik and Áron Zarnóczki (Budapest: ELTE, 2011), 44–59.

10 Some of the exceptions include F.S. Rodrigo, M.J. Esteban-Parra, and Y. Castro-Diez, "On the Use of the Jesuit Order Private Correspondence Records in Climate Reconstructions: A Case Study from Castille (Spain) for 1634–1648 A.D.," *Climatic Change* 40 (1998): 625–645, and Francisco Zamora Rodríguez, "'Quando el Agua llegare aquí Sevilla…' La avenida del río Guadalquivir en 1626 según un documento de la Biblioteca da Ajuda (Portugal)," *Historia, instituciones, documentos* 41 (2013): 407–431.

11 When discussing the letters in the Batthyány archive, the loss of material has to be taken into an account as well. Despite the fact that the archive of this family is one of the best preserved among the archives of the Hungarian nobility it has also suffered significant destruction of material during the last few centuries. In the nineteenth century, numerous letters perished in a fire in the building that housed the family archive. Moreover, thousands of letters were destroyed during the Second World War. Before 1945, Béla Iványi, the archivist of the collection at the time, wrote several treatises in which he used private letters that have not come down to

The largest group of individual items of evidence consulted for this chapter was the letters sent from the town of Körmend to the central administration of the Batthyánys. Körmend was the only major settlement in the domain, and thus the whole administration – both civil and military – was stationed in the town. After the peace treaty of Zsitvatorok left the above-mentioned fortification of Kanizsa in Ottoman hands in 1606, Körmend became an important point in the Hungarian defense system opposing the Ottoman stronghold.[12] This contributes significantly to our knowledge of the time on military matters, as many of the letters did not concern local farming but rather the frontier defense. From this town alone, in the first six decades of the seventeenth century more than 3,000 letters are extant. Only a few letters were sent from the other villages of the domain, but they were also included in the material consulted for this present book (see Appendix 1).

Another major group of letters survived from the nearby fortified settlement of Csákány, a few kilometers upstream from Körmend. The settlement, which despite its proximity to Körmend never belonged to the domain of Körmend but to the district of Strem (in present-day Austria), nonetheless was similarly owned by the Batthyánys. The economic role of Csákány was limited compared to that of Körmend, which had relatively extensive plowlands and significant grain production. Csákány, however, was fortified, just like Körmend, and the local military administration frequently exchanged news with their landlords, the Batthyánys, who were closely involved in the military affairs of the region in the Ottoman war period. The approximately 1,000 letters sent from Csákány functioned as good control material to see how much the problems occurring at Körmend were local or at least to some extent representative of a longer stretch of the Rába valley and the frontier. Also, while the letters provided less data on farming problems or natural phenomena (esp. floods), the problem of border defense was much more dominant in their content. Apart from these major parts of the Batthyány family's letter collection, letters sent from several other riverside settlements were included (see Fig. 3.2) to see whether the hydrological changes mentioned in the sources from these two settlements can also be found in other rivers in the same region. I decided to include some of the settlements at the Zala and Mura rivers as well as

us. This is especially important, as some of the letters discussed here are only known from his monographs and source editions. Most of all: Iványi, *Képek Körmend.*

12 Körmend along with Sárvár was planned to be the center of the border defense against Kanizsa (*gegen Kanischawärts ligende Grenze, confinia Canisae pposite*). See: Pálffy, "A török elleni," 204–205 and MNL OL P 1313 Memorabiliák [Memorabilia], no. 237.

Gyöngyös Stream to test the validity of the findings related to the Rába in a broader context.

Settlement	Nearest river(s)
Csákány	Rába
Gasztony	Rába
Körmend	Rába, Pinka
Lövő	Zala
Eberau	Pinka
Murska Sobota	Mura
Rakičan	Mura
Sárvár	Rába
Szecsőd	Rába
Szentgotthárd	Rába, Lapincs
Szentpéter	Rába
Szentgrót	Zala
Szőkefölde	Gyöngyös Stream
Vép	Gyöngyös Stream

Figure 3.2 Geographical coverage of the private letters studied

Two different administrative systems existed at Körmend during the seventeenth century, one military and the other civil. The defense of the town lay in the hands of the captains – or rather the vice-captains, as in most cases the latter were the office-bearers – who resided in the town. They were responsible for maintaining the fortifications and the moat. They had to organize supplies for the mercenaries living in Körmend at the time, most of whom were of German origin and were responsible for guarding and controlling the riverbank, protecting the hinterland from the Ottomans. The issue of border defense was, however, closely connected to the problem of the local water mill and its dam, as was already hinted at above. Therefore, vice-captains had to work together with the *provisors* and vice versa. This caused frequent problems, as major floods could destroy the bridge, the earth and wood fortification, as well as the mill, and the two officials had different priorities when it came to reconstructing the buildings (e.g., in September 1641 during the largest flood event of the seventeenth century – discussed in section 3.3.4.).

Despite the large number of surviving sources from within the period discussed here, the number of letters available differs significantly from year to year (see Fig. 3.3). There are seasons and years of which the sources allow analysis on a daily or weekly basis, such as the quarter of a century

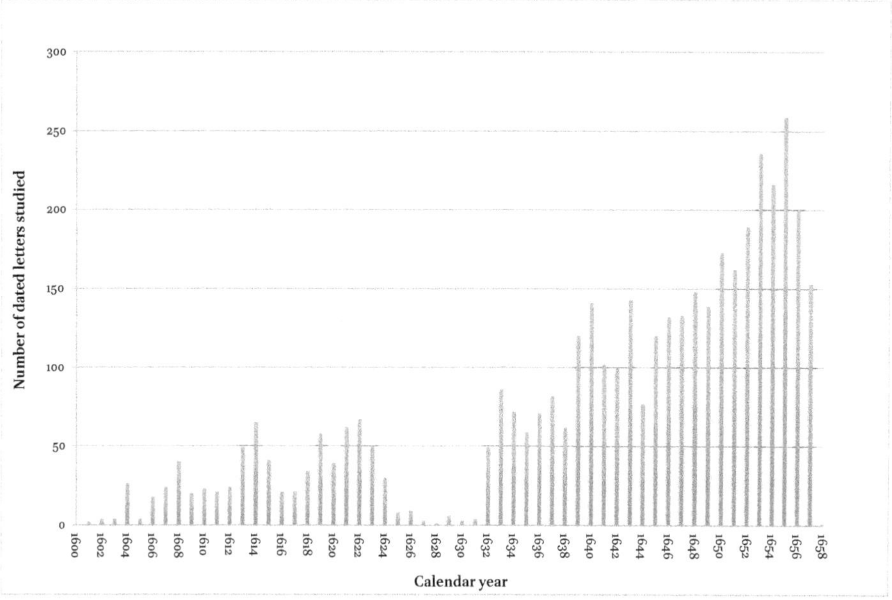

Figure 3.3. Annual distribution of letters studied from Körmend and Csákány (1600–1659)

starting with 1633. The evidence also provides sufficient material to study the environmental conditions between 1607 and 1625. However, in the period from 1600 to 1607 and from 1626 to 1633, there are far too few sources (less than 20 letters per year) to draw any far-reaching conclusions.

Another important feature of the letters is that the preserved ones are almost always the ones sent from the administrators of these settlements to the landlords, and not the other way around. Most of the letters sent to the Batthyánys at their residence at the time in Güssing (in present-day Austria) informed them of different issues. These letters seem to have always been deposited in the family's archive, which was systematically managed from Ádám's time onwards.[13] Letters that were sent from the landlords as answers to the local administrators, however, have rarely been preserved.

The letters are the most important sources, but they are not the only type of documentary evidence consulted for this chapter. Different

13 Already in the early sixteenth century, a registry was made on the archival materials of the family: *Registrum litterarum per dominum Benedictum de Bathyan Buda domino Balthasari de eadem per arbitrativam dispositionem proborum superinde inter eos factam datarum et assignatarum* (1511). See: Béla Iványi, "Gróf Batthyány Ádám, a levéltárrendező" [Ádám Batthyány, the archivist], *Levéltári Közlemények* 20–23 (1942–1945): 291–292.

administrative and economic history sources were produced on the settlements belonging to the domain of Körmend as well as Csákány. The most important of these were the terriers, already discussed in Chapter 2.[14] Every few years the settlements of the Batthyány complex were surveyed, and terriers described the state of the different buildings, the population, the different services they owed the landlords, and the taxes the settlements paid annually.[15] Furthermore, account books as well as instructive documents also contributed at certain points to a complex understanding of the environmental problems in these settlements.[16] Maps, discussed in the previous chapter, were also used in the analysis. Besides a unique map of the town of Körmend from the 1570s, there are no maps directly depicting the area until the 1770s. From the late eighteenth century onwards, however, there are relatively many available maps, some of which were drawn expressly to help the planning of the major water management works on the Rába river at the time.[17]

Some of the above-listed sources are rarely found among the sources used in early modern environmental histories. Nonetheless, the large quantity and the density of environment-related information enable us to provide a detailed analysis of the environmental conditions along the Rába valley in the first half of the seventeenth century, with a special focus on the problems related to floods and the water levels of the river, which, as I shall argue, were most probably directly attributable to the militarized nature of the river.

14 There are four major collections that preserved terriers from early modern Körmend and Csákány: MNL OL P 1313, and P 1322; MNL OL E 156 and the Batthyanyisches Herrschaftsarchiv Güssing [Family Archives of the Batthyány Family at Güssing] (BHG).

15 Ferenc Maksay, "Urbáriumok" [Terriers], in *A történeti statisztika forrásai* [Sources of historical statistics], ed. József Kovacsics (Budapest: Közgazdasági és jogi könyvkiadó, 1957), 119–144, Vilmos Mihalik Béla, Dániel Kálmán and Áron Zarnóczki, *A veszprémi káptalan 1727. és 1755. évi urbáriumai* [Terriers of the chapter of Veszprém from 1727 and 1755] (A Veszprémi Egyházmegye Múltjából, 24) (Veszprém: Veszprémi Érseki és Főkáptalani Levéltár and Veszprémi Érseki Könyvtár, 2012); see most importantly the introductory essay of the authors to the source edition.

16 MNL OL P 1322 Instrukciók [Instructions]. On the role of instructive documents in historical research, see: István Kenyeres, "Kamarai uradalmak igazgatása a XVI. században" [Administration of the domains of the Hungarian Chamber in the 16th century], in *XVI. századi uradalmi utasítások. Utasítások a kamarai uradalmak prefektusai, udvarbírái és ellenőrei részére*, 2 vols. [16th-century instructions. Instructions to the prefects, provisors and checkers of the Chamber domains], (Fons Könyvek, 2), eds. István Kenyeres and Péter Kis (Budapest: Szentpétery Imre Történettudományi Alapítvány, 2002), vol. I, 15–90, esp. 81–83.

17 Maps concentrating on individual domains were made mostly from the 1770s. E.g., *Mappa du territoire de Körmend et des villages qui endependent situé en basse Hongrie et relevé l'année 1771* (1771). The manuscript map is kept at: OSZK T 1659.

3.2 The Rába at Körmend

Little information is available on the environmental conditions along the Rába in the surroundings of Körmend for the period before the arrival of the Ottomans in the Carpathian Basin. The first written source that concerns this section of the river dates to 1255. In a privilege charter issued by Béla IV, the king endows a certain *comes* Dés of the Hermán clan with three plots of land at an estate called Torvaj. The donation was meant to help *comes* Dés finish building – and then maintain – his tower at Körmend. Although it is now lost, the location of this tower (fortification) can be identified precisely because it was depicted on the previously mentioned first map of the town from the early 1570s (see Fig. 3.4).[18]

This charter also refers to a more important feature than the tower, a certain river that either surrounded or flowed by the structure. The source of water of this river, also called Körmend, could only have been the Rába. As has been demonstrated on several occasions, sections of rivers in the Middle Ages were frequently named after the settlements they flowed by, thus it is likely that the Körmend River is identical to – or is a former branch of – the Rába itself.[19]

This branch lay relatively far from the present-day channelized bed of the Rába River. Other sources also indicate that the Rába had many branches and oxbow lakes around Körmend. It is confirmed by several hydronyms such as *Holt-Rába* (meaning 'dead Rába', probably an oxbow lake), *Sár* (meaning 'mud'), or *Ó árok* ('old ditch')[20] mentioned in many perambulations in the seventeenth century. Even though these branches of the Rába may have existed for a long time, there is no evidence that their water was used for milling or fishing, suggesting that they were filled with

18 On the earliest map of Körmend (*Kirment*), see LBW GK Planbände Bd. XIII. fol. 11. GI. 178. Published in: György Kisari Balla, *Karlsruhei térképek a török háborúk korából / Kriegskarten und Pläne aus der Türkkenzeit in Karlsruher Sammlungen* (Budapest: Kisari Balla György, 2000), 434. For the dating of the map, see Pálffy, *Die Anfänge der Militärkartographie*, 56–63.

19 On this question, see Bándi, *Körmend a középkorban*, 17; Gábor Kiss and Balázs Zágorhidi Czigány, "Víznevek névalkotó szerepéről – Vas megyei példákon" [On the role of hydronyms in forming place names – on examples from Vas County], *Helynévtörténeti Tanulmányok* 3 (2008): 111–120.

20 For the former, see e.g., Körmend város urbáriuma [Terrier of the town of Körmend] 1646. MNL OL P 1313 Majoratus Lad. 1. No 32. 10/a, 12/b p, A körmendi majorság és örökségföldek összeírása [Conscription of the manorial lands of Körmend], 1649. MNL OL P 1313 Majoratus Lad. 1. No 37. 290/b p, SZEK. 1.c. Can. vis. [Kazó-féle canonica visitatio (1697)] [Visitation of Kazó]. Available on microfiche: MNL OL X 767 (box: 52) 415–416, 418, and 420. p., for the latter, see e.g., MNL OL P 1313 Majoratus Lad. 1. No 32 7/b, 8/a, 8/b, 9/b. p.

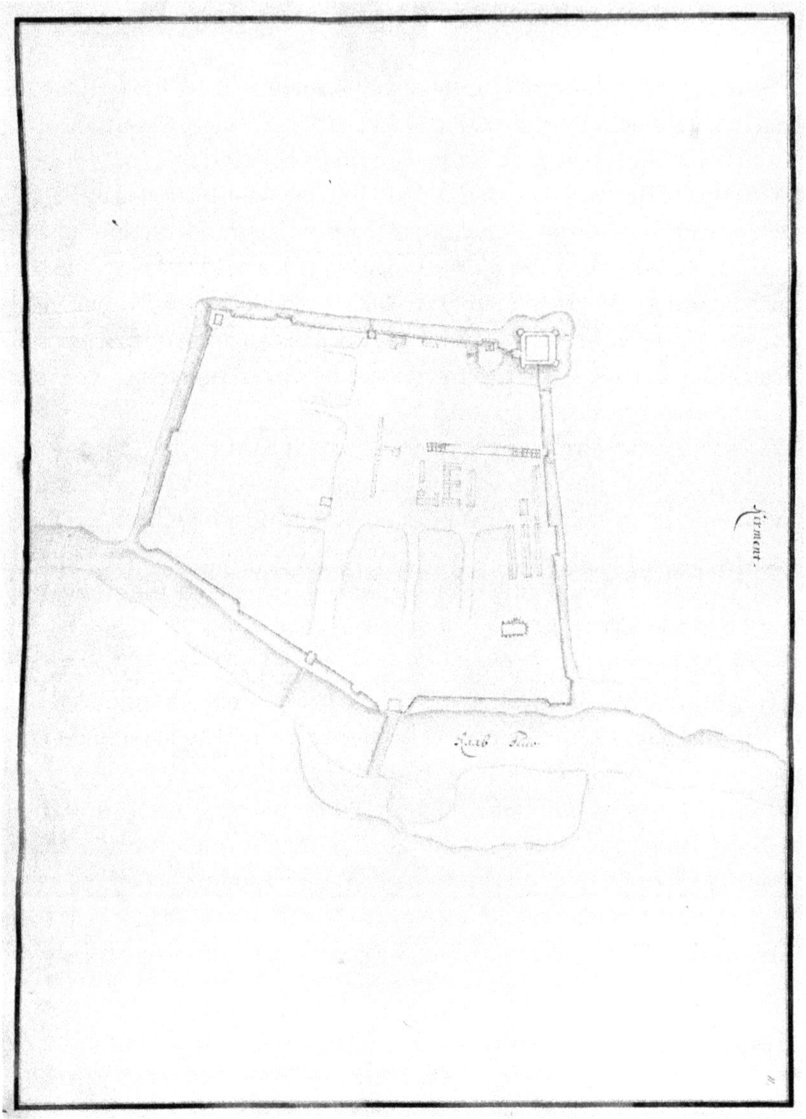

Figure 3.4 Körmend in the 1570s

water only temporarily. This, of course, was not true of the river's entire flow in this section of the Carpathian Basin. On the border of Püspöki, some 20 kilometers downstream of the Rába, an oxbow lake of the river was nourished by the floods in the springtime and was used systematically for fishing.[21]

21 See, e.g., MNL OL E 156 a Fasc. 119. no. 6. 34r (59. p.), the terrier from 1592.

According to the sources consulted for this chapter, the Rába changed its bed several times before systematic water management work started in the late eighteenth century by Körmend.[22] Not even the flow of the main riverbed was evident in the early modern period. In a letter written in 1637, the vice-captain of Körmend, András Hidasy, mentions that the Rába had eroded a section of the right bank and endangered the pier. According to Hidasy, the collapse of that section of the river bank could have easily sent the river to a new bed, by-passing the town's mill.[23] A few years later, in 1641, Hidasy again states that the Rába was about to move to a new bed just upstream of the previously mentioned area.[24] A few days later, in another letter, he also mentions that newly deepened river branches were seriously affecting the viability of roads and had to be filled if anything were to be transported.[25] In a letter dated 1646, the captain of Körmend, Gáspár Francsics,[26] mentions a similar problem: the spring floods had collapsed the riverbank above the town's bridge. This was probably due to the structure of the bridge hindering the undisturbed flow of the flooding river. The letter informing the landlord of the event also warns of the threat that the

22 Cf. Gellén, "A Rába védelmében," and Károlyi and Somogyi, "Felszíni vízfolyások."

23 *Ez mostanj ár majd az Raba az malom es híd között az Raba tulso felen nagj szakasztast tett, ugi annira hogi hordani nem veszik ugian keleötte … ha ária lenne, féleö hogi epenseggel el keruli az ki az molnat, S mostis bizonj feleö másikat veszen el* (This flood by the Rába did a lot of damage between the mill and the bridge on the other side of the Rába, so much that if no one takes care of filling it [probably the pier] in case of a flood it might find a new bed by-passing the mill) – letter of András Hidasy to Ádám Batthyány, 16 April 1637 (MNL OL P 1314 no. 19 159).

24 *Hanem it az Ar Víz anjra vagyon hogy az híd elöt igen kj vetek az Raba, el akaria az hidat kerülny masut akarja magat kj szakasztanj. Nagod küldene onnent föliül segetseget hogy mivel tudnank megh oltalmaznj mert igen nagy karos leszen ha megh nem jobotjuk Az malomak is igen nagy artasara leszen* (The flood is so serious here that above the bridge the Rába found a new bed and tries to by-pass the bridge. My lord, you should send down some help from up there [Güssing] to help in the rebuilding otherwise it causes a huge loss to the mill) – letter of András Hidasy to Ádám Batthyány, 3 March 1641 (MNL OL P 1314 no. 19 233).

25 *…az Hídon kjvül olj nagy arkokat mosot hogy sem lovas sem szekeres at nem mehet migh megh nem töltik, s ha egy az víz leszen, megh az malomnak is nagy kart teszen* (above the bridge it [the river] dug a bed so deep that neither a horseman nor a carter can cross until it is filled up, and in case of a flood it could cause a huge loss to the mill) – letter of András Hidasy to Ádám Batthyány, 10 March 1641 (MNL OL P 1314 no. 19 235).

26 For more details about Francsics, see Ildikó Ladányi-Benedikt, "Mindennapi élet a körmendi várban – Francsics Gáspár körmendi kapitány levelei 1648–1650-ben" [Everyday life in the castle of Körmend: the letter of Gáspár Francsics, captain of Körmend, 1648–1650], in *A Batthyányak évszázadai. Tudományos konferencia Körmenden 2005. október 27–29.* [Centuries of the Batthyány family. Conference held at Körmend, 27–29 October 2005], ed. Zoltán Nagy (Körmend and Szombathely: Körmend Város Önkormányzata, 2006), 205–216.

river may be moved to a whole new bed.[27] Just a few years later, in 1650, it was now the turn of the *provisor* of the domain, István Nemsem, to urge his landlord to take measures to fill a new riverbed to continue to provide water for the manorial mill.[28] These few sources demonstrate clearly that, despite the presence of some local water regulation and manipulation, the bed of the river was much less stable than in the late eighteenth or nineteenth centuries. Also, they reveal that changes in the environment upstream or even downstream could seriously affect local conditions and change the hydrography, the economic possibilities, and even the accessibility of the town, which was an important transportation hub dating back to the Middle Ages.[29] The transforming waterscape in the surroundings of the town and the changes in the environmental conditions in the wider area are reflected in the problem of floods causing significant damage to the built environment along the river, such as the manorial mill, the mill dam, the bridge, and the earth and wood fortification of the town.

Some works by Béla Iványi, a previously mentioned archivist of the Batthyány collection and enthusiastic researcher of medieval and early modern Hungary, and other historians from the late twentieth century refer to the problem of floods on the Rába, but none of them have posed the basic question of what the reason might have been for these events. Here I argue that the water-related problems at this settlement and the neighboring Csákány can be understood in the context of the new political situation in Transdanubia and the formation of the frontier between the Ottoman and Habsburg Empires in the surroundings of the Rába River from the mid-sixteenth century and early seventeenth century onwards. The primary focus here is on the water-related infrastructural elements

27 *Az itvalo Hidon alol tull az Raban ollj nagj árkoth assoth az Ar víz, ha mindjart eleieth nem keltik eppen az Raba az folliassabull kj megjen es az malom szarazon marad...* (Upstream from the bridge, the flood of the Rába dug a channel so deep that without taking care of it the Rába would find this bed by-passing the mill, leaving it dry) – letter of Gáspár Francsics to Ádám Batthyány, 13 March 1646 (MNL OL P 1314 no. 15 085).

28 *Mivel az Raba igen megh aradot, hogy Csötörtökön, s, Penteken az hidon tul az ép földben oly nag szakasztast tett, hogy ket szaz szeker agal es, negyven szal vórogh feniö fa karoival alig tudgyuk véghéz vinny, s, ha mostan segetsegel az csinalasan nem leszűnk az molnat el kerülj az Raba...* (The Rába heavily flooded, on Thursday and Friday above the bridge it left its bed flooding plowlands. We barely could take care of that with 200 cartloads of branches and 40 logs of larch. Unless we can take care of it now, the Rába would by-pass the mill eventually) – letter of Gáspár Francsics to Ádám Batthyány, 19 February 1650 (MNL OL P 1314 no. 33 770).

29 Szende, "Towns Along the Way," 197–199 and Magdolna Szilágyi, "Városok, utak, kereskedelem. Az úthálózat szerepe Vas megye városi fejlődésében a 13–14. században" [Towns, roads, trade. The role of the road-network in the urban development of Vas County in the 13th–14th centuries], *Savaria* 36 (2013): 223–241.

and the problem of border defense along the Rába discussed in the sources listed above.

3.2.1 Fortifications, Bridges, and Mills – Körmend's Waterscape

There is a relative scarcity of information on the fortifications at Körmend in the Middle Ages. A source from the early sixteenth century mentions the Saint Martin's Church outside of the walls, indicating the existence of some kind of fence or wall.[30] The first relatively precise map of Hungary, first printed probably in 1522 and then in 1528, also depicts the town being encircled by a wall.[31] These two items of circumstantial evidence suggest that some kind of fortification protected the town already in the Late Middle Ages. Gyula Siklósi, one of the archaeologists who conducted excavations in the town, hypothesizes that there was an earth and wood fortification around the town as early as the fourteenth century, but there is neither written nor archaeological evidence to confirm this assumption.[32] In the early modern period, the wooden palisade, if it did indeed exist, was significantly transformed. From the map from the 1570s and a written source from 1580, it is clear that a moat surrounded the town, probably constructed in the 1540s not independently of the changing political circumstances in these years as discussed in Chapter 2. From 1546 onwards, the sources usually distinguish between the town and its suburb (*suburbium, exterius oppidum,* and *kwlsew waras*, the latter meaning 'suburb' in Hungarian), which testifies to the town having been surrounded by a wall, most probably an earth and wood fortification.[33] The moat could only take water from the Rába, which suggests that floods not only endangered the Rába itself but also parts of the moat and the wooden palisade. Moreover, the south side of the fortification bordered directly on the Rába, which made the choice of construction material of vital importance, as different building materials reacted differently to the natural elements. According to sources from the seventeenth century, at least two different materials were used in building the fortifications. Some parts, such

30 Gyula Siklósi, *A középkori Körmend védelmi rendszere* [The defense system of medieval Körmend] (Testis Temporis, 15) (Körmend: Körmend Város Önkormányzata and Siklósi Gyula, 2006), 7.

31 Lazarus Secretarius, *Tabula Hungariae ad quatuor latera.* Manuscript map: OSZK K App. M. 126. For the history of the map, see *Lazarus Secretarius: The First Hungarian Mapmaker and His Work*, ed. Lajos Stegena (Budapest: Akadémiai, 1982).

32 Siklósi, *A középkori Körmend*, 7.

33 Tibor Koppány, *Körmend városának építéstörténete* [Construction history of the town of Körmend] (Körmend: Körmend város tanácsa és végrehajtó bizottsága, 1986), 27.

as those between the Upper Gate and the castle, were built of brick, while other walls, which were seriously affected by the Rába and required significant and recurrent rebuilding work, were of earth and wood (see section 3.3.1).

As with the fortifications, limited information is available on the bridge at Körmend even though it was an important crossing point for thousands of years because of its strategic location – at the crossroads of the north-south and northeast-southwest transit routes in the Carpathian Basin. The Roman Amber Road crossed the Rába near the settlement and turned north towards the Roman town of Savaria (at the site of present-day Szombathely) and then further to the north.[34] The location of the crossing point and of the settlement of Arrabona (not to be confused with the Roman settlement at the site of present-day Győr) have until recently been debated.[35] In the mid-twentieth century, Iványi mentioned the so-called "remains of the Roman road" (*a római út nyomdoka*),[36] which led to the Rába close to Körmend downstream from the town at the village of Katafa, basing his statement on a later lost map kept in the Batthyány archives. This localization can be confirmed by a late eighteenth-century map of the Rába valley.[37] This fairly detailed map was ordered as part of the planning involved in the Rába water control works. At that time, the map showed a Roman ruin, the "remains of the Roman-age bridge" (*vestigia antiqua Pontis tempore Romanorum*) in the river close to Katafa.[38] Some questions arise about this identification. First, how

34 Endre Tóth, "Körmend vidéke római kori történetéhez" [On the history of the Roman period of Körmend and its neighborhood], *Vasi Szemle* 33 (1979): 342–346, Gábor Kiss and Balázs Zágorhidi Czigány, "A Lapincs–Rába-vonaltól délre eső terület Árpád-kori történeti földrajzához: a megyetörténet műhelyéből. I. rész" [On the Árpádian-period historical geography of the area south of the Lapincs–Rába-line – from the workshop of county history, I], *Vasi Szemle* 64 (2010): 711–721.

35 See the *Tabula Peutingeriana*. Cf. András Mócsy and Mária Szilágyi, "Úthálózat" [Road network], in *Pannónia régészeti kézikönyve* [Archaeology handbook of Pannonia] (Budapest: Akadémiai, 1990), 119. See also: Vajk Cserményi and Endre Tóth, "Der Abschnitt der Bernsteinstrasse in Ungarn," *Savaria* 16 (1982): 288, and Endre Tóth, *Itineraria Pannonica. Római utak a Dunántúlon* [Roman roads in the Transdanubia] (Budapest: MNM, 2006).

36 Iványi, *Körmend története*. See also: Koppány, *Körmend*, 8, Tóth, "Körmend." For the different identifications, see Cserményi, and Tóth, "Der Abschnitt der Bernsteinstrasse," 288, and Tóth, "Itineraria," 21–22.

37 Klára Dóka, "A Rába-szabályozás kérdése 1786-ban" [The question of the Rába-regulation in 1786], *Soproni Szemle* 30 (1976): 55–60. *Mappa fluvii Arabonis ex mondato ex Consilii Locumtenentialis Regii* [from the late eighteenth century] – MNL OL S 12 Div XI no. 0125. [1/3]. See also: Szilágyi, "The Árpád Period Communication Networks."

38 *Mappa fluvii Arabonis ex mondato ex Consilii Locumtenentialis Regii* [from the late eighteenth century] – MNL OL S 12 Div XI no. 0125. [1/3]. See also: Szilágyi, "The Árpád Period Communication Networks."

were the ruins identified as Roman by eighteenth- and nineteenth-century map makers? Second, how had these structures survived floods for almost two millennia? If they were wooden bridges, as most similar constructions dating to the Roman period were, this is highly unlikely. If they were made of stone, how did they manage to survive for 1800 years and then perish in the past two centuries? It is also a mystery why, if the remains of a bridge were there in the Middle Ages, the medieval settlers of the region chose not to build their new town – Körmend –, and its bridge aligned to this already existing important infrastructural element. In many places throughout the Carpathian Basin, parts from a significant continuation of the Roman road network have been identified by both archaeologists and historians, but this case is an exception.[39] Although, as far as is known, the Roman road did not cross the Rába at present-day Körmend, from the Árpádian period onwards, the town certainly was an important crossing point of the river. It was recently revealed that the town, founded in the aftermath of the Mongol invasion of 1241–1242, lay at the intersection of important trade routes. This is clear from the relatively large market square compared to the town's size and its early market privileges,[40] although there is no direct written evidence from the Árpádian period on the existence of a bridge here.[41] Important indirect evidence – the name of the settlement – suggests, however, that Körmend had a bridge from its foundation. The most recent monograph on the town argues that among the possible etymologies of the name "Körmend", the most probable one is connected to the ancient Turkish word for a bridge (or ford).[42] From the early modern period onwards, several written sources testify to a bridge being in almost uninterrupted use at Körmend. The bridge was functioning both in 1544 and 1594 when the two river surveys discussed in Chapter 2 were carried out. Also, a source from the year 1548 referred to the duty of Körmend's inhabitants to maintain the bridge and the town's mill.[43]

39 For more details on this issue, see Lajos Glaser, "A Dunántúl középkori úthálózata, I–II." [Medieval road-network in the Transdanubia], *Századok* 63 (1929): 138–167, and 257–285, Szilágyi, "The Árpád Period Communication Networks," and eadem, "Római utak a középkori Dunántúlon. Az utak nevei és szerepük a középkori térszervezésben" [Roman roads in medieval Transdanubia. The name of the roads and their role in medieval spatial organization], *Történelmi Szemle* 56 (2014): 1–25.

40 Szilágyi, "Városok, utak, kereskedelem," 226–228, and Szende, "Towns Along the Way," 196–198.

41 Glaser, "A Dunántúl középkori," and Szilágyi, "The Árpád Period Communication Networks."

42 István Palkó, "Körmend névtudományi vizsgálata" [Onomastic study of Körmend], in *Körmend története*, 12.

43 MNL OL P 1313 Úrbéri iratok [Urbarialia] 203. rsz. 90. p. (1548)

In the seventeenth century, the bridge regularly appeared in the sources, both in the letters and in the terriers consulted. By this time, not only Körmend but also the neighboring villages of Szecsőd, Hídvég (certainly dating to the Middle Ages), and Csákány had a bridge. Data preserved in letters about the bridge at Körmend refer almost exclusively to its reconstructions. There is no detailed description of the bridge at Körmend nor of any other bridge built over the Rába. Still, based on what the sources say about its reconstructions, we know that the bridge, probably including its pillars, was made of wood. Bridges of similar structure were subject to destruction by floods, especially the ice floods that recurred every few years on the rivers in this region. The bridge here was probably similar – but of course smaller – to the one built over the Traun in Austria at the town of Wels. The latter bridge is of particular importance, as there are continuous records on its reconstruction and will be used as a parallel example.[44] Christian Rohr, an Austrian environmental historian, has given a detailed analysis of the more than 10,000 folios of preserved records written by the bridge masters of Wels. The accounts show that the fast current and the flooding of the river made some work inevitable every single year. Ice floods as well as the floods connected to the snowmelt in the Alps, which feeds the Traun, were expected annually, so much that the bridge masters usually ordered some amount of wood well before the arrival of the floods.[45] Rohr managed to point to a clear seasonality in the rebuilding works there (see Fig. 3.5). To some extent, a similar seasonality of bridge repairs can be identified at Körmend, but even more in the case of the repairing of the dam of the manorial mill. Seasonality was less expressed at Körmend, probably because the Traun's spring floods were much more disastrous and also because the two rivers had somewhat different water regimes. The floods on the Rába were less directly connected to the snowmelt, as the river's water came from both the Alps and their foothills.[46]

Two important points are apparent in the letters that refer to the bridge in the seventeenth century. First, just as at Wels, the bridge masters coordinated

44 Christian Rohr, "Measuring the Frequency and Intensity of Floods of the Traun River (Upper Austria), 1441–1574," *Hydrological Sciences Journal* 51 (2006): 836.

45 Rohr, "Measuring the Frequency," 834–847.

46 A similar problem, although on a much bigger river, the Danube is reflected in the account books of Bratislava. The bridge here had to be regularly rebuilt due to the floods, especially the ice floods and the icing of the river: János Király, *A pozsonyi nagy-dunai vám- és révjog története* [The history of the customs and toll at the Danube by Bratislava] (Bratislava: Drodtleff, 1890), 44–104. See also: Vadas, "'A Dunára én bizony nem megyek.'"

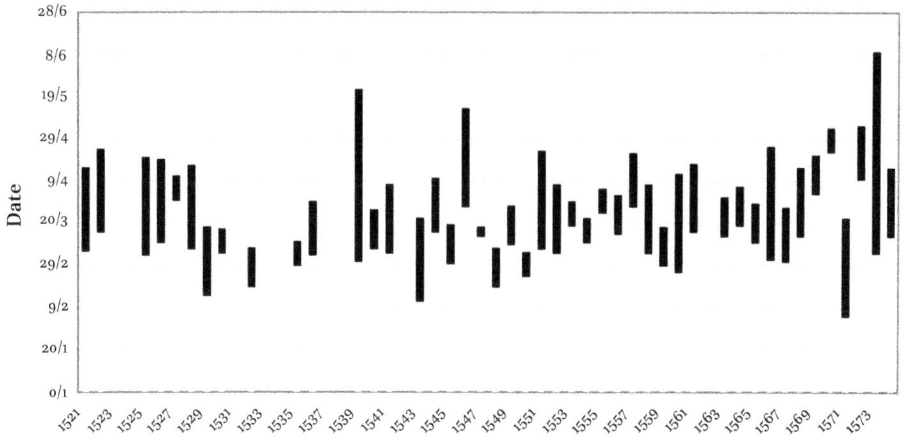

Figure 3.5 The beginning and length of the reconstruction of the Traun bridge at Wels in the light of the bridge masters' accounts (1521–1574)[47]

all reconstructions.[48] Second, there are at least five years in the period discussed in this chapter when major work on the bridge was necessitated by floods,[49] but probably some work was also needed on the bridge when the mill dam and the mill were harmed by floods. There is no direct evidence on the structure of the bridge at Körmend, certainly no etching or other visual source. The above-mentioned map from the 1570s depicts the bridge, but the only thing visible is that it had many pillars, probably for strengthening the bridge during floods. The bridge, however, seems to have been less costly than the mill or other buildings. It was nonetheless an important point on the frontier and on the Rába, which is evidenced by the fact that the sources recurrently refer to its state and importance in the communication network of western Transdanubia.

Most likely the mills of Körmend and especially the manorial mill were more exposed to destruction by the Rába than the bridge or other infrastructural elements. The history of the manorial mill goes back to the Middle Ages. The first time mills (or a mill with multiple wheels) are mentioned in Körmend was in 1358.[50] There is no information on whether this mill stood at the same site as the later manorial mill, but in 1458 an

47 Rohr, "Measuring the Frequency," 840.

48 Letter of László Bozay to Ádám Batthyány, 23 April 1637 (MNL OL P 1314 no. 7588).

49 1612 (MNL OL P 1314 no. 40 803), 1635 (MNL OL P 1314 no. 7531), 1637 (MNL OL P 1314 no. 7588), 1641 (MNL OL P 1314 no. 19 244, no. 16 076), 1650 (MNL OL P 1314 no. 33 771). See also: MNL OL P 1313 Úrbéri iratok [Urbarialia] 203. rsz. 90. p. (1548).

50 Document issued by the town of Körmend: 8 May 1358 (MNL OL DL 91 514).

abandoned mill is described as one being by the Rába.[51] In 1499, the last will of the owner, János Ellerbach, mentions this and another mill in Körmend, both by the Rába.[52] The same document also refers to a mill by the river Pinka as well as the first mill downstream from Körmend on the Rába at Szecsőd. Letters also frequently contain references to this latter mill, as the village of Szecsőd also fell under the authority of the manorial complex of Körmend. The value of the manorial mill is reflected in the fact that, according to a 1499 document, 10 percent of the income the mill brought in was enough to sustain a chaplain who was supposed to serve mass each day commemorating Ellerbach at the Saint Elisabeth's Church of Körmend.[53] The medieval origin of the manorial mill has been confirmed by recent archaeological excavations, as the remains of an early modern mill complex with medieval foundations were unearthed by the former pier of the Rába, but the detailed report of this excavation has yet to be published.[54] Just as in the case of the bridge, limited information is available on the actual structure of the mill building. Based on the excavations and some of the materials provisioned for reconstruction works, most of it was made of wood, but parts of the foundations were of stone. In the lists of provisioned

51 *...in quarum faciebus quedam domus lapidea existere dicuntur ac uno loco molendini, similiter deserto, in fluvio Raba sito* – MNL OL DL 100 755. See: Iványi, *Körmend története*, 25–26 and MNL OL P 1313 Majoratus Lad. 6. no. 5. 269. p. The document issued 11 May 1462 has been preserved however: MNL OL DF 259 127.

52 *...item iterum legamus decimam partem omnium proventuum nostrorum tam de molendinis nostris in opido Kermend in fluvio Raba infra scolam ac in possessione nostra Zechewd similiter in fluvio Raba decurrentem nobis evenire debentium altari beate Marie virginis in ecclesia parochiali Zenthelsebeth vocata in eadem Kermend fundata constructo sicuti eciam prius genetrix nostra carissima legaverat ante obitum suum, ita videlicet quod rector altaris predicti vel plebanus eiusdem ecclesie, qui scilicet ipsorum ipsam decimam partem dictorum proventuum molendinorum ipsorum percipiet cottidie debeat et teneatur cantare unam missam in predicto altari beate virginis, insuper similiter decimam partem proventus nostri de molendino circumspecti Stephani literati in eadem Kermend habito nobis evenire debentem legamus eidem rectori altaris eiusdem, qui scilicet dictam missam celebraverit, hoc tamen adiecto, quod si plebanus pretactus percipiet dictos proventus, extunc debeat tenere capellanum ad serviendum Deo et celebrandum super dictum altare...* – MNL OL DL 19 510. Partly edited in: Bándi, *Körmend*, 72 note 108. The date on the document is 1489, but as was demonstrated by Zsuzsanna Bándi (eadem, *Körmend a középkorban*, 38), it was issued indeed in 1499.

53 Dénes, "Körmend," 66.

54 Bettina Bajdó, "Előkerült a régi malom" [The old mill has been discovered] [interview with the archaeologist, Gyula Siklósi]. Online document: http://www.vasnepe.hu/fokusz-ban/20070615_elokerult_a_regi_malom/print (last accessed: 1 May 2020), and Gyula Siklósi, "Körmend, Mónus Illés u. 15." [Körmend, 15 Mónus Illés Street] in *Régészeti kutatások Magyarországon 2007* [Archaeological excavations in Hungary, 2007], ed. Judit Kisfaludi (Budapest: Kulturális Örökségvédelmi Hivatal, 2008), 239–240.

materials for the mill, however, metals were the only material other than wood included. They must have been used to attach the wooden parts.[55]

The mill itself is mentioned in many of the sources consulted, but the dam of the mill is referred to even more frequently. Mill dams are relatively simple constructions that back up the water in rivers to control the water level, providing better opportunities for milling, fisheries, and other activities. Of course, as was discussed in the previous chapter of this book, they also created an obstacle to the flow of the water, which was a problem that dated back to the Middle Ages by the Rába. The earliest source that tells of mill dams being destroyed when their existence was not in others' interests dates to as early as 1247.[56] However, until the late medieval period, these cases were usually settled within a short period.[57] The case of the mill dam of the manorial mill at Körmend, however, is more complex, as it had additional roles. As noted above, it was instrumental in providing constant water flow into the moat that surrounded the fortifications of Körmend.[58] This role was not exceptional, though; several major urban moats in Central Europe[59] as well as the moat at the neighboring Csákány had a similar water supply system.[60] The structure of the mill's dam was

55 For a similar list of material provisions, see András Vadas, "Vízépítés és munkaszervezési formák a késő középkori és kora újkori Magyar Királyságban" [Water construction and labor organization in late medieval and early modern Hungary], in *Techné: a mesterségbeli tudás átadásának lehetséges színterei az ó- és középkorban* [Techné: the scenes of knowledge transfer in the Antiquity and the Middle Ages], eds. Márta Munding, Kornél Szovák, and László Takács (Piliscsaba: Avicenna Közel-Kelet Kutatások Intézete, 2018), 477–510.

56 MNL OL DL 318. Edited in: *Urkundenbuch des Burgenlandes und der angrenzenden Gebiete der Komitate Wieselburg, Ödenburg und Eisenburg*, 3 vols. (Publikationen des Instituts für Österreichische Geschichtsforschung, 7), eds. Hans Wagner and Irmtraut Lindeck-Pozza (Graz and Cologne: Böhlau, 1955–1979), vol. I, 218–219.

57 See: István Tringli, "A magyar szokásjog a malomépítésről" [Hungarian customary law of mill construction], in *Tanulmányok a középkorról* [Studies on medieval history] (Analecta Medievalia, 1), ed. Tibor Neumann (Budapest and Piliscsaba: PPKE, 2001), 251–267, and András Vadas, "Some Remarks on the Legal Regulations and Practice of Mill Construction in Medieval Hungary," in *Wasser in der mittelalterlichen Kultur / Water in Medieval Culture. Gebrauch – Wahrnehmung – Symbolik / Uses, Perceptions, and Symbolism* (Das Mittelalter. Perspektiven mediävistischer Forschung, 4), eds. Gerlinde Huber-Rebenich, Christian Rohr, and Michael Stolz (Berlin: De Gruyter Verlag, 2017), 291–304.

58 József Kelenik, "Körmend a hadtörténelemben 1526–1711" [Körmend in military history], in *Körmend a hadtörténelemben* [Körmend in military history], eds. László Veszprémy et al. (Körmend: Önkormányzat, 1992), 63–64 note 64.

59 András Vadas, "Városárkok és vízgazdálkodás a késő-középkori Közép-Európa városaiban" [Urban moats and water management in the late medieval towns of Central Europe], *Urbs. Magyar Várostörténeti Évkönyv* 10–11 (2015): 323–353.

60 See Chapter 4.

simple: it was mostly made of wood, but certainly some amount of stone was also built into the foundation to strengthen the structure. The stone component, however, was seldom affected by flooding, as only one letter mentions acquiring stone to rebuild the foundations of the dam.[61] The dozens of letters referring to reconstructions suggest that large fir stakes stood on the stone foundation and that between the stakes, wattle was crammed, consolidated, and strengthened by mud. River flooding, driftwood, and especially ice clearly affected similar dams along the Rába. The dams at the river mentioned in the previous chapter may have all had virtually the same structure, and thus were all probably affected by these events to some extent. According to the sources consulted, minor work was necessary each year.[62] The *provisors* demanded numerous working days from local tenant peasants specifically for maintaining the dam and mill each year.

The reconstruction of the dam proved to be problematic on many occasions. First, the quantity of wattle necessary for the work had to be brought in, usually from the Őrség area (some twenty kilometers west of Körmend). Transportation from the Őrség was frequently halted due to problems in the communication networks, sometimes caused by the same floods as those that had damaged the dam at Körmend. The other problem with rebuilding was always the shortage of laborers. This shortage in unpaid mandatory labor is noteworthy, as from the mid-sixteenth century onwards, apart from the usual 40 days annually, the inhabitants of Körmend were also obligated to work three days a year exclusively on rebuilding the mill dam.[63] With the arrival of the Batthyánys, these 43 days were raised to 52 (one day every

61 *Tudositanam Nagodath, hogy micsoda munkasok kellenenek az malom gatra ... gialogok kellenenek az kik az Gatba karokat ... köveket hordani* (I wanted to let you know my lord that we need tenant peasant to pole logs and carry stones for the dam) – letter of Mátyás Gerdákovics to Ádám Batthyány, 4 December 1640 (MNL OL P 1314 no. 16 040).

62 *Ngodnak akaram ertesre adnom ... nagy szakasztasokat tett az Raba, mind az Malom gattian, es azon kivőlis, melynek megh csinalasanak es megh toltesenek mostan vagion ideje ... Az Varast kólól keritennÿ Palankkal, annakis most volna ideje...* (I wanted to let you know my lord that the Rába caused major damage on the mill's dam and elsewhere the rebuilding of which we are currently working on as this is the time of its [of the structure of the dam] filling. The town had to be surrounded with the fence as this is the time to do that) – letter of László Bozay to Ádám Batthyány, 11 May 1637 (MNL OL P 1314 no. 7593).

63 Terrier to the tenant peasant of Farkas Tarnóczy, 1561. MNL OL P 1313 Úrbéri iratok [Urbarialia] 203. rsz. 394–395. p. On the duties of the tenant peasant slightly earlier, in 1548, see: MNL OL P 1313 Úrbéri iratok [Urbarialia] 203. rsz. 90. p. Amongst the duties of the local tenant peasant, help in rebuilding of the mill dam was listed even in 1700: MNL OL P 1313 Majoratus Lad. 2. no. 119. 88–91. p. On the document, see János Varga J., *Jobbágyrendszer a magyarországi feudalizmus kései századaiban, 1556–1767* [The tenant peasant-system in Hungary in the late feudal period] (Budapest: Akadémiai, 1969), 299. On other duties, see: Tóth, "Körmend a kora újkorban," 117–121.

week) in the early seventeenth century, and a significant number of these days were devoted to rebuilding the mill's dam. From the 1600s onwards, moreover, the burghers of Körmend also had to face the problem of providing quarters and supplies for a rapidly growing number of soldiers.[64]

A couple of times during the period studied, floods and other natural phenomena damaged the mill, the dam, or the bridge, and tenant peasants were called on to fell trees, bring wattling, and strengthen the structure by adding mud where necessary. Several times, however, other, more pressing issues appeared, for instance agricultural work such as sowing or harvesting had to be carried out instead. In these cases, work on the structures was either deferred or tenant peasants were ordered to come to Körmend from the Őrség, and in some cases from Güssing, even further away. At times when the work was halted, the mill could not function, which in some cases caused a significant loss of income to the domain and the landlord.[65]

One other aspect of rebuilding needs to be emphasized, namely that this activity required certain environmental conditions. If water levels on the Rába were high or the weather was particularly wet, it was impossible to access the structures.[66] Most of the damage occurred during winter ice floods,

64 In a letter dated 28 February 1649, István Nemsem mentions half-day unpaid labor (MNL OL P 1314 no. 54 886), but this probably refers to unpaid labor with animals, as letters both from 1641 and 1649 mention one day a week of unpaid labor.
(MNL OL P 1314 no. 16 083 and no. 33 726). On the problem of supplying the military personnel, see: Tóth, *Jobbágyok, hajdúk, deákok*, 21–27.

65 Amongst others: MNL OL P 1314 nos. 16 123, 16 183, 33 643, 33 644, and 33 750. For the income provided by the mills in domains, see István Kenyeres, "The Economy of Castle Estates in the Late Medieval Kingdom of Hungary," in *The Economy of Medieval Hungary* (East Central and Eastern Europe in the Middle Ages, 49), eds. Balázs Nagy et al. (Leiden and Boston: Brill, 2018), 394–416.

66 E.g., *...Az malomhozh valo segitseget Ngod mostan rendellie megh, might egieb dolgok eleó jónnek, hogÿ az gatat megh tolthessek es Gialogh emberek Kik azt be rakjak, mivel mostan az Rábán kicsin, mostan munkalkodhatnanak raita...* (My lord, you should order the help to the mill now before anything else comes up because the tenant peasant could work on the filling of the dam as the Rába is low at the moment; it is possible to do the works) – letter of László Bozay to Ádám Batthyány, 9 June 1637 (MNL OL P 1314 no. 7595). Both factors – the low water level of the Rába and the good weather – appear in a letter by István Nemsem to Ádám Batthyány, 20 September 1649: *...Kegmes uram Nagodnak csak azzon dologh felöl akaram megh talalnom mivel mostan az Raba kicsin, s, Istennek hala az idois io. kerem Nagodath mint Kegmes Uramath, az iövő hetre rendelien Nagod az Ujvarj tartomanybul 15 avagy 16 szekeret hat hordatnok agot az itt valo malom gat toltesere, mert most szinten io ideie volna...* (My lord I am writing you because the Rába at the moment is low and thank God so is the weather; therefore, I ask you my lord that you would order from the domain of Güssing down 15 or 16 carts to help filling the dam because this is right the time to carry out the work) – MNL OL P 1314 no. 33 748. In the latter case, help arrived belatedly, as a letter from 13 October still refers to the mill as being in ruins: *...Kerem ezeert Nagdath mint kegmes uramat parancsollia megh S[i]monicsnak hogy jövö hetre*

and in the springtime, the harm done to the mill was connected to melting snow in the mountains in present-day Austria. Reconstruction works usually fell between the period after the spring flood and the second half of spring or the beginning of summer, which is generally the busiest period of the agricultural year.[67] Even if there had been no major floods, some work was necessary on these wooden structures on an annual or bi-annual basis. The parts standing in water rotted rapidly,[68] and these structures began to sag unless they were renewed regularly.[69] The stakes that stood in the water had to be of fir to withstand the water as long as possible, but acquiring wood was usually circuitous, as special permissions were required to cut trees for any purpose in the local manorial forests (the so-called 'forbidden forests', as noted above) and there were no fir trees in the domain.[70]

Maintaining the fortifications, the bridge, the mills, and their dams seems to have been a major problem for the *provisors*, consuming huge amounts of money, and requiring immense efforts made by the local peasantry. The size of the medieval source material is somewhat insignificant compared to the early modern sources on the settlements along the river, as they almost entirely lack references to the problem of the mills being abandoned or the bridge being out of use. In the early modern period, however, these problems often appeared in the surviving sources, and in most of the cases, the main reason behind them was the frequency and the magnitude of floods, which is discussed below. The main task is to find the reason for the numerous floods.

küldgie ala az szekereket mert igen kar illien kerge ideöben, az malomnak pusztan allany... (I am asking you my lord that you would order Simonics to send down the carts next week because it is pity that the mill is wasting in this brisky weather) – MNL OL P 1314 no. 33 750.

67 See e.g., MNL OL P 1314 no. 15 974 (letter of Mátyás Gerdákovics to Ádám Batthyány, 13 June 1638), no. 16 010 (letter of Mátyás Gerdákovics to Ádám Batthyány, 7 July 1639), no. 16 178 (letter of Mátyás Gerdákovics to Ádám Batthyány, 19 June 1644), no. 54 775 (letter of Mátyás Gerdákovics to Ádám Batthyány, 28 March 1643), 43 767 (letter of András Somogy to Ádám Batthyány, 26 May 1620), no. 33 640 (letter of István Nemsem to Ádám Batthyány, 3 May 1646).

68 *Az minemö Ágall valo töltesth tsinaltunk volt, az Raba mellett, es az Palank közben, imar mind el avult, es el rothattak...* (the pier we filled with branches by the Rába as well as close to the fortifications all became obsolete and did rot by now) – letter of Gáspár Francsics to Ádám Batthyány, 10 March 1651 (MNL OL P 1314 no. 15 301).

69 *...Az itt valo malom gattia annÿra el sülliedt, hogÿ czak aligh megien az vÿz az malomra, az kövek ninczenek egÿ teniernÿ vastagok ugÿ elvastak...* (the dam of the mill here is so low that there is hardly any water that reaches the wheels, and the grinding stones are barely only about a palmful thick) – letter of György Falusy to Ádám Batthyány, 27 July 1658 (MNL OL P 1314 no. 13 280).

70 See, e.g., the letter of István Nemsem and Ádám Batthyány, 21 March 1651 (MNL OL P 1314 no. 33 801).

3.3 Floods – The Unintended Consequences of Dams and Mills?

In an important but not contextualized study by Béla Iványi, and in a list compiled by another historian following in Iványi's footsteps, Zsolt Gellén, numerous floods that affected the Rába valley were collected. The data emphasize that some of the floods had serious consequences for the town of Körmend. Iványi suggested that "from the mid-1630s to the end of the 1650s, the mill at Körmend fell victim to the Rába multiple times every year."[71] This gives grounds for some further considerations. First, although they both listed several floods, they failed to highlight that it indeed was such a frequently occurring phenomenon in the early modern period, and second, and much more importantly, if their assumption is correct, and there was a high frequency of major floods in the seventeenth century, especially in the above periods, than why was that the case?

As already pointed out in the previous chapters, the Rába has one of the most changeable water levels in the Carpathian Basin even in modern times. To mention only a few events as examples, major floods occurred on the river in 2009, followed by more damaging ones in 2012, 2013, and 2014. It is unlikely, however, that a settlement founded in the mid-thirteenth century was established at a terrain directly endangered by floods and that the buildings connected to the river were built at places where the floods could potentially endanger profitable operations. Changes in the flood regime of a river can be attributed to numerous factors. Changes in the land-use patterns and the forest cover of the catchment area could certainly affect the water regime of a river. In other parts of Central Europe, as mentioned above, changes in the water regime and the water table have been attributed to climatic processes.[72] Floods, of course, can be caused by direct manipulation of the riverbed and river banks, such as building up the flood plain or by constructing dams and levees.

One must, of course, also keep in mind the possibility that there was no change in the water regime of the Rába, and that the seemingly large number of floods in the first half of the seventeenth century, which will

71 Iványi, *Képek Körmend*. For the quotation, see Gellén, "A Rába védelmében," 249.

72 The reconstruction of Lajos Rácz published in 2001 did not presume a particularly wet period in the studied decades, according to his work the first half of the seventeenth century can rather be characterized by a relatively dry period: Rácz, *Magyarország éghajlattörténete*, 297–298.

be discussed below, is no more than a product of the better-documented nature of the period. Floods of the Rába have been recorded since the eleventh century; however, the scarcity of sources up to the 1600s allows limited insight into the flood regime of the river.[73] A few scattered data on floods from the period up to 1700 come from legal evidence and narrative sources in addition to the numerous seventeenth-century private letters. With the growing number of written sources available, the number of such references increased from the mid-sixteenth century. In 1543, a letter refers to a flood delaying the operation of the mill at Ikervár, mentioned also in the above-discussed survey from the same year.[74] Similar sources refer to the problem of crossing the Rába because of flooding,[75] which does not seem to have had an extraordinary impact. But as early as the mid-sixteenth century, a letter refers to a village in the river valley being recurrently endangered by floods. This happened in May 1556 when, according to a letter written to Tamás Nádasdy – the same landlord who had ordered the survey of the river some ten years before – the tenant peasants of Csány had to take extraordinary measures to protect their settlement from the water.[76] This was probably a serious flood, but hardly any other information is available on this event. The growing military importance of the river combined with the new management system discussed above led to the significant growth in the amount of information available on all kinds of matters related to manorial management. The letters allow a nuanced discussion of the most significant environmental problem: floods. Below I outline how the domain of Körmend was affected by floods and how the local administration responded to this recurring problem.

73 Andrea Kiss, "Floods and Long-Term Water-Level Changes in Medieval Hungary," PhD-diss. (Central European University, 2011), and based on that: eadem, *Floods and Long-Term Water-Level Changes in Medieval Hungary* (Springer Water) (Cham: Springer International Publishing, 2019).

74 Elemér Mályusz, "Az Országos Levéltár Nádasdy-levéltárának magyar levelei: IV. közlemény," [Hungarian language letters of Nádasdy archive of the Hungarian National Archive] *Levéltári Közlemények* 3, no. 1 (1925): 76 (no. 69). For the original: MNL OL E 185 [letters of Lőrinc Farkas to Tamás Nádasdy (1531–1543)] 38–41. p.

75 E.g., the letter of Márk Horváth to Tamás Nádasdy, 1 May 1557. MNL OL E 185 [letters of Márk Horváth to Tamás Nádasdy (1548–1561)] 115–116. p. edited in: *Négyszáz magyar levél a XVI. századból, 1504–1560.* [400 letters from the 16th century], ed. Ágoston Szalay (Pest: [Landerer és Heckenast], 1861), 231 no. CCXLVII, and the letter of Ambrus Bejczy to Boldizsár Batthyány, 21 April 1569. The letter is lost, but was referred by Béla Iványi, *Képek Körmend*, 101–102.

76 *500 magyar levél a XVI. századból*, 2 vols. [500 letters from the 16th century], ed. Sándor Őze (Budapest: MNM, 1996), vol. I, 224–225 (no. 124).

3.3.1 The Impact of Floods on Communication Networks and the Local Economy

There is no point in analyzing individually each of the flood reports that have been preserved from the period between 1600 and 1659.[77] Here I give an overview of the different spheres of the economy that were impacted by the floods, including communication problems, farming difficulties, and changes in the built environment of Körmend. I will also discuss how forest resources from the area were used to rebuild after a flood.

Numerous letters address the problem of the river's impact on communication networks, most of all the operation of the bridge. The first such reference is from November 1603, when a terrier notes that the bridge at Hollós, the first village with a bridge downstream, was destroyed by floods. There can be little doubt that the same flood impacted the bridge at Körmend as well.[78] In a letter dated 20 March 1624, one of the officials of the Batthyány family, Kristóf Hagymásy, informed his landlord that even if he tried, it was unlikely he would be able to cross the Rába at either of the bridges (Csákány, Hollós, or Körmend) because of flooding.[79] The year 1637 saw multiple floods. This time it was probably the spring flood in the middle of April that destroyed the bridge.[80] The *provisor*, László Bozay, wrote to Ádám Batthyány a week after Hagymásy, excusing himself for not having written earlier but that he first had to discuss the losses with the miller and the bridge master.[81] The rebuilding did not go as fast as hoped, as on 11 May he again had to urge sending a "steward" (called *sáfár* in Hungarian) from Güssing to Körmend so that they would be able to start work on the bridge. Several letters are preserved describing similar cases from the 1640s and

77 Vadas, *Egy határfolyó környezettörténete*, 134–162, and András Vadas, "The 'waters leave their beds frequently' – A Western-Hungarian Town and the Flooding of the Rába/Raab River in the Seventeenth Century (1600–1659)," *Water History* 5 (2013): 267–286.

78 *Habetur hic Pons Super fluvio Raba, a quo dum currus transferunt, Soluntur a Singulo denarius 4 qui nunc propter inundationem fluvii destructus est* – MNL OL P 1313 Majoratus Lad. 1. no. 16. 27. p and MNL OL E 156 Fol. 11. no. 27. 33. p. (original pagination).

79 Letter of Kristóf Hagymássy to Ferenc Batthyány, 20 March 1624 (MNL OL P 1314 no. 108 673). The letter is referred to in: Iványi, *Képek Körmend*, 101–102, and Gellén, "A Rába védelmében," (the latter erroneously dating the letter to 1634).

80 MNL OL P 1314 no. 19 159. See above note 186.

81 *Az mjnemő szakajtast az Raba tetth, az bizzonios Ngos Uram, hogj igen nagj munkat szerzet, oka az, hogj addigh Ngodk nem jrtam, akartam elsőben, az Molnar, es Hidmesterekel felőle szollanj es discurralnj...* (the breakage that the Rába caused my lord certainly made major work necessary. The reason that I did not write you earlier is that first I wanted to talk and discuss with the miller and the bridge masters) – letter of László Bozay to Ádám Batthyány, 23 April 1637 (MNL OL P 1314 no. 7588).

the 1650s alike; they show that the flooding of the Rába had a considerable impact on the viability of the roads and communication between the two banks of the river. This was a major problem for Batthyány, as much of his land lay on the right bank of the river (see Fig. 3.1).

Other letters concern the different economic difficulties caused by floods and changes in the Rába's water level in general. Some of them refer to the problem of the flooding of plowlands, meadows, and gardens. For instance, a letter from 1614 mentions that instead of lush meadows, the cattle had to be driven to wastelands. The lack of meadows and hay was an ever-growing problem at Körmend, as the arrival of significant military forces at the settlement meant that their horses needed huge amounts of hay.[82] In other cases, floods endangered the beginning of the harvest. In a letter dated 10 July 1611, for instance, András Somogyi, the *provisor* of Körmend during the lordship of Ferenc Batthyány, explains that they cannot begin the harvest due to flooding. The date of the letter is as important here as the information about the planned beginning of the harvest. In this area in the first half of the seventeenth century, the beginning of the harvest was usually some ten days later, but according to data from other Central European areas, much of Switzerland, and the Czech lands, the weather was particularly warm from the spring into June of 1611.[83] The letter does not specify how the floods endangered the harvest, and thus it is not clear whether it was the accessibility of the fields or the flooding of the fields themselves that caused problems. Nevertheless, Somogyi notes that it was not only the floods that caused a potential threat but also the rain, which could rot the grain on the stems.[84]

Besides endangering the plowland and the gardens recurrently, flooding also affected livestock on some occasions. Either for pannage or for grazing, both pigs and cattle were driven out onto the floodplains, which were beneficial lands. On some occasions, however, floods arrived so quickly

82 *Kegielmes uram Nagisagodat kellietek megh talanom, az kathonak az mi kivesen vadnakis et keőrmenden nem tudnak holot fiuelnie, mivelhogi az polgarsaghnak az hol effele parragok volt, mind fől kőlt iobbara immar, raita veszőttek marhaiokkal a sok az vizek miat Nagisagod rendelne valahol Egÿ darabot a ki is kótózhetnenek* (My lord, I have to write you because no matter how few soldiers are here in Körmend, they have no place [for their horses] for grazing because the fallows that the burghers own have already been peeled. The cattle peeled it because of the high waters. I beg you my lord that you would provide them with a piece of land where they could move [their horses there]) – letter of István Blaskovics to Ferenc Batthyány, 11 May 1614 (MNL OL P 1314 no. 6834).

83 Dobrovolný et al., "Monthly and Seasonal." For the raw data, see ftp://ftp.ncdc.noaa.gov/pub/data/paleo/historical/europe/dobrovolny 2010temperature.txt (last accessed: 17 February 2020).

84 Letter of András Somogy to Ferenc Batthyány, 10 July 1611 (MNL OL P 1314 no. 43 755).

that livestock was stranded between the branches of the rivers, and on two occasions, losses of significant numbers of pigs were described in letters.[85] Besides livestock, another area of animal exploitation was affected by the water regime of the river, that is, fishing. The role of fishing in the pre-modern Hungarian economy was important, as fish was part of the everyday diet. Medieval and early modern average fish consumption may have been multiple times the present-day annual amount of approximately five kilograms per person in Hungary, and it was even more important during Lent.[86] Flooding, according to several sources, hampered fishing and in some cases made it completely impossible. This was especially true for the larger species, as nets could not be used during major flooding.[87]

Another problem related to local agriculture was the impact of floods on forest resources. In this context, at least two issues must be considered: the impact of floods on the forest resources themselves and the timber supply for reconstruction work after floods. As it is clear from the river survey of 1543–44 discussed above as well as terriers from the manorial complex, the only major forest in the immediate surroundings of Körmend was the so-called Dogobói Forest. This forest was right at the confluence of the Rába and the Pinka,[88] one of the larger tributaries, and thus the forest was affected by floods of both rivers. The trees – mostly oaks – could withstand temporary water cover in the forest, but the lack of acorns, which were the basis of pig foraging for a good part of the year, caused problems in the economy. Since the forest was one of the two forbidden forests (*silva prohibita*) at the manor, its main function was to provide not timber but acorns (for pannage). This problem is emphasized, for instance, in a letter from 1646 in which it is explained how the local *provisor*, as was usual for issues related to local farming and provisioning, warns the landlord that the acorns left in the woods after the flood would only support the pigs for

85 See section 4.3.4 on the floods of 1641.

86 Anon., "A haltermelés és -fogyasztás alakulása a világon és Magyarországon," [The fish production and consumption in the world and Hungary] *Statisztikai Tükör* 6, no. 84 (2013): 1–4. Online document: https://www.ksh.hu/docs/hun/xftp/stattukor/halaszat.pdf (last accessed: 3 April 2020)

87 For similar problems at Rába, see: *Halat Ngodnak kűltem volna de sokbul szeret nem tehetem, mivel nagy árr víz vagion, minden szenankat el vitte az víz…* (I would have sent you fish my lord, but I could not find much because of the floods, it also took all our hay) – letter of István Nemsem to Ádám Batthyány, 9 July 1648 (MNL OL P 1314 no. 33 705). For the same problem by other rivers, see Vadas, "Floods in the Hungarian Kingdom," 92 and 98 (nos. 45 and 49).

88 *Item penes eandem possesionem Nadallÿa silvam dolorosam et glandiferam ad prefatum castrum Kermendh pertinentem interfluvios Raba et Pÿnkwa in territorio dicti dominÿ Kermend existendi sitam* – MNL VML XII.1. Jegyzőkönyvek [Registers] 1589. no. 85. 120. fol.

about a month.[89] This estimate may have been cautionary – or intentionally overexaggerated to prompt a reaction from the landlord – but by the end of the year, there was no mast left on the domain.[90]

The forest was not used intensively in the local wood supply, as the timber used for re-building the buildings was usually not cut there. For instance, after a major flood in early 1648 that required all the wooden parts of the manorial mill to be replaced,[91] the wood was provisioned from outside the domain.[92] Even though the Dobogói Forest was not used systematically for wood supply, it had to be maintained regularly, especially after floods. A letter from 1655 leaves no doubt of this. After a flood in February 1655, the accountant of the castle of Körmend orders tenant peasants from two nearby villages, Boldogasszony and Győrvár (both belonged to the manorial complex for only a short period), to collect the trees fallen in the forest. The job was not completed for a while, as a few days later the Rába inundated the forest again and the tenant peasants had to be sent home.[93] Eventually, the fallen trees were collected in the forest, and a letter dating from ten days later tells of timber processed there being prepared for delivery to the mill's dam.[94] This letter indicates the most frequently occurring problem related to flooding along the Rába: its impact on the built environment, most importantly on water-related constructions.

One of the recurring problems was the erosion of the town wall. The multiplied speed and amount of water and drift carried by the river during floods frequently resulted in small landslides by the piers, which meant that parts of the mostly wooden town wall would begin to collapse either into

89 Letter of István Nemsem to Ádám Batthyány, 28 September 1646 (MNL OL P 1314 no. 33 647).

90 Letter of István Nemsem to Ádám Batthyány, 22 December 1646 (MNL OL P 1314 no. 33 651).

91 *Mivel az Nagod parancsolatia szerint az Uyvarj álcz ala iűt volt az itt valo malom nyzetni, azt mondgya az aálcz, hogy itt az minymeö faia vagion az malomnak, az mind oda vagion...* (as ordered by you my lord, the carpenter from Güssing came down here to see what happened with the mill here. The carpenter says all the timber of the mill is busted) – letter of István Nemsem to Ádám Batthyány, 3 January 1648 (MNL OL P 1314 no. 33 676).

92 Letter of István Nemsem to Ádám Batthyány, 20 January 1648 (MNL OL P 1314 no. 33 677).

93 *Kegmes uram ha Ngod io valia Dobogoi Erdöben nemmÿ dűlt fa vagion Györvarÿ gialogokat, Boldogaszony falviakat ra haitom, es hasogatast csinaltatok, s, az vár, elöt valo keret be keritetem ne legien az vár kert nélkűl, be is iűtek volt az gialogok, de nagy Viznek miatta haza boczatam űketh...* (My lord, as there are many fallen trees in the Dogobói Forest in case you gave permission, I could order the tenant peasants of Győrvár and Boldogasszony to chop the trees, and I will have the garden in front of the castle surrounded by a fence so that the castle would not be without a garden. The peasants came for the work, but I had to let them go because of the high waters) – letter of Mihály Szokoly to Ádám Batthyány, 12 February 1655 (MNL OL P 1314 no. 48 128).

94 Letter of Mihály Szokoly to Ádám Batthyány, 22 February 1655 (MNL OL P 1314 no. 48 121).

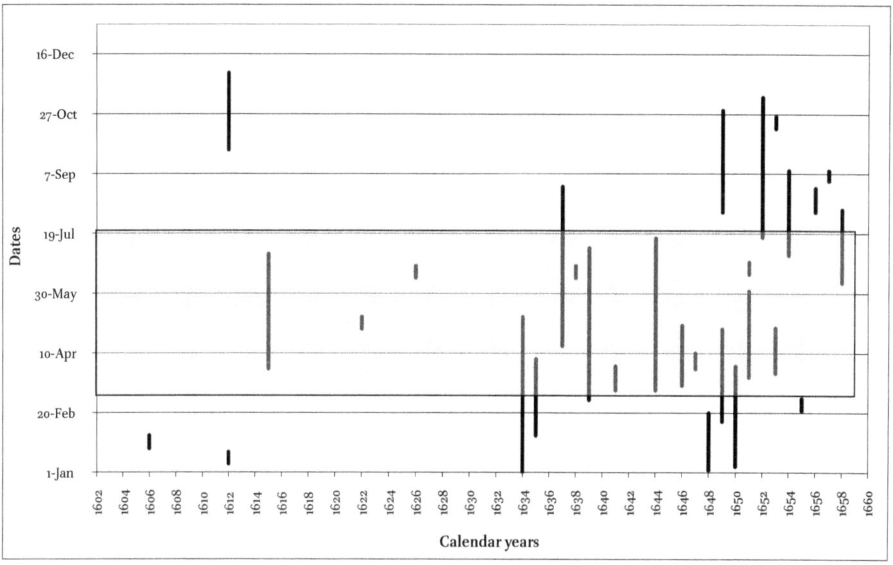

Figure 3.6 The dates of building work on the mill dam and Körmend's fortifications by calendar years (the period marked in gray is the most frequent period)

the Rába or the moat that surrounded Körmend. Thus, each year after the spring floods, the *provisors* tried to have the earth and wood fortifications and the mill dam re-mudded so that the constructions would be consolidated as soon as possible. A letter from May 1620 nicely demonstrates this by mentioning that work should be done, as this is the "usual time to do so."[95] Although the letters and *terriers* give information on the more diverse scheduling of similar works, clearly May and June – before the harvest – were most frequently the period for rebuilding the dam and the fortifications (see Fig. 3.6). As noted above, this time of the year was probably the busiest during the agricultural year, but nonetheless this timing was the most appropriate, similar to that of other works.

While sowing or harvest could hardly be postponed, works on the fortifications and dam were postponed frequently.[96] Several letters reveal that the construction on one of the buildings was so belated that a new flood arrived before the previous rebuilding had been completed or had even begun. In 1637, for instance, after a great flood in April, no major work was

95 Letter of András Somogy to Ferenc Batthyány, 26 May 1620 (MNL OL P 1314 no. 43 767).

96 Letter of László Bozay and András Hidasy to Ádám Batthyány, 30 May 1634 (MNL OL P 1314 no. 19 312), and letters of László Bozay to Ádám Batthyány, 11 May 1637 (MNL OL P 1314 no. 7593) and 9 June 1637 (MNL OL P 1314 no. 7595).

started until mid-June despite the joint efforts of the *provisor* and the vice-captain of Körmend. Because of the frequent postponement of rebuilding work, the landlords were often stung into action by letters with dramatic wording. After the flood in March 1639, the recently appointed *provisor*, Mátyás Gerdákovics, wrote to Ádám Batthyány that unless they started work immediately on the rebuilding, the whole mill would be destroyed.[97]

According to the letters, the construction sometimes required major efforts from the local administration, as huge amounts of material had to be transported and then built into the structures. In 1644, the same Gerdákovics wrote again to Batthyány that a flood in March had broken the mill dam at two points and carried away its parts. In the letter, he informed the landlord of the necessity of thousands (!) of cartloads of sticks for wattling. This seemingly immense amount is not unheard of in the early modern period, but this particular problem may have been exceptional, for the provisor added that without intervention, the mill might not be able to operate at all in that year. This probably did not happen in the period covered by letters and terriers consulted for this chapter, although out of the five or six wheels of the manorial mill, in most cases only two or three were in operation at any one time.[98] Despite the warning, the landlord took no action, as far as can be ascertained based on a letter from later that year.[99] A rebuilding of similar size was urged only two years later, similarly after a flood. It is certainly noteworthy that this time it was not the *provisor* but an accountant of the captain of the castle who was trying to put pressure on the landlord to have the mill dam rebuilt as soon as possible.[100] It seems

97 *Kegmes uram az mostani telbelj viznek sebesege es jege az malom gatian sok kart tett az mely maid el keltene pusztulni ha Ngod Gondiat nem visely…* (My lord, the current and the icing of the river in the winter these days caused major damage at the mill which could be destroyed unless you my lord took care of it) – letter of Mátyás Gerdákovics to Ádám Batthyány, 2 March 1639 (MNL OL P 1314 no. 16 003). Letters with similar wording: MNL OL P 1314 no. 7603, 13 285, 16 056, 16 178, 16 183, 16 197, 16 201, 33 642, 33 643, and 33 725.

98 *Az mastani Ár Víz Nagyságos Uram szántalan sok kárt tett, az malom gátnak két darabját elszakasztván elvitte, az malom alól is elhozta a töltést, ugyannyira, hogy egy nihány ezer szekér ágh kivántatnék* (This flood, my lord, caused a number of damages, it broke two parts of the mill's dam, carried away the pier from the surroundings of the mill; therefore we would need a few thousand cartloads of branch) – letter of Mátyás Gerdákovics to Ádám Batthyány, March 1644. The letter is lost but is referred to in: Iványi, *Képek Körmend*, 104–105. On the problem of the mill's wheels, see Vadas, *Körmend és a vizek*, 40–44.

99 Letter of Mátyás Gerdákovics to Ádám Batthyány, 7 June 1644 (MNL OL P 1314 no. 54 786).

100 *Ma kapitan Uram eleiben hivattam az Molnarokath, es vegere mentem tűlők mennÿ szeker Ágh kivantatik az gatolashoz azt mondgyak, hogy Ezer szekerel io ha erik be vele, az szaz karoval az mint vagion penigh izoniu dologh ha az Víz megh apad czak egy fordulast sem tehet az malom. Mostan ha egy Parzazh agh lenne raita egy hetig, talan megis eleit venni. Azon kívűl az Malom*

that the dam was just as important for the military defense as for the local economy. According to the millers, a few hundred cartloads of wattling were first needed to avoid even greater damage. Consequently, thousands of cartloads of wattling and a hundred stakes would be required for the rebuilding process. A month later, the dam was still unrepaired. Although by then the landlord had probably taken measures to rebuild the structure, the expected helping hands from Őrség did not arrive.[101]

To adjust the riverbed after floods, sometimes not only the dam but also the piers of the river had to be strengthened by driving stakes down into the alluvium. For instance, after a flood in February 1650, the *provisor*, at that time István Nemsem, ordered 200 cartloads of wattle and 40 fir stakes to restore the pier. According to him, the order was necessary, for otherwise the main riverbed would eventually have changed its channel to bypass the manorial mill.[102] Apart from the significant amounts of timber that were ordered on several occasions, the workforce caused another problem because building a thousand cartloads of wattle required many days of forced labor. The days that the peasantry of Körmend were required to work were far fewer than the number of days that would have been needed. For instance, in 1651, again during a rebuilding of the fortifications, an additional 60 to 70 tenant peasants were asked to come down to Körmend from the center of the Batthyány domains in Güssing to help the local population. The letter with this request for the additional tenant peasants also implies that, despite the high number of mandatory unpaid labor days, they still could not finish all of the rebuilding work.[103] A letter from a week later leaves no doubt that the process was not completed, as the *provisor* again draws attention to the need to drive cartloads of wattling to the mill.[104]

The military and civil administrations of the town sometimes had different priorities, but their interests coincided when it came to the importance of rebuilding the mill dam or the fortifications. This common interest is clearly reflected in a case from 1651. The newly appointed captain of

silibeit zugoianak deszkait mind el vitte az Víz. (Today I asked the millers to come to see the captain to get to know how many cartloads of branches are necessary for the dam. They said they would need thousands of cartloads at least, along with a hundred logs. However, until the water level decreases the mill cannot work at all. Now if there were a few hundred [cartloads of] branches this would help for about a week. The decks of the mill race leading to the sluice were also carried away by the water.) – letter of Mihály Szokoly to Ádám Batthyány, 14 March 1646 (MNL OL P 1314 no. 48 109).

101 Letter of Gáspár Francsics to Ádám Batthyány, 19 April 1646 (MNL OL P 1314 no. 15 103).

102 Letter of István Nemsem to Ádám Batthyány, 19 February 1650 (MNL OL P 1314 no. 33 770).

103 Letter of Gáspár Francsics to Ádám Batthyány, 29 April 1651 (MNL OL P 1314 no. 15 306).

104 Letter of István Nemsem to Ádám Batthyány, 29 April 1651 (MNL OL P 1314 no. 33 816).

Körmend, István Svastics, informed Ádám Batthyány of the need to rebuild the section of the fortification wall that bordered the main flow of the Rába. The letter urges that the work be carried out soon, as the water level was low at that time, which was a precondition for completing the task.[105] This time, there was a new element to the work: the planned workforce. The captain enlisted the help of troops from the army in charge of defending Vas County, something that had never occurred in the sources before. Svastics most likely had more influence than previous captains. This is also an indication that the fortification was considered an important site for defending the broader hinterlands and not just the manor itself.

The numerous water-related problems were not disregarded by the local administrators. A few sources demonstrate that measures were taken to moderate the losses caused by flooding. One of the major problems was the huge amount of drift that was carried downstream by floods. Because of the large number of dams and bridges that stood in the river upstream, each flood carried many parts of wooden structures of various kinds. To defer losses, stakes were driven into the alluvium somewhat upstream from the bridge and the dam in the Rába to stop or deflect at least some of the drifting wood.[106] As discussed in Chapter 2, there is little evidence in this period of any flood-control system similar to what was explained by Takács.

The number of problems listed above indicate clearly that water, especially the floodwaters of the Rába, played a key role in the local economy. Significant efforts were required by the local and central administration of the Batthyány family to protect Körmend from recurring inundations and major losses to plowland, gardens, and meadows. To understand exactly how frequently this problem reoccurred, below I give an overview of the flood frequencies of the Rába in the period covered by the sources here. I then try to answer the most relevant question: what is the reason behind the frequency of these events?

3.3.2 The Frequency and Magnitude of the Floods

As a starting point for studying the monthly and seasonal distribution of the reported flood events, it is crucial to discuss the distribution of letters sent from Körmend by month and by season separately. It is important to note that the source material is not evenly distributed, especially not in the number of letters surviving from different years (see Fig. 3.3). This does not,

105 Letter of István Svastics to Ádám Batthyány, 18 April 1653 (MNL OL P 1314 no. 45 055).
106 Letter of Gáspár Francsics to Ádám Batthyány, 14 June 1645 (MNL OL P 1314 no. 15 049).

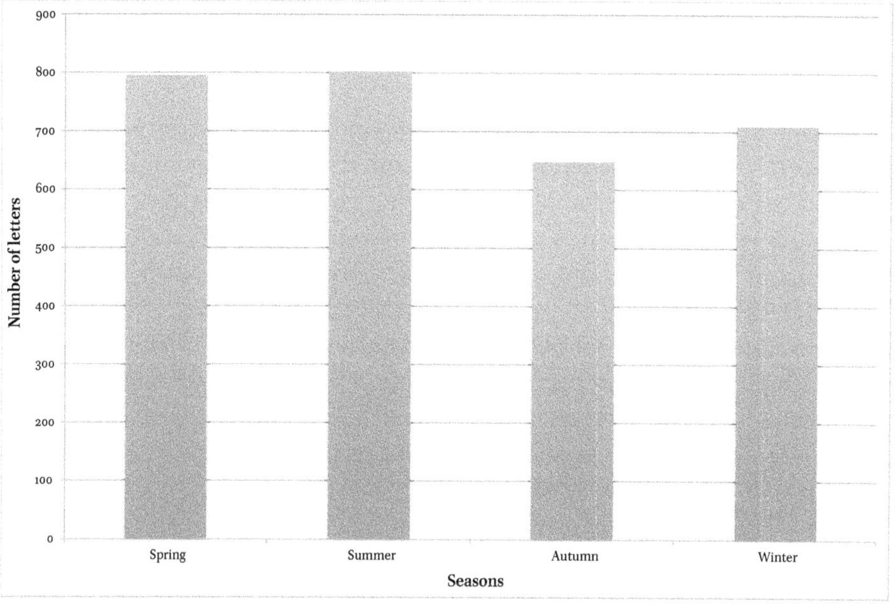

Figure 3.7 The seasonal distribution of letters sent from Körmend (1600–1659)

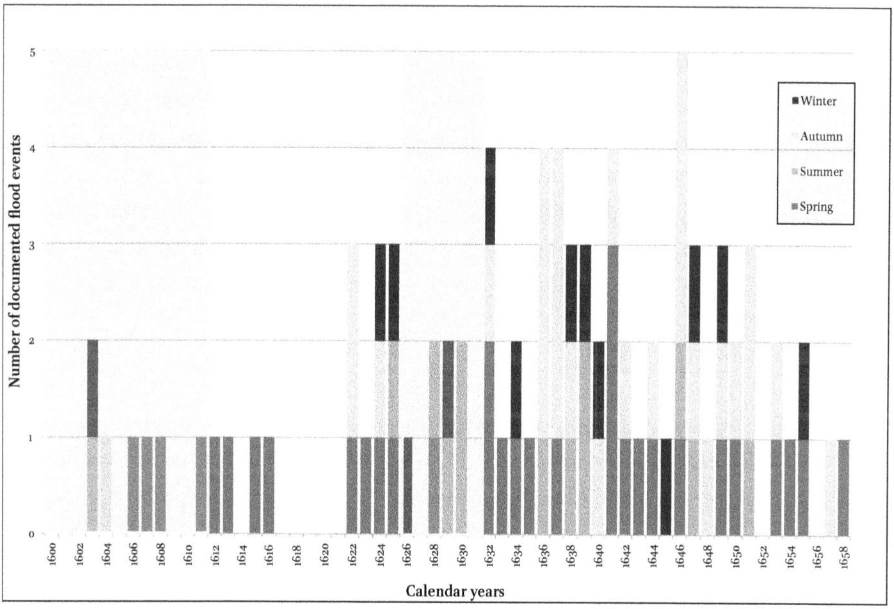

Figure 3.8 The number and seasonality of floods on the Rába River (1543–1658)
(Gray background indicates more than 20 letters per year)

however, mean that this group of evidence cannot be used to conclude the seasonality of flooding in the early modern period. While there are years from which there are virtually no surviving letters, while in other years there are more than 200 letters, the difference in the number of surviving letters between the best-documented May and the least-documented February is less than 50 percent. This makes it possible to assume that the monthly distribution of flood events reflected in the letters is certainly closer to reality than in the case of the annual distribution. This is especially true for the seasonality of floods because the seasonal distribution of letters is more even than for the distribution of letters from individual months. There are only 20 percent fewer letters from the least documented autumn than from the summer, which is the richest in terms of surviving letters (see Fig. 3.7). This supports the contention that the study of seasonality is statistically well-founded.

During the 60 years covered in the relevant source material, there are fewer years without major floods on the Rába at Körmend than years with floods (see Fig. 3.8). Two short intervals in the studied period had elevated flood frequencies, especially the eight years between 1634–1641 and between 1645–1653. During the first period, there was one year with four individual flood events (1641, see below), one year with three major floods (1634), and in three consecutive years (1637 to 1639) there were two floods annually. The other period with a large number of inundations was between 1645 and 1653, in which at least one flood event can be documented in each year. In two years, there were four floods (1645 and 1646); in three other years, three flood events (1650, 1651, 1653); and in two of the years (1647 to 1648) two floods occurred each year on the Rába. Among these, some were more serious, causing damage to the harvest, domestic animals, and to the buildings lying back from the river.[107]

Besides these periods with higher flood frequencies, floods were documented in most years from 1610 onwards. These events most often occurred in spring, but from time to time the letters also record ice floods on the river. The summer and autumnal floods occurred, with rare exceptions, in the two periods mentioned above (1634–1641 and 1645–1653). This phenomenon has not yet been explained, nor is there a similar pattern on any other Central European river in the same period.

It is equally important to consider the magnitude of floods that occurred during the examined period. Literature provides us with different systems that can be used to evaluate the magnitude of flood events on a particular

107 For each of the flood reports, see Vadas, *Körmend és a vizek*.

Magnitude	Event	Primary indicator	Secondary indicator
1	Small, regional flood	Little damage, e.g., fields and gardens close to the river, wood supplies that were stored close to the river are moved to another place	Short flooding
2	Above-average or supra-regional flood	Damage to buildings and structures related to the water like dams, weirs, footbridges, bridges, and buildings close to the river such as mills; water in buildings	Flood of average duration; severe damage to fields and gardens close to the river, loss of animals and sometimes human lives
3	Above-average or supra-regional flood on a disastrous scale	Severe damage to buildings and structures related to water, i.e., dams, weirs, footbridges, bridges, and buildings close to the river such as mills; water in buildings. In part, buildings are destroyed or torn away by the flood	Duration of the flood: several days or weeks; severe damage to fields and gardens close to the river, extensive loss of animals and people; morphodynamic processes such as sand sedimentation cause lasting damage and change in the surface

Figure 3.9 Classification of the intensity of historical floods in Western and Central Europe[108]

river. They vary in sensitivity and specificity. Some of the systems consider the spatial extent of the floods and what destruction was caused by the downstream flow of water;[109] others are more specific and focus on the damage at certain points on the rivers.[110] In Hungarian historiography, Andrea Kiss has developed a three-scale system, mostly based on legal evidence she consulted, to differentiate the spatial extent and the magnitude of the flood events (see Fig. 3.9).[111] The sources analyzed here do not contain data suitable for a study of the spatial extent of the flood events, as letters seldom refer to settlements further away from Körmend and its close environs.

108 Sturm et al., "Hochwasser in Mitteleuropa," 15; present graph from Kiss, "Floods and Long-Term Water-Level Changes," 103.

109 Katrin Sturm et al., "Hochwasser in Mitteleuropa seit 1500 und ihre Beziehung zur atmosphärischen Zirkulation," *Petermanns Geographische Mitteilungen* 145, no. 6 (2001): 14–23.

110 See, e.g., the system created by Rudolf Brázdil et al., "Flood Events of Selected European Rivers in the Sixteenth Century," *Climatic Change* 43 (1999): 239–285 and the one used by Rohr, "Measuring the Frequency." For an overview of the literature on the topic, see also: Christian Rohr, *Extreme Naturereignisse im Ostalpenraum. Naturerfahrung im Spätmittelalter und am Beginn der Neuzeit* (Cologne, Weimar, and Vienna: Böhlau, 2007), 204.

111 Kiss, *Floods and Long-Term Water-Level Changes.*

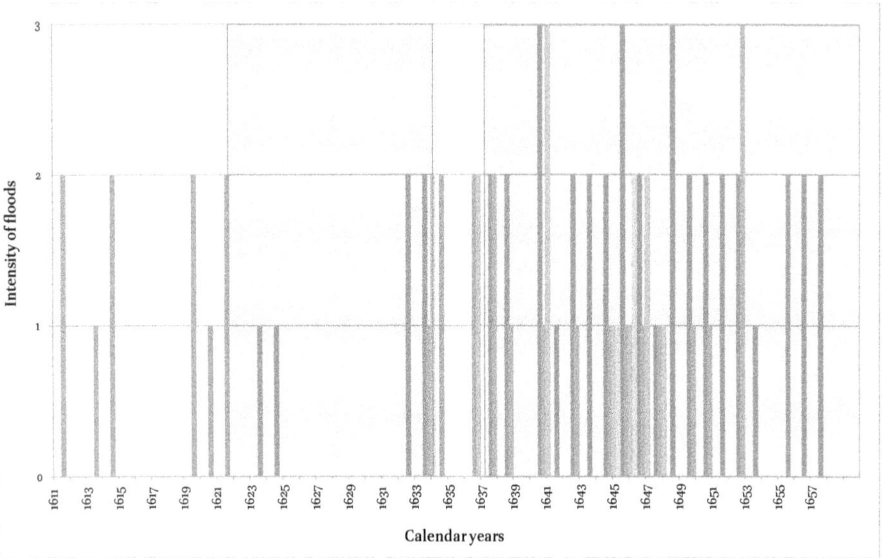

Figure 3.10 The magnitude of the floods on the Rába between [1543] 1600 and 1659[112]
(Gray background indicates more than 20 letters per year)

They do, however, yield outstanding insight into local environmental and weather circumstances. It is therefore possible to estimate the magnitude of the floods based on the primary indicators described by Sturm and his colleagues.[113] The sources provide information on the destruction of buildings connected to the river and are thus the most important tools according to Sturm's system for quantifying the magnitude of floods. It is not possible to estimate the discharge of water based on the available sources because of fundamental modern changes in the environment of the Rába River around Körmend and the almost complete disappearance of the constructions mentioned by the letter writers.[114]

The magnitude of the floods in the studied period has a high variability (see Fig. 3.10). The flood events classified as first and second levels of intensity occurred with roughly the same frequency (26 and 27 cases, respectively). Even in the third group, there were at least five events within these 60 years. In these years, 1641 must be emphasized as well, for it had two disastrous

112 Based on the system of Sturm et al., "Hochwasser in Mitteleuropa."

113 Sturm et al., "Hochwasser in Mitteleuropa."

114 On the example of the Rhine, see e.g., Oliver Wetter et al., "The Largest Floods in the High Rhine Basin since 1268 Assessed from Documentary and Instrumental Evidence," *Hydrological Sciences Journal* 56 (2011): 733–758.

(third-category) flood events that caused serious losses in material wealth and among livestock, overwhelming the people. The period after 1633 is not only important because of the large number of documented floods but also because of the magnitude of these events. Between 1634 and 1641, almost every individual case can be classified in one of the above-average categories (second or third level). After 1649, although the frequency of flooding was lower than in the preceding decade, the magnitude of the flood events increased significantly. Because of this, one may submit that there was a lasting change in the water regime of the Rába. Earlier research has also addressed the problem by suggesting a lasting decrease in the capacity of the manorial mill.[115] Although several of the source letters recommended a response to the changing conditions,[116] it is unlikely that the mill did not function relatively efficiently, as otherwise the first major intervention in the building complex would certainly have come earlier than the mid-eighteenth century, when the whole building was rebuilt and the flow of the Rába was brought under control.

3.3.3 The Seasonality of Floods – The Early Modern Period Compared to Modern Times

Several studies have dealt with the flooding of the Rába River in recent decades, which makes it easy to compare the present-day situation with conditions in the seventeenth century. Like most of the rivers on the Western periphery of the Carpathian Basin, the Rába River derives most of its water from the Eastern Alps. In the catchment area, the lowest precipitation over the year occurs in January and February, although these two short periods of higher-than-average temperatures (and partial snowmelt) can result in severe ice floods.[117] In modern times, flood frequency increases greatly in March compared to January and February, as the snow melting in the Alps raises the water levels, which must have been the case in historical times as well.[118] From spring to the beginning of autumn – especially on the Hungarian section of the Rába – drought periods occur with relatively

115 Gellén, "A Rába védelmében," 249–250, and Iványi, *Képek Körmend*, 105.
116 *…az itt valo malomnak igen keves vize vagion mind az gat alat megien* (the little water that reaches the mill runs under the dam) – letter of István Nemsem to Ádám Batthyány, 27 October 1654 (MNL OL P 1314 no. 33 943).
117 For the most comprehensive treatment of the modern floods of the Rába, see: Bergmann et al., *Hydrologische Monographie*.
118 See: Kiss, *Floods and Long-Term Water-Level Changes*, and Vadas, "Floods in the Hungarian Kingdom."

high frequency, something that can be deduced from the early modern source material as well.[119] Despite the presence of drought periods, the flood frequency in June and July is higher than the average for the twelve months. In the twentieth century, apart from the average water levels in January, the lowest values were measured in September, with a very low probability for flood events to take place in this month. The average water levels of the Rába, and the corresponding chance of floods, start to increase in October, although the probability of floods is still significantly lower than in March or June–July.

When studying the seasonality of floods in the early modern period – in this case, the first 60 years of the seventeenth century – both parallels and differences can be seen. Because of the relatively low number of cases in the early modern period, however, it is impossible to come to any conclusions about the changes in flood frequency or the average water levels of one month in the seventeenth century. Thus, I will discuss here only the comparison between the seasonal flood frequency in the seventeenth century and that of modern times. In the period from 1600 to 1659, 40 percent of the documented flood events occurred in the spring, which is similar to the information coming from modern instrumental measurements.[120] What is surprising, though, is the relatively even distribution of floods in the other three seasons in the examined period. This picture is fundamentally different from twentieth-century flooding on the Rába River. The relatively high number of documented floods in the autumn, however, is connected to two years – 1645 and 1646 – when three individual flood events occurred on the Rába. This represents half of all known floods occurring in the autumns of the studied period. The eleven known flood events in the winter were, without exception, ice floods. Although in modern times, ice does form on all the rivers in the Carpathian Basin, human intervention (such as water regulation, drainage) has changed the ice regime of the rivers drastically, including the Rába, with the result that ice floods are no longer common on the rivers of Transdanubia.[121] The flood regime of the river was probably somewhat different in early modern times; however, mostly because of the

119 E.g., in 1637 [MNL OL P 1314 no. 7595], in 1649 [MNL OL P 1314 no. 33 748], in 1652 [MNL OL P 1314 no. 33 845], in 1653 [MNL OL P 1314 no. 45 058], and in 1656 [MNL OL P 1314 no. 13 293]).

120 Bergmann et al., *Hydrologische Monographie*.

121 Katalin Takács, Zoltán Kern, and Balázs Nagy, "Impacts of Anthropogenic Effects on River Ice Regime: Examples from Eastern Central Europe," *Quaternary International* 293 (2013): 275–282, and Katalin Takács and Zoltán Kern, "Multidecadal Changes in the River Ice Regime of the Lower Course of the River Drava since AD 1875," *Journal of Hydrology* 529 (2015): 1890–1900.

major interventions to the river's flow, it is now rather difficult to pinpoint the reasons why.

3.3.4 A Year of Severe Floods: A Case Study of 1641

The year 1641 was one of the richest in floods, and the magnitude of the flood in September of this year must have been one of the greatest. For this reason, and also because of the extensive documentation, the floods of 1641 will be examined here in detail to show the potential of qualitative research on floods based on private letters and to demonstrate how fundamental fighting with floods was in a frontier settlement. The aim here is not only to discuss the details available on the flood events but, if possible, to study their weather backgrounds. Because of the clear parallels with the floods described in letters from the aforementioned Csákány, I am referring to the situation there as well.

The magnitude of the first 1641 flood connected to the snowmelt is well documented in a letter written by András Hidasy, the vice-captain of Körmend.[122] According to his account of the event, the bridge of Körmend changed the flow of the river so drastically that it moved to a new bed, diverting the water in another direction and jeopardizing the water supply of the manorial mill. As discussed, a change in the flow of the Rába was hardly unique, but it still shows the magnitude of this flood event. A week later, Hidasy asked for help in rebuilding the original channel, which was again something that was to happen recurrently. Even when the flood peak moved away from Körmend, it was impossible to get close to the River Rába with either horses or wheeled vehicles.[123] The reconstruction work following this massive flood event required workers from outside the manorial complex, as in many of the cases mentioned above. However, the aid did not arrive as had been hoped. The *provisor* of the estate, Mátyás Gerdákovics, wrote a letter pointing out the likelihood that the mill had entirely been destroyed. The work done by the millers and their assistants was not sufficient to reconstruct the building, according to Gerdákovics.[124] Although the reason for his visit is unknown, Ádám Batthyány went to see Körmend in the middle of April of that year, which seldom happened during his long lordship. A few days later, a letter from Hidasy mentioned that while his lord had been in the town, he could see the flood subsiding, but then

122 Letter of András Hidasy to Ádám Batthyány, 3 March 1641 (MNL OL P 1314 no. 19 233).
123 Letter of András Hidasy to Ádám Batthyány, 10 March 1641 (MNL OL P 1314 no. 19 235).
124 Letter of Mátyás Gerdákovics to Ádám Batthyány, 29 March 1641 (MNL OL P 1314 no. 16 056).

the water level started to increase again, which endangered the bridge and the mill.[125] Floods occurred on other rivers as well. Hidasy mentions that when Batthyány left Körmend for Keszthely (a town some 70 kilometers to the southeast, by Lake Balaton), he sent an envoy with a message. The envoy reached the town of Keszthely, but he could not follow Batthyány's path further because of flooding. Although Batthyány's itinerary has yet to be compiled, it is likely that the envoy's passage was blocked not by the Rába – or not only by that – but rather by the Zala River and other rivers in the western Transdanubia.

The flood related to the snowmelt was similarly serious at Csákány. Just as at Körmend, a mill controlled the local water level, and its dam was similarly destroyed by the flood. As the Dobogói Forest was closer to Csákány than to Körmend, the local administrator (technically a *provisor*), Ferenc Gencsy, asked for permission to fell trees in the forest for rebuilding.[126] Without this work, the mill could not grind grain at all. When finally, in May, the landlord gave the order to bring timber to Csákány, trees were not cut in Dobogó but at Ivánc in the Szenterzsébet Forest.[127] I could not determine the location of the Szenterzsébet Forest with certainty, but Ivánc is some five kilometers south of Csákány on the right bank of the Rába. The magnitude of the spring flood here is evidenced by the fact that it inundated not only meadows but also some plots that were probably used as plowland and were leased to peasants at the time.

Although the spring floods in 1641 may have caused damage, the flood in September was far more destructive than the preceding ones. It seems to have had the greatest magnitude in the whole of the studied period. The sources tell of this event in more detail than about any other case from the seventeenth century, since both Hidasy and Gerdákovics wrote about the

125 Letter of András Hidasdy to Ádám Batthyány, 17 April 1641 (MNL OL P 1314 no. 19 244).

126 *Kegielmes Uram akaram Nagodnak meghirnom, hogj az telen nem tudom mj okbol el mulek az Nagod chakanj Molnanak faianak megh hozatasa melj Nagodnak nagj karaara következik hiszen kegielmes Uram az Dobogon avagj az Füzesi erdőn megh talatok volna minden fajat, mert most az Nagj arviz mind fejfaiat s mind penigh az kin az fej fak allotak azokat az szegeket ell törtő az arviz...* (My lord I wanted to remind you that for some reason the provisioning of the timber for the miller at Csákány was halted in the winter which causes major loss to you, my lord. We could find the wood either at Dobogói or at the Füzesi Forest but now there is a big flood that broke the top of the mill as well as its foundations under the logs) – letter of Ferenc Gencsy Ferenc to Ádám Batthyány, 15 March 1641 (MNL OL P 1314 no. 15 832).

127 Letter of Ferenc Gencsy to Ádám Batthyány, 12 May 1641 (MNL OL P 1314 no. 15 835). On the constructions see also: MNL OL P 1322 IV. Memorialék. Összeírások. Vegyes iratok (Számadások, leltárak, készpénzkimutatások, malomépítéshez anyagösszeírások) [Memorabilia. Conscriptiones, miscellanea (Accounts, inventories, cash accounts, and material provisions for mills)] 1641 no. 49.

event in numerous letters. The main body of the first letter on the issue sent to Ádám Batthyány on 1 September does not refer to a flood; Hidasy added that information in an extensive post script. He must have had his message written down earlier by one of his scribes, so the post script refers to the most recent events. The vice-captain mentions that in the morning when the letter was written, the water level was not particularly high, but then the river started to rise rapidly. The fact that cattle were driven out to the pastures as well as the fact that their herders were marooned between the branches of the Rába demonstrate the sudden onslaught of the flood. When this first letter was written, nothing was known about the destruction because the current of the Rába was so fast that no one could get close to the river itself. The water levels must have been higher than ever before in the study period, as it not only affected buildings connected to the river but inundated the town's market square, which was very likely to have served its function at that site since the mid-thirteenth century at the latest. The river also flooded Hidasy's grain stocks. The terriers are not explicit about the location, but storage buildings were undoubtedly constructed in a place that was ordinarily deemed safe from floods, just like market squares. The letter also notes that the water reached houses, including one that was owned by György Falusy, one of the leading administrators and the *provisor* of the town.[128] The Falusy family owned three buildings in Körmend according to an account of the houses compiled somewhat later in 1649. The one on Kapitány Street (close to the market square) was seemingly where the family itself lived because in the account it was called the house belonging to the orphans of György Falusy.[129]

Although it happened a decade later, a flood report from 1653 shows striking parallels with the events described above. In this case, both the *provisor* and the captain, at that time István Nemsem and István Svastics, wrote about a flood in their letters. They reported problems similar to the ones that occurred in the previously discussed instances. The *provisor* emphasized that the flood covered the plowland, probably with the newly sown grain in the soil. The most interesting parallel between the flood events in the two consecutive decades is the speed of the rise in the water level and their consequences. As in September 1641, the water of the Rába rose so quickly, according to Nemsem, that some livestock were marooned between the branches of the river. This affected the horses Batthyány kept

128 Letter of András Hidasy to Ádám Batthyány, 1 September 1641 (MNL OL P 1314, No. 19 263).
129 Bálint Ila, "Körmend város 1649. évi összeírása," [Conscription of the town of Körmend from 1649] *Ethnographia* 78 (1967): 556–568.

at Körmend and most of the pigs that had been driven to the gallery forests for pannage. According to the *provisor*, only the older pigs were able to survive the flood.

The next letter informing Batthyány about the September flood of 1641 was written two weeks after the first report. According to Hidasy, the water swept away the river pier. The vice-captain also draws attention to the necessity of re-building the bridge quickly, asking permission from his landlord to cut timber in the Dobogói Forest. The letter also relates that water levels were still high, which suggests a long-lasting flood event. According to Hidasy, the wheels of the mill were still beneath the waters of the Rába. He reports to Batthyány that the location of the cowherd and the herder was still unknown, and as no later message refers to them, they were probably killed in the flood.[130] This is one of the very few reports before the modern period in Hungary that points to human victims in a flood event.

Gerdákovics, the *provisor*, referred to high water levels in a letter dated 19 September. Although the flood was not yet over, some of the damage it had caused was already clear.[131] By that time, it was evident that the structure of the mill was partly destroyed, but a few days later Gerdákovics and Hidasy each wrote of the damage being somewhat less than they had previously supposed. Gerdákovics thought that the bridge would have to be re-built first and then the mill. According to his estimates, around 500 cartloads of wattling and at least 100 fir stakes were needed.[132] These are large quantities, but as discussed above, they were not unheard of, not even within the 60 years discussed here. A day later, on 23 September, Hidasy again reported a lower loss than previously estimated. As revealed at that time, only a part of the dam's floodgate had been destroyed by the flood, and the building itself had not been washed away by the river. Although there were losses to the dam and its timber, the Rába had not washed out the stakes comprising the foundations of both the mill and the bridge. Hidasy still wrote about the need to repair the dam by filling the gaps between the stakes with wattle and mud. According to his letter, the water was in flood and the weather was rainy as well. This reflects another problem that was partly related to floods and partly to rain: forage for the animals had been destroyed, and

130 Letter of András Hidasy to Ádám Batthyány, 16 September 1641 (MNL OL P 1314, No. 19 265).

131 Letter of Mátyás Gerdákovics to Ádám Batthyány, 19 September 1641 (MNL OL P 1314, No. 16 075).

132 Letter of Mátyás Gerdákovics to Ádám Batthyány, 22 September 1641 (MNL OL P 1314, no. 16 076).

no seed or grain remained in town to plant for the following year.[133] Only a week later, Hidasy asked for forage to overwinter the animals.[134]

The sources from Csákány are less numerous for this period. Only one letter has survived, from September 1641, but it does report the same flood wave. The captain of Csákány, János Keczer, who later became chief-commandant of the army of Transdanubia, wrote that all his grain had been carried away by the flood, and he could not even feed his horses.[135] The letter only mentioned the losses to his own harvested grain and fodder, but three months later, in December, Keczer informed Ádám Batthyány of the need to send grain to Csákány to sustain the settlement.[136]

According to the sources from Körmend around 1641, there was major flooding along the Rába as well as elsewhere, but little is known about floods on rivers other than the Rába in the Carpathian Basin or the wider region. There are some data that can help to clarify the weather situation during the period of this major flood. The most important source for the weather situation in August and September 1641 in Central Europe is an extremely detailed weather diary kept by Herman IV, landgrave of Hesse-Rotenburg (1607–1658). He observed the weather for four hours each day at different times, taking notes on the temperature (only by sensing, however, not measuring), precipitation, how clear the sky was, etc. His notes were taken in this period at Rotenburg an der Fulda. According to his notes, the first half of August was particularly warm, while from 24 August until 2 September, continuous heavy rain prevailed.[137] Scholarship in Germany recognizes a very rainy summer in Germany based on both historical evidence and tree-ring data.[138] According to the weather records of the tanner master Michael Stüeler in Krupka (in the northwestern part of the present-day Czech Republic), the

133 Letter of András Hidasy to Ádám Batthyány, 23 September 1641 (MNL OL P 1314, no. 19 266).
134 Letter of András Hidasy to Ádám Batthyány, 29 September 1641 (MNL OL P 1314, no. 19 267).
135 Letter of János Keczer to Ádám Batthyány, 16 September 1641 (MNL OL P 1314 no. 24 367).
136 *mivel hogÿ az mostanÿ Ar vizek az mi keves gabona termet volt it chakanban azt mind el vitte* ("the little crop we had here at Csákány was carried away by the flood") – letter of János Keczer to Ádám Batthyány, 4 December 1641 (MNL OL P 1314 no. 24 369).
137 See: Walter Lenke, *Klimadaten von 1621–1650 nach Beobachtungen des Landgrafen Hermann IV. von Hessen* (*Uranophilus Cyriandrus*) (Offenbach a. M.: Dt. Wetterdienst, 1960). The data were added to The Climate and Environmental History Collaborative Research Environment database. Available at: http://tambora.org (last accessed: 2 May 2021). See also for the temperatures in August and September 1641. Dobrovolný et al., "Monthly and Seasonal."
138 Rüdiger Glaser, *Klimageschichte Mitteleuropas – 1200 Jahre Wetter, Klima, Katastrophen* (3rd ed. Darmstadt: Primus, 2013), 148 and Ulf Büntgen et al., "Combined Dendro-Documentary Evidence of Central European Hydroclimatic Springtime Extremes over the Last Millennium," *Quaternary Science Reviews* 30 (2011): 3947–3959.

summer and the autumn of 1641 were extremely cold.[139] This cold humid mass of air could easily have influenced the precipitation in the catchment area of the Rába. The sources from elsewhere on the domains of the Batthyánys and from other settlements in Transdanubia do not tell of major floods in the first days of September, but as mentioned above, rainy weather was described in one of Hidasy's letters from mid-September. This was not only noted at Körmend; a chronicle compiled by members of the Payr family at Sopron (some 75 kilometers north of Körmend) also reported continuous rain between 13 and 15 September, as a consequence of which there were catastrophic floods on the Rába and Rábca Rivers (one of its tributaries before modern river water control works).[140]

One other river showed significant flooding at this time: the aforementioned Mura River running south of the Rába. The Mura, which also originates in the Eastern Alps, reaches the lowlands at the western borders of the Kingdom of Hungary. The Batthyánys had lands in this part of the country as well, and their *provisor* at Rakičan, in a letter dated 27 September, left one of the most detailed accounts of a flood in the early modern sources consulted for this book. The letter explains that the Mura may find a new riverbed – a phenomenon that frequently occurred in discussions about the Rába as well – as the pier had been destroyed. In this case, the situation was highly delicate, as according to the letter the changing riverbed could result in pieces of land being attached to Styria.[141] The source is one of the few mentions of the role alluvium played in identifying the borders of estates and, in this case, countries.[142] The rains mentioned in the Payr family's

139 Rudolf Brázdil, Hubert Valášek, and Oldřich Kotyza, "Meteorological Records of Michel Stueler of Krupka and their Contribution to the Knowledge of the Climate of the Czech Lands in 1629–1649," in *Czech Geography at the Dawn of the Millenium*, ed. Dušan Drbohlav (Olomouc: Palacký University, 2004), 95–112.

140 Heimler Károly, *Payr György és Payr Mihály krónikája 1584–1700* [The chronicle of György Payr and Mihály Payr] (Sopron: Soproni Városszépítő Egyesület, 1942), 28–29. See also: Vadas, "Floods in the Hungarian Kingdom," 93.

141 *Nagodnak aztis akaram megh jelentenÿ, hogy ez mostanti el mult napokban, hogy az vizek megh arattanak volt. Régéden alol Setyncz nevő falunal az Mura annira ide kÿ vetette magat, hogy erre az Nagtok Tartomaniara főlőte nagy karokat tett, és ha Ngotok ÿdejen nem prevenial, lehet az Mura vize éppen kÿ szakaztia magat, es az mura mellett eppen kett birosagott Stÿriahoz el foglalia*... (I wanted to inform you my lord that in the past days the waters flooded. Downstream from Régéd [?] at the village of Setnycz [?] the River Mura left its bed so much that it caused huge losses and unless your lord intervenes the Mura may find a new bed and attach two villages [?] to Styria) – letter of Balázs Temlin to Ádám Batthyány, 27 September 1641 (MNL OL P 1314 no. 48 773).

142 On the issue, see the fundamental work of the fourteenth-century jurist, Bartolus de Saxoferrato (*Tractatus de fluminibus seu Tyberiadis*). See also: András Vadas, "Border by the River – But

chronicle could not have caused the flooding of the Rába at the beginning of September, but the presence of this body of air in Central Europe may have affected the catchment area of the Rába and may have contributed in particular to the duration of the flood as well as to its geographical extent. Although the data on the weather in the summer and the autumn of 1641 is not sufficient to draw major conclusions, it is likely that the significant flood event in September was not merely a consequence of human activity but can be connected to the weather situation. The very rapid rise in the water levels identified at the Rába, however, may not have been independent of the different obstacles – bridges, dams – in the river.

This short case study demonstrates that private letters can be used as valuable sources in the study of the impacts of individual flood events on local economies and environments. It is also worth noting that similarly detailed flood reports are only known in greater numbers from the nineteenth century onwards for the Carpathian Basin. Flooding seems to have caused significant damage locally, and although based on less comprehensive sources, similar tendencies can be observed elsewhere along the Rába. Therefore, it is interesting to examine why no attempts were made to control these floods and why there is no evidence of flood management systems similar to those that may have functioned in the Árpádian period. To answer such questions, one has to look at what was according to the sources an even bigger threat than high waters, that is, low water levels.

3.4 Low Water – An Even Less Fortunate Event?

Flooding, as clearly reflected in its frequency at Körmend and the events of 1641, had a fundamental impact on the town's economic opportunities in the seventeenth century. Floods, however, had an identifiably positive impact; high water, as noted above, could prevent plundering raids by the Ottomans from the right bank of the Rába. The letters clearly indicate that the local administrations considered this a positive impact of floods. In a letter from 1644, the local *voivode* mentions to Ádám Batthyány that he

Where is the River? Hydrological Changes and Borders in Medieval Hungary," *Hungarian Historical Review* 8 (2019), 336–360; Bence Péterfi, "Debates Concerning the Regulation of Border Rivers in the Late Middle Ages: The Case of the Mura River," *Hungarian Historical Review* 8 (2019): 313–335, and Renáta Skorka, "On Two Sides of the Border. The Hungarian–Austrian Border Treaty of 1372," *Hungarian Historical Review* 8 (2019): 290–312.

had heard from one of his men that the Ottomans had attempted a raid but could not cross the Zala River – some 20 kilometers south of the Rába – due to the floods. The letter refers to waters in the plural; it is not exceptional for multiple rivers in western Transdanubia to flood at the same time.[143]

Two years later in another letter, Gáspár Francsics mentions that because of the high water level of the Rába, the mills (probably referring to the manorial mill at Körmend and that of Szecsőd downstream on the river) could not operate.[144] As captain of the town of Körmend, he also added that at least the "dogs" – referring to the Ottomans (who were frequently denoted this way in the early modern period) – would not be able to cross the river. In a letter a year later, Francsics also recounted the flooding of the Rába and mentioned that the Ottomans would find it difficult to cross, adding that they were nonetheless watching the river.[145]

Just as high water provided a better situation for defending the hinterland, low water levels signaled a potential threat. The plundering was not one-sided, however; periods with low water levels were also times when Hungarian and/or German troops conducted military actions on the areas controlled by the Ottoman Empire.[146] The problem of low water levels also appears in several letters. György Falusy, in the context of the mill dam of the village of Szecsőd, clearly describes the problem:

> My lord, as I also wrote you in the past days about the mill of Szecsőd, it is in a dangerous situation, as its dam got broken. One could cross the Rába anywhere, as the water is not higher than a paddle; we can tell that with my lord, Captain András Fülöp. Therefore, I beg you, my lord, that you send help, so that the poor are not threatened by the Pagans, and you have no further loss because of the mill not being able to run.[147]

143 Letter of Benedek Borz to Ádám Batthyány, 10 January 1644 (MNL OL P 1314 no. 7440).

144 *Az Raba most is ollÿ nagÿ hogÿ az molnak sem őrölhettnek, az vizek karok nagiok, most nem felö, hogÿ az eb altal vihete az Raban…* (The Rába is so high now that the mills cannot grind, the losses are great, but we do not have to fear that the dogs would cross the Rába) – letter of Gáspár Francsics to Ádám Batthyány 25 November 1646 (MNL OL P 1314 no. 15 145).

145 *It az Raba meleth nagj Sarok es vizek vadnak, mi mindazonaltal … vigiázásban leszünk…* (here by the Rába there is a big mud and water all around, but nonetheless we will be guarding) – letter of Gáspár Francsics Gáspár to Ádám Batthyány, 11 November 1647 (MNL OL P 1314 no. 15 202).

146 See most of all the studies in the volume: *Ransom Slavery along the Ottoman Borders* (*Early Fifteenth–Early Eighteenth Centuries*) (The Ottoman Empire and Its Heritage: Politics, Society and Economy, 37), eds. Géza Dávid and Pál Fodor (Leiden and Boston: Brill, 2007).

147 *Nagos Uram ez el mult napokbanis irtam vala Nagodnak az Szecződj malom felől, az bizonj igen veszedelmes alapattal vagion hogi az gattia el szakadot, mivel eővedzőig valo vizben mindenüt altal mehetnÿ az Raban … Annak okaert kerem alazatosan Nagodat valahonnand rendellien*

István Svastics, the captain of Körmend, described the low water level caused by the lack of a dam in similar terms to those quoted above. He draws attention to the growing Ottoman threat at the Rába in the area of the village of Hollós due to the dam breaking at the village.[148]

To understand the complexity of the impact of the river on border protection and the economy, the drought periods with low water levels are also to be considered. Droughts and low water levels on rivers in Europe have in general received considerably less attention than floods in the scholarship. Fewer sources refer to low water levels than to floods, simply because the impacts are usually less visible.[149] This is, however, certainly not true for areas such as Africa or the Middle East where extensive irrigation prevailed in local economies and where the water levels fluctuated between greater extremes than in Europe.[150]

I discovered about a dozen sources that address the problem of low water levels, while more than a hundred sources refer to floods and high water levels, which indicates a difference in the research potentials. In most cases, the low water levels on the Rába were not problematic because of the river's economic exploitation as a source of energy but they were a problem from the point of view of frontier defense. Lasting low water levels on the river often required the above-discussed river cutting, which similar to

Nagod segetseget, hogÿ az szegenisegnekis valamj veszedelme az poganjsagh miat, s Nagodnak töb töb kara ne kővetkezek az malom nem forgasa miat (My lord I wrote you about the mill at Szecsőd already. That is in a very dangerous state, because one could cross the Rába anywhere as its water is only as deep as a paddle. I humbly ask you my lord to order some help because the poor people are threatened by the pagans, and you lord should not have more loss from the mill wasting) – letter of György Falusy to Ádám Batthyány, 9 May 1656 (MNL OL P 1314 no. 13 193).

148 *Holosra menienek az két vaÿda, mivel hogj gát el szakadot, az víz igen kíczin malom felot, azert szorgalmatos vigiazas kől* (the two voivodes went to Hollós, as because of the breakage of the dam their water level is very low above the mill, there is a need for care taking) – letter of István Svastics to Ádám Batthyány, 4 May 1656 (MNL OL P 1314 no. 45 164).

149 E.g., Ulf Büntgen et al., "Five Centuries of Southern Moravian Drought Variations Revealed from Living and Historic Tree Rings," *Theoretical and Applied Climatology* 105 (2011): 167–180 or based exclusively on historical sources: Fernando Domínguez Castro et al., "Reconstruction of Drought Episodes for Central Spain from Rogation Ceremonies Recorded at the Toledo Cathedral from 1506 to 1900: A Methodological Approach," *Global and Planetary Change* 63 (2008): 230–242, Rudolf Brázdil et al.: "Documentary Data and the Study of Past Droughts: A Global State of the Art," *Climate of the Past* 14 (2018): 1915–1960. For Central Europe and the Carpathian Basin, see Andrea Kiss, "The Great 1506–1507 Drought and its Consequences in Hungary in a (Central) European Context," *Regional Environmental Change* 20 (2020) no. 50. doi: 10.1007/s10113-020-01634-5.

150 See, e.g., Alan Mikhail, "Oriental Democracy," *Global Environment* 7 (2014): 381–404, idem, *Under Osman's Tree: The Ottoman Empire, Egypt and Environmental History* (Chicago: University of Chicago Press, 2017), 19–33. See also Husain, "Changes in the Euphrates River."

floods and post-flood activities, had a more or less typical annual rhythm.[151] Interestingly, the rhythm of these works at Körmend was more regular compared to what was recorded at Csákány, perhaps because the Pinka joined the Rába by Körmend, providing a more stable water level than at Csákány. In most cases, low water levels happened in winters and summers and were less common in springs and autumns.

If low water levels persisted, more and more fords opened that could provide an easy crossing for troops on both sides. In these cases, the Ottomans went along the river to look for fords that were not watched from the other side of the river.[152] A letter by the captain of Csákány in the 1620s or the 1630s explains the problem clearly:

> The Rába is so small that they [the Ottomans] can cross wherever they wish. I tried to follow them up as far as Tarnak, but I could not catch up with them. Lord Hidasy also came, accompanied by nine others, but there are so many fords on the Rába that we cannot see them, even if we watched as we did well enough this last night.[153]

Low water level situations were among the few instances where the military force of Vas County was systematically mobilized to control this area. In some cases, however, by the time they arrived the water level had risen, and they were immediately released.[154] A recurring problem with the county's forces was their accommodation. It was hard on the local population if the forces had to be present for any length of time, which happened several times during the studied period. Therefore, whenever soldiers' services

151 *ide mynduntalan sok gonosz hirek erkeznek, valamynt akarja szinte ugÿ jöhet az Raban* (bad news keeps arriving here, they [the Ottomans] can cross the Rába wherever they wish) – the letters of István Keserű to Ferenc Batthyány, 24 December 1602 (MNL OL P 1314 no. 26 390) and 29 August 1605 (MNL OL P 1314 no. 26 395).

152 E.g., the letters of István Potyondi to Ádám Batthyány, 1637 [Monday] (MNL OL P 1314 no. 38 283) and 1637 [Monday] (MNL OL P 1314 no. 38 284).

153 *az Raba is kichin valahol akar ot jöhet altal, az mi kérésére voltam utanok mentem akoris Tarnakig, de nem erhetek ell őket, Hidasÿ uramis ide iőtt volt kilenczed magaval de sok az kelő az Raban, nem vehetők iesre őket, az el mult eÿel, eleget lestők őket* – Undated letter of István Potyondi to Ádám Batthyány (MNL OL P 1314 no. 38 296).

154 *Az Ngod paranczolattia szerent az katonakat ki szalatottuk ide az Raba melle, mível azoth az ár vizek mint itt valo letelűnk talan tovab nem kivantatik, akarunk Ngodtul ordinatiat venj ha haza boczassuk az katonakat avagi, mit czelekedgiűnk velok iria megh Ngd* (As ordered by you my lord, we sent out the soldiers to the Rába; however because of the floods their presence is not required any more. Please my lord let us know whether we should send them home or what shall we do about them) – letters of Ádám Hertelendi to Ádám Batthyány, 17 February 1646 (MNL OL P 1314 no. 18 718). See also 23 December 1655 (MNL OL P 1314 no. 18 759.).

did not seem necessary, the local military leadership would immediately release them.[155]

Low water was usually connected to summer drought periods, but winters also tended to bring lower water levels due to the large quantities of precipitation in the form of snow in the catchment area of the river. Low water levels, which made the water flow slower, led to the Rába freezing, which, as discussed in Chapter 2, raised a potential threat during the war period.[156] The sources consulted for this chapter contain only a few mentions of the Rába in this specific context,[157] but one is worth emphasizing, as it tells of the importance the military leadership of the region attributed to the Rába, even if the threat was probably somewhat exaggerated. A letter mentions that if there was a lack of help in defense, it would bring "things that are harmful or that affect the whole of our country."[158]

Even based on the few preserved references to low water levels and the related problem of the Rába freezing over, one can assume that from a military point of view this was a major threat. These circumstances may have affected the local economies less, but because of the reactions required from the civil and military administration – such as watching longer stretches of the river, breaking the ice, 'cutting the river,' and so on – this also proved to be a burden on the local tenant peasants and in some cases even for those who were exempt from unpaid labor.

3.5 Was Körmend an Exception? Some Conclusions

This chapter has aimed to present in detail the economic and environmental problems along a rather short stretch of the Ottoman-Hungarian frontier zone. As is clear from Chapter 2, the river had a fundamentally different role from the sixteenth century onwards. The question this part of the book has addressed is whether the changing political situation affected the

155 Letter of Ádám Hertelendi to Ádám Batthyány 23 December 1655 (MNL OL P 1314 no. 18 759.)

156 On the problem of the frozen River Rába and the border defense, see the numerous letters of Gáspár Francsics and István Svastics, captains of Körmend: MNL OL P 1314 no. 15 150, 15 152, 15 153, 15 202, 15 207, 108 523, 45 041, 45 058, and 45 164.

157 *nagodnak tutara vagion az raba igen kicsin felö az dolog ha meg fagi anal roszab lészen* (you well know my lord, that the Rába River is very low, and it could be even worse in case it freezes over) – letters of János Keczer to Ádám Batthyány, 10 January 1654 (MNL OL P 1314 no. 24 659). See also the fairly similar letter of Benedek Hollósi to Ádám Batthyány, 23 October 1652 (MNL OL P 1314 no. 19 575).

158 Letter of Bernát Csány to Ádám Batthyány, 9 July 1655 (MNL OL P 1314 no. 8799).

environmental conditions on a local scale or not. Many economic hardships connected to the Rába River have been identified. While in the Late Middle Ages, the river seems to have primarily served the local economies, from the arrival of the Ottomans in the region it took on strategic importance. As the previous chapters I hope have shown, the water-related environmental problems were significant in this period. The main question that arises is whether it was indeed the river's changing role that exerted considerable environmental pressure on the local economies.

Before summarizing how the water-related economic pressure and the frontier were related, one other question must be addressed, that is, how representative this case is. Were the economic difficulties experienced in the seventeenth century specific to Körmend and the Rába valley, or were they also experienced along other rivers in the region? To understand the extent to which the case study of the domain of Körmend points to a more general problem, as indicated in the introduction to this chapter, sources from Csákány as well as from settlements in the same region lying on smaller or similar-sized rivers (listed in Fig. 3.2) were considered. There is no space to give a detailed analysis of the case of Csákány. There are clear parallels in the locations of the two settlements, most importantly that they both had castles that were protected by the Rába River through their moats. While Körmend lay close to the river, Csákány was somewhat more protected from the floods, as it was connected to the water system of the Rába through a small tributary called Vörös Stream.[159] A dam close to the mouth downstream on the Rába must have backed up the water of the river to provide water for the moat of the local castle. Despite Csákány's considerably minor role in the border defense system compared to that of Körmend and even than that of the nearby Szentpéter (present-day Őriszentpéter), the local captains and provisors were also in continuous contact with the Batthyánys and with the captains and provisors of Körmend. Similar to Körmend, Csákány had a manorial mill on the Rába, and thus the main environmental conditions are certainly comparable. Fewer floods were recorded at Csákány – 30 in

159 *...az it valo napokban holt rabakon es folio rababanis arja mind el apad imar ... az chastély arokait is igen meg halasithatia bar czak az ujvarj tizt viselöknek irion nagod hogi meg vegiek az halonak valo szerszamott* (these days the water both on the Dead Raba [oxbow lake] and the River Rába also started falling ... you can now use the moat of the castle as a fish pond. Please write to the officials at Güssing that they bought the tools for the net) – letter of János Keczer to Ádám Batthyány, 27 March 1650 (MNL OL P 1314 no. 24 465). The ditches on the nineteenth-century cadastral map are still filled with water and can be well-identified even nowadays. For the sheet of the cadastral map in question, see: MNL OL S 78 (Térképtár. Kataszteri térképek) 264. téka Nagycsákány, no. 8.

total compared to the almost 90 known in Körmend – but this can probably be explained by the different number of letters that have come down to us from the two settlements. The same problems recorded at Körmend seem to have prevailed upstream to some extent as well.

It is more relevant in the light of similar problems at Körmend and Csákány to determine whether floods and water-related environmental problems occurred with similar frequencies along other Transdanubian rivers. Most of the sources consulted to understand another river's environment originate from the valley of the aforementioned River Mura. The river's basic hydrogeographical conditions are similar to those of the Rába. It also brings water from the Eastern Alps, but two differences are to be noted. First, because of its higher discharge and more extreme water level fluctuations, it had different mills than those by the Rába, mostly ship mills operated on the Mura.[160] Second, it received somewhat less attention in the frontier protection of the country because of its geographical position.[161] The more than 500 letters consulted from the region referred to very few flood events – mostly those that were considered major at Körmend. These, of course, can be associated with the weather situation rather than the direct manipulation of the river's course.

Disregarding the fact that the written sources that were consulted pointed to fewer floods in other rivers of the region, the changing precipitation, its distribution throughout the year, and the lower temperatures usually associated with the Little Ice Age (as discussed in Chapter 2) could potentially explain the very high number of flood events on the Rába. To argue for the existence of a long-term change in the climatic conditions of the catchment area, there is a need for high-resolution reconstructions (temperature and precipitation) for the lower region of the Eastern Alps in the period. No high-resolution historical data-based reconstruction is available for the catchment area of the Rába, but the Central European temperature reconstructions do include Austria. here is no clear shift in the temperatures in any of the seasons in the examined period, or the changes were not sufficient to explain the great number of extreme floods on the Rába.[162]

160 On ship mills, see Tibor Sabján, "Ship Mills in Historical Hungary," in *Ruralia 5: Water Management in Medieval Rural Economy / Les usages de l'eau en milieu rural au Moyen Âge*, ed. Jan Klápště (Prague: Institute of Archaeology, Academy of Sciences of the Czech Republic, 2005), 242–250.

161 Nonetheless, see the strategic importance Miklós Zrínyi attributed to the region: Hausner and Németh, *Zrínyi-Újvár.*

162 Dobrovolný et al., "Monthly and Seasonal." On flood frequencies on other rivers, see Rüdiger Glaser et al., "The Variability of European Floods since AD 1500," *Climatic Change* 101 (2010): 235–256.

Besides reconstructions based on historical sources, precipitation reconstructions based on tree-ring data are available for the broader neighborhood of the catchment of the Rába. Giorgio Strumia has studied tree rings of the Vienna Basin area for the period of 1436 to 1998. He found a precipitation increase in the summers during the period 1602 to 1651. It is noteworthy that his records show that the summer of 1634 brought the most rainfall in the previous half millennium. A major summer flood occurred at Körmend that year. Precipitation in the spring-summer periods was also high in the 50 years starting in 1617, within which the years between 1625 and 1634 seem to have been the wettest of the previous half millennium. More recent scientific reconstructions, however, point to less evident changes in the region. According to other reconstructions of summer precipitation, the period does not show any noteworthy change.[163]

Although the results published by Strumia identify some critical years that may have contributed to floods on the Rába, these in themselves do not explain the flood frequencies in the case study area. The long-term change in the forest cover of the catchment area may have been another factor that contributed to higher flood frequencies and problems in the fluctuation of water levels. There is no detailed reconstruction of the forest cover in present-day Austria, the area where most of the catchment area upstream from Körmend lies, but studies have not discovered major clearances or crises related to a lack of forest resources until the modern period. In Chapter 4, I will argue that the situation was similar in the Kingdom of Hungary.[164]

Finally, recent reconstructions of Central European flood frequencies in the past half a millennium also do not point to an extremely flood-rich period in the first half of the seventeenth century (and certainly not up to 1640).[165] The large number of floods that caused material losses at Körmend and its neighboring settlements have to be understood in the light of the above. The lack of changes in climate and climate-related events and

163 Giorgio Strumia, "Tree-Ring Based Reconstruction of Precipitation in Eastern Austria," (PhD diss., BOKU, 1999), 85 and 95. I am thankful to Zoltán Kern for drawing my attention to this work. See also: Robert J.S. Wilson, Brian H. Luckman, and Jan Esper, "A 500 Year Dendroclimatic Reconstruction of Spring–Summer Precipitation from the Lower Bavarian Forest Region, Germany," *International Journal of Climatology* 25 (2005): 611–630.

164 Hannes Mayer, *Wälder des Ostalpenraumes; Standort, Aufbau und waldbauliche Bedeutung der wichtigsten Waldgesellschaften in den Ostalpen samt Vorland* (Stuttgart: Gustav Fischer, 1974). For a rather general overview from an environmental history perspective, see Gerhard Weiss, "Mountain Forest Policy in Austria: A Historical Policy Analysis on Regulating a Natural Resource," *Environment and History* 7 (2001): 335–355.

165 Gönter Blöschl et al., "Current European Flood-Rich Period Exceptional Compared with Past 500 Years," *Nature* 583 (2020): 560–566.

vegetation do not explain what happened in the riverine environment. As has been demonstrated through the example of the Danube, mostly in the context of Vienna, a river can be considered a socio-natural site where the arrangements (dams, bridges, and other structures) and practices (regulation works, energy production) have to be discussed together to understand the long-term changes in the fluvial landscape.[166] The case study in Chapters 2 and 3 suggests that the direct influence of the societies along the course of Rába, combined with natural river bed changes, significantly impacted the societies of the settlements near the river. The material discussed above points to an unstable environment that can probably be attributed to the region's role as a frontier. The river took on a new function: it became a complex system that had to serve both economic and military purposes, and in many cases, these two purposes heavily conflicted with each other.

Numerous studies in Western Europe – and some in Hungary – have also treated water as a subject of conflict of interests.[167] As manipulations of watersheds or even the very courses of rivers harmed the economic interests of other actors of the same area, many lawsuits were lodged, providing important sources for historians in particular.[168] Similar conflicts also applied in the case of the Rába River, although with somewhat different

166 See most importantly: Verena Winiwarter, Martin Schmid, and Gert Dressel, "Looking at Half a Millennium of Co-Existence: The Danube in Vienna as a Socio-Natural Site," *Water History* (2013): 101–119. See also the thematic volume of *Water History* 5, no. 2 (2013).

167 E.g., Stefania Barca, *Enclosing Water: Nature and Political Economy in a Mediterranean Valley, 1796–1916* (Cambridge: White Horse Press, 2010), Matthew Evenden: *Fish versus Power: An Environmental History of the Fraser River* (Cambridge: Cambridge University Press, 2004), and *Eaux et conflicts dans l'Europe médiévale et moderne: actes des XXXIIes Journées Internationales d'Histoire de l'Abbaye de Flaran, 8 et 9 octobre 2010*, eds. Sandrine Lavaud and Patrick Fournier ([Toulouse]: Presses Universitaires du Mirail, 2012). Similar problems with an ethnological approach in Hungarian context: *"Áldás és átok a víz" – Tudományos emlékülés a Mirhó-gát megépítésének 200. évfordulóján* ['Blessing or curse is water' – Scientific conference on the 200th anniversary of the construction of the Mirhó Dam], ed. Albert Tóth (Kisújszállás: Városi Tanács, 1987), Miklós Szilágyi, *Halászóvizek – halásztársadalom – halászati technika. A tiszai halászat történeti-néprajzi elemzése* [Fishing waters – fishing society – fishing techniques. Historical-ethnological analysis of fishing at Tisza] (Debrecen: KLTE Néprajzi Tanszéke, 1992), and Tibor Bellon, *A Tisza néprajza. Ártéri gazdálkodás a tiszai Alföldön* [the ethnography of Tisza. Flood plain economy at the Tisza in the Great Hungarian Plain] (Budapest: Timp Kiadó, 2003). See most recently: András Vadas, "Who Stole the Water? The Control and Appropriation of Water Resources in Medieval Hungary." (PhD diss., Central European University, 2020).

168 E.g., B. Dénes Jankovich, "Adatok a Körösvidék középkori vízrajzához és a vizek hasznosításához" [Data on the medieval hydrography and water management in the Körösvidék], *A Békés Megyei Múzeumok Közleményei* 16 (1996): 305–349, Tringli, "A magyar szokásjog," Kiss, *Floods and Long-Term Water-Level,* 106–107 and 322, Vadas, "Some Remarks," 291–304, and idem, "Terminológiai és tartalmi kérdések a középkori malomhelyek körül" [Questions regarding the

actors, as water management was carried out in the interest of diverse actors along the river.

First, a constant and relatively high water level on the Rába was important for military strategy to prevent incursions by Ottoman troops. Different means were used to raise the water level both temporarily and in the long run, such as river-cutting (*bevágás*, discussed in Chapter 2) and relying on the mill dams (as discussed in Chapters 2 and 3). By the late medieval period, these constructions were so frequent along the river that they provided a good opportunity to base the defense system on their sequence. In some places, the need for higher water levels can also be attributed to the diversion of water to surround fortifications with moats, as at Körmend and Csákány.

Second, in general, controlling water levels contributed significantly to effective farming in Körmend. In cases of high water levels, however, the otherwise beneficial mills frequently caused problems, as the sluices could seldom divert the water surplus. The millers also had to consider that too much water sent through the sluices could cause floods downstream.[169] In the case of Körmend, this was particularly salient, as the first village downstream was Szecsőd, which also belonged to the same manorial complex and to the same landlord. The high water levels in many cases caused losses by not providing enough meadows for the livestock, or it led to the inundation of the gallery forests, resulting in a lack of mast for pigs herded in these forests.

It was not in the interest of the landlords – in this case, the Batthyánys – to endanger their settlements and their populations. Frequent rebuilding work required huge amounts of labor and material, as Chapter 3 has shown. The buildings built into the floodplain – such as the mills and the bridge – were close enough to the water that certain frequency of loss was counted in. The losses in the first half of the seventeenth century, however, went far beyond this. The 90 floods and related rebuilding projects affected the domain of Körmend and the neighboring lands of the Batthyány estate complex, from the Őrség to Güssing. From time to time, unpaid labor had to be provided from relatively distant domains.

The river was a new challenge that the town of Körmend and the local administrations had to face. Of course, floods were a recurrent phenomenon in the Middle Ages, but nothing shows that this was a major economic problem in the life of the settlement. From the sixteenth and seventeenth

terminology and meaning of the term *locus molendini* in the Middle Ages], *Történelmi Szemle* 57 (2015): 619–648.

169 On the problem of the suite of mills by the same section of a river, see Vadas, "Some Remarks."

century and the onwards, however, the frequent maintenance of the river and its environment provided the left bank and the hinterland with relative safety from Ottoman plundering. The major change in the environmental conditions of the river can most likely be understood as an indirect cost of the Ottoman war periods and, as such, the most significant footprint of the war period in the region.

4 From Endless Forests to Meadows and Wastelands?

What Happened to the Forests Along the Border?

Abstract
The chapter analyzes the relationship between the Hungarian-Ottoman wars and the loss of forests in Transdanubia (western Hungary). It provides an estimate of the most important ways in which forest resources were used in order to investigate whether or not the forest loss can be linked to the Ottoman presence in the Carpathian Basin. The most important spheres of forest-related consumption – fortification works and the production of war materiel – are considered and contrasted with other types of forest-resource use.

Keywords: Environmental history, Kingdom of Hungary, forest history, earth and wood fortifications, war materiel

As argued in Chapters 2 and 3, the waterscape of the frontier area between the Ottomans and the Hungarians in the Carpathian Basin was in all likelihood significantly transformed in the sixteenth and seventeenth centuries. However, as is clear from the sources used in Chapter 2 as well as from the existing scholarly literature, water was not the only natural resource heavily used in wars in pre-modern times – so were forests. The main question addressed in the present chapter is how forest resources in the Carpathian Basin were influenced by the Ottoman wars. As discussed in the introduction, the few works that do consider the impact of pre-modern wars on the environment refer to its severe impact on forest resources. Most of the scholars attribute much more importance to forest resources in pre-modern warfare than to rivers and waterscapes.[1] The rather traditional

1 Hupy, "The Environmental Footprint," McNeill, "Woods and Warfare," or Tucker, "The Impact of Warfare."

Vadas, A., *The Environmental Legacy of War on the Hungarian-Ottoman Frontier, c. 1540–1690*. Taylor & Francis Group, 2023
DOI: 10.5117/9789463727938_CH04

narrative of historians also follows this view. Below, I discuss the ways in which the use of forests in this period can be approached to test the validity of the assumption that forest resources in the lowland and hilly areas of the Carpathians became scarce in the period of the Ottoman wars. First, however, I provide a summary of the most important points that scholars have made about early modern war-related forest consumption in the Carpathian Basin

4.1 Forest Resources in the Carpathian Basin in Pre-Modern Times – Endless or Scarce?

The political change of the sixteenth century discussed in section 1.3 significantly transformed the spatial and economic structure of the Carpathian Basin. It deeply impacted the settlement network and farming on the Great Hungarian Plain as well as in Transdanubia. Although by the late medieval period, cattle on the hoof were already the most significant export product of Hungary, it was the early modern period that saw a tremendous increase in the number of animals driven to German towns and towards markets in northern Italy.[2] This had a significant impact on the land use of the areas where extensive herding was practiced, as these parts of the central plains supported a significantly smaller agricultural population than intensive agriculture. This, combined with Ottoman military pressure on the villages of the occupied territories, resulted in the transformation of the settlement network and the vegetation around these villages. While many villages were abandoned in the Great Hungarian Plain and to a lesser extent in Transdanubia, there was significant growth in the populations of the market towns in the lowland areas of the Basin.[3] This transformation process corresponds to the decline narrative put forth by scholars, as explained in

2 E.g., László Bartosiewicz, *Animals in the Urban Landscape in the Wake of the Middle Ages: A Case Study from Vác* (BAR International Series, 609) (Oxford: Tempus Reparatum, 1995). See furthermore as an overview: Ian Blanchard, "The Continental European Cattle Trades, 1400–1600," *The Economic History Review* [New Series] 39 (1986): 427–460, and more recently, with an emphasis on the trade with Italy: Andrea Fara, "An Outline of Livestock Production and Cattle Trade from Hungary to Western Europe in Late Middle Ages and Early Modern Period (XIVth–XVIth Centuries)," *Crisia* 45 (2015): 87–95, and idem, "Il commercio di bestiame ungherese verso la Penisola italiana tra tardo Medioevo e prima Età moderna (XIV–XVI secolo)," *Mélanges de l'École française de Rome – Moyen Âge*" 47, no. 2 (2015) [Online journal]. doi: 10.4000/mefrm.2709.

3 Most of all, see the recent book: Sárosi, *Deserting Villages* including an exhaustive research history of the topic.

the introduction to this book. In more recent literature, this transformation is seen as less of a crisis and more as a string of changes in the economy necessitated by the presence of the Ottomans. Part of this decline narrative is the loss of forests in the Great Hungarian Plain and Transdanubia and the extension of marshes in the lowlands – particularly the frontier, where in many cases the literature assumes that a total deforestation occurred. Changes in the function of woodlands in this area are clear thanks to several sources discussed below, but a change and moreover a decrease in the extent of the woodlands is hard to demonstrate based on the archival sources available for either the pre- or post-Ottoman period. This chapter will address how the war may have impacted the region's forest resources, both directly and indirectly, by looking at the most important forms of wood consumption in the Ottoman period both in the immediate frontier zone and to a lesser extent in the hinterlands.

In Hungary, research into forest cover is approached from at least two directions: from the natural sciences (such as palynology) and from the social sciences (such as history). The Carpathian Basin has clearly been under strong human influence for millennia before the arrival of the Hungarians or the Ottomans in the region. One must therefore be cautious when trying to give exact numbers for the extent of woodland in the last millennium based exclusively on pollen records. Nonetheless, palynology has proven to be an important tool in identifying shifts in vegetation and land-use patterns.[4] Another approach in forest history research uses written sources. As far back as the nineteenth century, forest scientists were keen on comparing historical woodland management to that of the modern period, and thus major efforts were made to collect data on medieval and early modern forest management.[5] Most historians, however, have been preoccupied with the extent of forests in historical times, as a result of which until the 1990s studies usually tried to provide numbers for the whole of the Carpathian Basin. The discussion focused on the loss of woodlands, which corresponds particularly well to the above-mentioned narrative of the destruction of the Ottoman period.

The most important point that scholars have made is that the Ottomans either directly or indirectly are to be blamed for the destruction of the so-called "traditional" forested landscape of the lowlands of the Carpathian Basin. Such claims, however, are not new, similar allegations having been

4 See mostly the numerous works of Pál Sümegi and his colleagues.

5 See most importantly this tripartite cartulary: *Magyar erdészeti oklevéltár*, 3 vols. [Cartulary of Hungarian forestry], ed. Károly Tagányi (Budapest: Pátria, 1896).

made already by Renaissance humanist authors in Hungary. In the 1530s, Miklós Oláh, later archbishop of Esztergom, wrote his most influential work, *Hungaria*, which provides a description of the geography and history of Hungary. He presents a rich, prosperous country full of green meadows and rich forests in the lowlands, a landscape that certainly never existed in his lifetime.[6] This has proven to be an enduring narrative. When two prominent historians, Bálint Hóman and Gyula Szekfű, wrote a new *Hungarian History* in five volumes during the interwar period, the image they presented was only slightly less biased than Oláh's four centuries earlier. In a chapter Szekfű wrote about the seventeenth century, the Great Hungarian Plain is characterized in apocalyptic terms:

> The change [in the landscape] was the consequence of the Ottoman conquest. Villages were destroyed, both people and livestock left the area, either fled or were taken into Ottoman captivity. As we shall see, instead of twenty-five or thirty villages, only one mid-sized market town was left, the inhabitants of which possessed 2–300,000 acres of uninhabited wasteland [*puszta*]. With people leaving, the reign of grasses started, lands became covered in grass, here it became sand dune, and there it turned to salt marsh. The whole plain turned into a fallow, the remaining forest could no longer provide enough moisture, with the loss of moisture the wasteland started to expand again, which for centuries the Hungarians of the previous centuries had thought to have cast away for good from Hungarian soil.[7]

6 See Oláh's *Hungaria*: Nicolaus Olahus, *Hungaria – Athila* (Bibliotheca Scriptorum Medii Recentisque Aevorum. Saeculum XVI.), eds. Colomannus Eperjessy and Ladislaus Juhász (Budapest: K.M. Egyetemi Nyomda, 1938). See also: Emőke Rita Szilágyi, "Zur Überlieferungsgeschichte von Nicolaus Olahus' Hungaria," in *Wiener Archivforschungen: Festschrift für den ungarischen Archivdelegierten in Wien, István Fazekas* (Publikationen der Ungarischen Geschichtsforschung in Wien, 10), eds. Zsuzsanna Cziráki et al. (Vienna: Institut für Ungarische Geschichtsforschung in Wien and Ungarische Archivdelegation beim Haus-, Hof- und Staatsarchiv, Wien, 2014), 69–75, and Emőke Rita Szilágyi, "Az önkéntes száműzetés alakzatai és trópusai Oláh Miklós Brüsszelben" [Figures and tropes of voluntary exile. Miklós Oláh at Brussels], in *Börtön, exilium és szenvedés: Bethlen Miklós élettörténetének kora újkori kontextusai* [Prison, exile, suffer. The early modern Contexts of the life of Miklós Bethlen], eds. Anna Fajt, Emőke Rita Szilágyi, and Zsombor Tóth (Budapest: Reciti, 2017), 51–60, and Attila Restás, "A kora újkori Magyarország-toposzok történetéhez Paczoth Ferenc (1617) és Johann Fechner (1650) beszédei" [On the history of tropes in early modern Hungary: The speeches of Ferenc Paczoth (617) and Johann Fechner (1650)], *Irodalomtörténeti Közlemények* 122, no. 4 (2018): 490–507.

7 Bálint Hóman and Gyula Szekfű, *Magyar történet*, vol. 5 [Hungarian history] (Budapest: Királyi Magyar Egyetemi Nyomda, 1928), 6. (written by Gyula Szekfű)

The image of the settlements abandoned due to the Ottoman plundering and of forests and plowlands becoming wastelands was later criticized by ecologists, historians, and archaeologists; moreover, forestry experts were already critical in the interwar period.[8] A thorough re-evaluation of the landscape changes on the Great Hungarian Plain in the last millennium has yet to be completed,[9] but some important steps that have been taken are already worth noting. First, ecologists have demonstrated that the forest steppe that may have dominated the landscape in the Carpathian Basin before human influence was more or less gone by the Middle Ages, and certainly long before the early modern period.[10] In the late medieval period, the areas in the Great Hungarian Plain best suited for extensive cattle herding became more valuable than before, which probably reduced the size of the already

8 István Vági, "Van-e hazánkban ezeréves puszta, vagy azt a török hódoltság okozta. Megváltozott-e a Nagy-Alföld éghajlata a török hódoltság miatt aszályosabb irányban, továbbá a talajok is alig javíthatóan megromlottak-e a valóságban" [Was there a millennial puszta in Hungary or was it caused by the Ottomans. Did the climate of the Great Hungarian Plain became drier as a result of the Ottoman Empire and did soils indeed devastate to a barely amendable state], *Erdészeti Lapok* 73 (1934): 670–682, János Weidlein, "A dűlőnévkutatás történeti vonatkozásai" [Historical applications of research into field names], *Századok* 69 (1935): 665–692 and Károly Kaán, *Alföldi kérdések. Erdők és vizek az Alföld kérdéseiben* [Question of the Great Hungarian Plain. Forests and waters in the questions of the Great Hungarian Plain] (Budapest: Stádium, 1939), 11–43.

9 Criticism of this view arose from different aspects: Ferenc Szakály, *Magyar adóztatás a török hódoltságban* [Hungarian taxation in Ottoman Hungary] (Budapest: Akadémiai, 1981), László Bartosiewicz and Erika Gál, "Ottoman Period Animal Exploitation in Hungary," in *Archeology of the Ottoman Period in Hungary.* (Opuscula Hungarica, 3), eds. Ibolya Gerelyes and Gyöngyi Kovács (Budapest: Magyar Nemzeti Múzeum, 2004), 365–376. Most recently: Sárosi, *Deserting Villages*, 57–94.

10 From a historian's angle, see Péter Szabó, *Woodland and Forests in Medieval Hungary* (Archaeolingua – Central European Series, 2 = BAR International Series, 1348) (Oxford: Archaeopress, 2005). For the ecologists' and botanists' results: *Alföldi erdőssztyepp-maradványok Magyarországon* [Forested steppe remains in Hungary] (WWF Füzetek, 15), eds. Zsolt Molnár and András Kun (Budapest: WWF Magyarország, 2000), *Lejtősztyepek, löszgyepek és erdőssztyeprétek Magyarországon* [Slope steppes, loess steppes and forest steppe meadows of Hungary], eds. Eszter Illyés and János Bölöni (Budapest: [MTA Ökológiai és Botanikai Kutatóintézete], 2007). Online document: http://www.obki.hu/publikacio/pdf_anyagok/ILLYES_BOLONI_2007_Lejtosztyepek_ loszgyepek_es_erdossztyepretek_Mon.pdf (last accessed: 7 May 2017), Marianna Biró, "A történeti térképekre alapuló vegetációrekonstrukció és alkalmazásai a Duna-Tisza közén" [The possibilities and applications of using historical maps in vegetation reconstructions in the Danube-Tisza Interfluve area], (PhD diss., Pécsi Tudományegyetem, 2006), esp. 69 and Zsolt Molnár et al., "A Duna-Tisza közi homoki sztyepprétek történeti tájökológiai jellemzése" [Historical ecological analysis of steppe meadows in the Danube–Tisza Interfluve], in *Talaj-vegetáció-klíma kölcsönhatások. Köszöntjük a 70 éves Láng Editet* [Soil – vegetation – climate. Honorary volume on the 70th birthday of Edit Láng], eds. György Kröel-Dulay, Tibor Kalapos, and Andrea Mojzes (Vácrátót: MTA Ökológiai és Biológiai Kutatóintézete, 2008), 39–56.

small areas where forest steppe still occurred. Geomorphologists using optically stimulated luminescence (OSL) have managed to date the major sand-movement periods in the Great Hungarian Plain in the Holocene. They cannot point to any major change with the arrival of the Ottomans in the region.[11] They suggested that the processes that created salt marshes and wastelands started much earlier than the arrival of the Ottomans in the region. Although these processes may well have been accelerated by the growing cattle trade and the environmental pressure caused by the arrival of the Ottomans, they cannot be connected to a single historical period or political change. One further aspect, raised mostly by the Ottomanist Gábor Ágoston, is worth considering when discussing the landscape devastation caused by the presence of the Ottomans. Ágoston draws attention to the simple fact that if the Ottomans had planned to stay in the Carpathian Basin for at least a few decades, integrating the central part of this area into the Ottoman administrative system, it would not have been in their interest to rapidly destroy the local economies by consuming their resources.

Another factor to be taken into account is the difference between wood-cutting and forest clearance, which both the Ottomans and Hungarians were well aware of. The latter form of forest exploitation was not typical, not even in the areas under the Ottoman authority and not even if there were fair prospects for cattle herding. As has been demonstrated by research, scorched-earth tactics that could potentially cause rapid deforestation was seldom used by the armies on either side in the Carpathian Basin.[12] This did not mean that trees were not cut down regularly in the early modern period – they certainly were – but this is not to be confused with

11 Diána Nyári, Tímea Kiss, and György Sipos, "Investigation of Holocene Blown-Sand Movement based on Archaeological Findings and OSL Dating, Danube-Tisza Interfluves, Hungary," *Journal of Maps. Student Edition 2007*, 46–57, Diána Nyári and Tímea Kiss, "Blown Sand Movement at Kiskunhalas on the Danube-Tisza Interfluves, Hungary," *Journal of Environmental Geography* 2, no. 3–4 (2009): 31–36 and Márta Tóber and Andrea Kiss, "Landscape History of the Medieval Sand Ridge Area in Central Hungary: Examples of Sand and Arboreal Vegetation in medieval Documentation Compared to the Results of Natural Scientific and Archaeological Investigations," *Siedlungsforschung. Archäologie – Geschichte – Geographie* 31 (2014): 247–269. On landscape changes in the area, see Sárosi, *Deserting Villages*, 35–56.

12 Pálffy, "The Origins." The lack of the widespread usage of scorched-earth tactics can be explained by economic reasons: Pálffy, "Scorched-Earth Tactics," 183. See also: Géza Pálffy, "Elképzelések a török hódoltság elpusztításáról a XVI–XVII. században. (A Habsburg Birodalom magyarországi hadszínterének néhány főbb sajátosságáról)" [Ideas on the destruction of the Ottoman Empire in Hungary in the 16th–17th centuries (Some specificities of the Hungarian military scene of the Habsburg Empire)], in *"Quasi liber et pictura" Tanulmányok Kubinyi András hetvenedik születésnapjára / Studies in Honour of András Kubinyi on his Seventieth Birthday*, ed. Gyöngyi Kovács (Budapest: ELTE Régészettudományi Intézet 2004), 387–403.

the intentional and direct deforestation suggested by earlier literature. Trees were cut down for both military and everyday purposes. The military used wood for the construction of earth and wood fortifications and the production of gunpowder, while everyday purposes of woodcutting included the building of dwelling houses, mills, ships, etc. The question is not so much how many trees were felled for these purposes as how quickly and how well the harvested woodlands could regenerate. Changes in farming practices and land use in the Great Hungarian Plain from the thirteenth and fourteenth centuries onwards may have had a major impact on this process. As parkland became more and more dominant because of cattle herding, there was little chance for reforestation.[13] However, the entire area, with its relatively dense settlement network, was not used for herding in the late medieval and the Ottoman period. One of the best-known settlements that was abandoned in the Ottoman period is Ete in Tolna County in the eastern part of Transdanubia. This late medieval market town was gradually depopulated in the sixteenth century. Rich legal evidence from the Middle Ages and extensive archaeological excavations allow us to reconstruct a complex picture of the town. Palynological research has demonstrated that intensive reforestation occurred in the centuries following the Middle Ages, indicating that this part of the central plains of the Carpathians was probably never transformed into an area of extensive cattle herding.[14] Ete's example is not unique; similar processes have been found in other places, also based on other source material in the central lowlands.[15]

13 Péter Szabó, "Erdők a kora újkorban: történelem, régészet, ökológia" [Forests in the early modern period: history, archaeology, ecology], in *Környezettörténet*, 137–156. See also: Kaán, *Alföldi kérdések.*

14 For Ete, see József Holub, "Ete város története. (Adalékok a Tolna megyei Sárköz település- és gazdaságtörténetéhez)" [The history of the town of Ete (Data to the settlement and economic history of the Sárköz in Tolna County)], *Történeti Statisztikai Közlemények* 2, no. 3–4 (1958), 28–46, and Zsuzsa Miklós, "Beiträge zur Siedlungsgeschichte des mittelalterlichen Marktfleckens," *Acta Archaeologica Academiae Scientiarum Hungaricae* 53 (2002): 195–254. For the reforestation here, see Pál Sümegi et al., "Middle Age Paleoecological and Paleoclimatological Reconstruction ins the Carpathian Basin," *Időjárás* 113 (2009): 293 and Pál Sümegi et al., "The Environmental History of Southern Transdanubia during the Medieval and the Ottoman Period in the Light of Palaeoecological and Geoarchaeological Research," in *"per sylvam,"* 37.

15 For instance, field names: Weidlein, "A dűlőnévkutatás." See also: András K. Németh, "Fák és erdők a középkori Tolna megyében" [Trees and woodlands in medieval Tolna County], *A Wosinsky Mór Megyei Múzeum Évkönyve* 38 (2016): 45 and most recently: András K. Németh and Gábor Máté, "Vázlat a pusztafalvak tájtörténeti kutatásához a Kapos menti Enyőd példáján" [Outlines of a research project on landscape history of deserted medieval villages on the example of Enyőd village], *Történeti Földrajzi Közlemények* 5, no. 1 (2017): 44–58, esp. 56.

In addition to forest cover in general, wood consumption by different military industries has been considered by research, which is crucial for this book, as it could be instrumental in understanding the pressure caused by the military activities on this landscape element. The above-mentioned scholar Géza Pálffy, who studied the gunpowder industry at Košice (in present-day Slovakia), one of the most important Hungarian industrial centers of the period, came to the conclusion that the fuelwood consumption of the military industry was not so significant, or at least not based on the detailed workshop account from 1592.[16] Similarly, when studying the iron-smelting unit of the manorial complex of Muránsky Hrad, Béla Sarusi Kiss registered relatively limited amounts of charcoal purchased.[17] The few studies that have addressed the problem thus far do not provide a basis for concluding that military-industrial consumption was a significant reason for changes in the forest cover in the Carpathian Basin.[18]

Recent research into the earth and wood fortifications along the frontier has also tried to shed light on the problem of timber and fuelwood consumption, but with little success so far.[19] Controversial conclusions have been drawn by different scholars on this issue, sometimes based on the same primary sources and datasets. Sándor Takáts, at the end of the nineteenth and the beginning of the twentieth century, published dozens of works on the history of the Ottoman period based on archival research he carried out in Vienna. In the absence of systematic research on this issue, works ever since have been using his data to demonstrate large timber and fuelwood consumption during the Ottoman period in the fortifications of the border defense systems, despite the highly scattered nature of the information he managed to gather.[20] The numbers provided by Takáts are

16 Géza Pálffy, "A főkapitányi hadiipari műhely kiépülése Kassán és nyersanyagellátó forrásai" [The formation and the raw material provisioning of the chief captainship at Košice], in *Végvár és környezet*, 194–198. The sources of the expenses on building timber are from here: MNL OL E 211 Series I. IV. Fasc. fol. 88-a.

17 Béla Sarusi Kiss, "Vasgyártás és vasgazdálkodás Murányban a XVI. században" [Iron production and iron economy at Muránsky Hrad in the 16th century], *Fons. Forráskutatás és Történeti Segédtudományok* 4 (1997): 79–98.

18 Ágoston, "Where Environmental and Frontier Studies Meet," 75–76.

19 In general, see Rácz, "The Price of Survival," idem, *The Steppe to Europe*, 137–140 and idem, "Környezeti változások a kora újkori Magyarországon – környezettörténeti vázlat" [Environmental changes in early modern Hungary – environmental history sketch], in *Környezettörténet*, 157–165.

20 R. Várkonyi, "Környezet és végvár," 17–19, Gábor Ágoston and Teréz Oborni, *A tizenhetedik század története* [History of the seventeenth century], (Magyar századok) (Budapest: Pannonica, 2000), 90–91, Ágoston, "Where Environmental and Frontier Studies Meet," 74. The source of most of their related data comes from this article: Sándor Takáts, "Dunai hajózás a XVI. és

difficult to interpret, as they always describe wood consumption (or rather provisioning) per one single fortification per one single year, which reveals very little about the overall consumption. István Sugár followed a different approach in the 1990s. He studied the timber acquisitions of the castle of Eger, one of the most significant frontier castles on the northern edge of the Great Hungarian Plain in the mid-sixteenth century that withstood a major Ottoman siege in 1552 eternalized by the popular Hungarian novel (Géza Gárdonyi's *Egri csillagok*, 'Eclipse of the Crescent Moon'). Sugár looked at the supply zones for different types of wood for the fortification. He pointed out that timber was usually brought to the castle from the neighboring forests, which is no surprise as the hilly areas around Eger, most importantly the Bükk Mountains, provided a good basis for that. However, the sources he used also testify to the acquisition of wood from as far away as Maramureș, some 300 kilometers east of Eger. This latter data cannot be taken as a sign of the lack of wood in the neighborhood, as in most cases the castle was supplied from within a radius of a dozen kilometers.[21]

These few and scattered data certainly do not make it possible to draw general conclusions on the wood consumption for the construction of the earth-and-wood, brick, or stone fortifications. To understand the scale of the wood consumption needed for constructing the small but dense network of earth-and-wood fortifications (also referred to in the literature as palisade, *palánk,* and *palanka*), a different approach seems to be more fruitful.[22] In a short but important study, an archaeologist, Gyöngyi Kovács, and a palynologist, Pál Sümegi, attempted to calculate the wood consumption of earth-and-wood fortifications based on the archaeological data. With some restrictions, I will argue in Chapter 4 that this is the best possible approach to understand the impact of the construction of the border defense system on the forest cover.

Understanding the impact of the Ottoman forts on the woodlands is of major importance, as they are one of the most frequently discussed factors

XVII. században. III. közlemény" [Navigation on the Danube in the 16th–17th centuries. Part III], *Magyar Gazdaságtörténelmi Szemle* 7 (1900): 218.

21 István Sugár, "Az egri vár építőanyagainak beszerzési helyei 1548–1564" [Origin of the building materials of the castle of Eger from 1548 to 1564], in *Végvár és környezet*, 177–178.

22 See recently: Gyöngyi Kovács and Pál Sümegi, "Palánkvárak, fák, erdők. Régészeti és környezettörténeti adatok a török kori palánkvárak faanyag-felhasználásához" [Earth and wood fortifications, trees, and forests. Archaeological and environmental history data to the timber consumption of Ottoman-period fortifications], in *Várak nyomában. Tanulmányok a 60 éves Feld István tiszteletére* [On the trail of castles. Studies in honor of István Feld on his 60th birthday], eds. György Terei et al. (Budapest: Castrum Bene Egyesület, 2011), 113–120.

in the transformation in the early modern landscape. No less importantly, a major role was attributed to forests in defending the border of the Kingdom of Hungary in the sixteenth and seventeenth centuries, as was demonstrated in Chapter 2. Rivers and gallery forests were common features of the defense system, and thus the sources consulted for this book frequently address the issue of forests and woodland in the frontier zone. The sources used in Chapters 2 and 3 do not allow for a reconstruction of the forest coverage in the frontier zone but give some hints of the vegetation in some areas, especially important features of how the vegetation was perceived. The next section discusses the possibilities of reconstructing the Ottoman-period forest cover on a regional scale.

4.2 Reconstructing Forest Cover in the Early Modern Period – Perspectives and Limitations

Research into early modern forestry has been preoccupied with finding out how much of the medieval woodland was lost and roughly what percentage of the lands was covered by woods in the different regions of the Carpathian Basin in the early modern period. However, to my knowledge, there is no systematic source material on this issue with the potential for assisting in answering these questions. Nonetheless, this chapter intends to make some points that can help us understand the environmental legacy of the Ottoman wars from a forest history point of view by offering a partly new approach to the problem. Any analysis on the changes during the Ottoman presence in the extent of woodlands in the Carpathian Basin requires surveys that cover at least a statistically representative part of the area before and after the Ottoman conquest. Even if the source base is better for the sixteenth and seventeenth centuries than for earlier periods, more attempts have been made to reconstruct the forest cover of the medieval Carpathian Basin. But the documentary evidence gathered by Péter Szabó, discussed above only supports limited conclusions. By studying land estimations (*aestimatio communis* or [*köz*]*becsü* in Hungarian) from the fifteenth and early sixteenth centuries, he managed to gather information on approximately one percent of the area of the Carpathian Basin. Even less information is available for the Great Hungarian Plain, which, as was discussed above, is considered to have been most significantly affected by the presence of the Ottomans in terms of forest resources. Luckily, there is somewhat more data on the area of Transdanubia, but this still does not allow us to come to any far-reaching conclusions. According to Szabó, the endless forests that allegedly covered

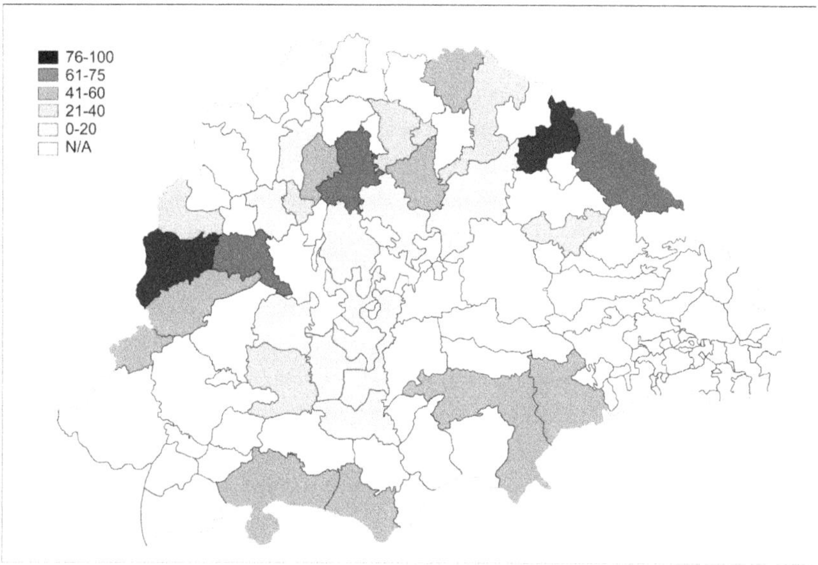

Figure 4.1 Forest cover in the Carpathian Basin in the Late Middle Ages and the frontier zone in the early modern period (after Szabó, *Woodland and Forests*)

the lowlands and hilly areas did not exist by the late medieval period, and probably never existed in the Holocene era. The 25 to 30 percent of overall forest cover calculated by Szabó based on the estimations and other information means that the Great Hungarian Plain and Transdanubia probably had less than 20 percent of woodland cover (and less than ten percent in the counties of the low plains; see Fig. 4.1).[23]

Despite or probably because of the seemingly better availability of sources – terriers, land conscriptions, and other documents – no attempt has been made to estimate the extent of woodlands in the sixteenth and seventeenth centuries on the scale of the Kingdom of Hungary, and even studies on counties or smaller units are missing.[24] The source availability again changes with the systematic military mapping of the country in the eighteenth century. The sheets of the aforementioned First Military Survey from the late eighteenth century provide a good research opportunity, but similar to the terriers from the early modern period, they have not been analyzed from the point of view of vegetation on the scale of the Carpathian Basin. In this case, at least the Great Hungarian Plain has been studied based on the maps of the First Military Survey. The results show sparse

23 Szabó, *Woodland and Forests*, 47–55.

24 Szabó, "Erdők a kora újkorban," 142.

forest cover (less than five percent) in the Danube-Tisza Interfluve and even lower in the Transtisza region (less than one percent). The numbers based on the survey from the 1780s reveal relatively little about the processes of the preceding centuries and therefore are not particularly useful for understanding the impact of the Ottomans on the vegetation of the plains. Nonetheless, they at least provide a snapshot of forest cover from a century after the Kingdom of Hungary recaptured the above-mentioned parts of the Carpathian Basin.[25] Two things seem to be clear at this point: first, based on the research carried out thus far, it is impossible to quantify the changes in the forest cover of the Carpathian Basin in the early modern period; and second, forests were probably never endless resources in the Hungarian Middle Ages nor in the following centuries. This may, of course, have been the case in the high mountains in the period of early statehood; however, in areas that were accessible to locals, the situation was probably already different by the Árpádian period.

The assumption that in the Middle Ages and the early modern times there was a lack of consideration for the finiteness of forest resources is a recurrent phenomenon in Hungarian scholarly works on the historical exploitation of forests. It has only been in the last 20 years that historians have devoted considerable attention to the problem of forest resources in the early modern period, and many works have argued that in various parts of Europe – and even outside of that – this is the period when this resource was understood as finite, which resulted in more conscious management of the remaining resources with significant involvement of royal administration and legislation.[26] It has been shown that legislation pertaining to the use of forest resources was promulgated in several countries from Spain, the Venetian Republic, the German areas, Poland, and the Ottoman Empire to as far as

25 Zsolt Molnár, "A Duna-Tisza köze és a Tiszántúl növényzete a 18–19. század fordulóján I. Módszertan, erdők, árterek és lápok" [Vegetation of the Danube–Tisza Interfluve and Transtisza Regions at the turn of the 18th and 19th centuries, I: methods, woodlands, floodplains, and fens], *Botanikai Közlemények* 95 (2008): 11–38, Mariann Bíró and Zsolt Molnár, "Az Alföld erdei a folyószabályozások és az alföldfásítás előtti évszázadban" [The forests of the Great Hungarian Plain in the century before the river regulations and the reforestation], in *Környezettörténet*, 169–206 and Pinke, "Alkalmazkodás és felemelkedés."

26 Cf. the so-called *Holznot* problem: Rolf-Jürgen Gleitsmann, "Rohstoffmangel und Lösungsstrategien: Das Problem der vorindustriellen Holzknappheit," *Technologie und Politik* 16 (1980): 104–154, Joachim Radkau, "Holzverknappung und Krisenbewußtsein im 18. Jahrhundert," *Geschichte und Gesellschaft* 9 (1983): 513–543, and Joachim Radkau, "Zur angeblichen Energiekrise des 18. Jahrhunderts. Revisionistische Betrachtungen zur 'Holznot'," *Vierteljahrschrift für Sozial- und Wirtschaftsgeschichte* 73 (1986): 1–37.

Japan.[27] The factors that influenced the change in the extent of the available forest resources differ in each case, as did the chronology of the process. But the problem itself, i.e., the understanding of the finiteness of the resource, was similar in the above polities. The control of natural resources became an increasingly important aspect of state-building processes. In addition, it served as a platform through which state control could be expressed.[28]

In Hungary, research has drawn attention to decrees and other documents that concern the protection of the forests from the early modern period onwards, attributing the appearance of these mandates to the scarcity of forest resources.[29] However, most of these documents – for instance, those published in the most important related source collection, the *Magyar erdészeti oklevéltár* (Cartulary of Hungarian Forestry) – do not mention the lack of woodland, nor do they refer to a scarcity of firewood or timber. Numerous documents fairly similar to this instruction by Ferenc Nádasdy (discussed above) have survived from the Ottoman period:

> We also wish that he [the local steward] in his position took good care of having the forests guarded, he would not allow the forests to be destroyed by not taking care of them, and without our order and wish no wood would be distributed for construction from the forbidden forests. Moreover, we wish that he, consulting with us, tried to acquire as much income, money both for building and for fuel as possible. Apart from that anyone who is found in our forests without giving a previous notice should, including themselves, their carts, animals, axes, and chains, be caught and brought

27 For a concise overview of the relevant legislative efforts in Europe at the time, see: John T. Wing, *Roots of Empire. Forests and State Power in Early Modern Spain, c. 1500–1750* (Brill's Series in the History of the Environment, 4) (Leiden and Boston: Brill, 2015), 19–28. For the different countries, see Wing, *Roots of Empire* and Felix Labrador Arroyo and Koldo Trápaga Monchet, "Forestry, Territorial Organization, and Military Struggle in the Early Modern Spanish Monarchy," *Environmental History* 23 (2018): 318–341 (for Spain and their colonies), Karl Appuhn, "Inventing Nature: Forests, Forestry, and State Power in Renaissance Venice," *The Journal of Modern History* 72 (2000): 851–889 (for the Venetian Republic), Paul Warde, *Ecology, Economy, and State Formation in Early Modern Germany* (Cambridge: Cambridge University Press, 2006) (for Germany), Mateusz Falkowski, "Fear and Abundance: Reshaping of Royal Forests in Sixteenth-Century Poland and Lithuania," *Environmental History* 22 (2017): 618–642 (for Poland), Alan Mikhail, *Nature and Empire*, 128–136, and White, *The Climate of Rebellion*, 28–31 (for the Ottoman Empire), Conrad D. Totman, *The Green Archipelago: Forestry in Pre-Industrial Japan* (Berkeley, CA: University of California Press, 1989) (for Japan).

28 E.g., Arroyo and Monchet, "Forestry, Territorial Organization."

29 Tagányi, *Magyar erdészeti oklevéltár*, vol. I, passim. See most recently: András Vadas, "For the Benefit of Generations to Come or for the Sake of Survival? Measures for Protecting Forests in Early Modern Hungary," *Historical Studies on Central Europe* 2, no. 1 (2022): 4–26.

to our houses, leaving the punishment to us, giving a third to those who caught and found them.[30]

These kinds of instructive documents were usually meant to regulate the use of local forests and should not be regarded simply as proof of resource scarcity. Instead, such documents should be treated as sources of an increasingly diverse administration of forests and conscious management of woodlands, which as mentioned was not unique to polities at the time, especially not ones involved in lasting military conflicts.[31] The sources gathered by Károly Tagányi for the above cartulary were collected predominantly from urban statute books which despite the more concentrated use of forest resources seldom refer to a scarcity. It is worth adding, however, that most sources originate from Upper Hungary (present-day Slovakia) or the western edges of the kingdom, which were rich in forests compared to the lowlands.

When sources refer to a scarcity of wood, this by no means supports the assumption that there was a complete lack of forests in a particular region. On the one hand, to ensure the long-term survival of forests, clearcutting was not common in the early modern period; on the other hand, in many cases only a specific type of timber or firewood – either a specific species or a specific size – was scarce. A short list with caps on the prices of timber and wood from the early seventeenth century from the domain of Eisenstadt (in Western Transdanubia in present-day Austria) supports the latter idea. The list includes more than 15 different prices for wood and timber for specific uses spanning from branches for wattling and stakes for mill dams to timber for the wheels of mills.[32] The list does not even differentiate between the different species used as timber. Most of the materials were probably not difficult to harvest in the nearby forests, but wood for specific

30 The instructions of Ferenc Nádasdy to the stewards at Lockenhaus, Deutschkreutz and Klostermarienberg (part of Mannersdorf an der Rabnitz), 20 January 1597. Edited in: Tagányi, *Magyar erdészeti oklevéltár*, vol. I. 270–271 (no. 182).

31 See, e.g., the decree of Sopron from 1541: Jenő Házi, *Sopron szabad királyi város története. II. rész, 2. kötet, Végrendeletek, közgyűlési jegyzőkönyvek, polgárkönyvi feljegyzések és különféle számadások 1400-tól 1541-ig* [History of the free royal town of Sopron. II/2. Last wills, town council minutes, town book entries and different account books] (Sopron: Székely és Társa Könyvnyomdája, 1931), 280. See also: István Csapody, "Sopron város (és volt úrbéres községei) erdeinek története (XII–XX. század)" [History of the forests of the town of Sopron (and its belonging villages), 12th to 20th centuries], *Erdészettörténeti Közlemények* no. 3–4 (1968): 3–16.

32 Lajos Merényi, "Kismartoni uradalmi faárszabás 1632-ből" [Price limitations of timber and wood from the domain of Eisenstadt from 1632], *Magyar Gazdaságtörténelmi Szemle* 12 (1905): 280. For a similar list, see: Anon., "Regéczi uradalmi erdőrendtartás" [Forest-use instruction from the domain of Regéc], *Magyar Gazdaságtörténelmi Szemle* 5 (1899): 462–464.

purposes, indicated by the varying prices, may have had to be acquired from further distances even in the case of the town, which was located in an area probably relatively wooded in the early modern period. Recurring problems in acquiring a certain material locally can be suggestive, but a systematic analysis of the available source material is required. However, such analyses have seldom been carried out in the context of early modern forest resources in the Carpathian Basin.[33]

As the above shows, there are many problems in trying to draw conclusions on changes in the forest cover during the Ottoman period, and quantifying these changes is virtually impossible. Still, the fact that at this point it seems unlikely that research will ever have a solid ground to suggest the exact percentage of forest cover in the early modern Carpathian Basin does not mean that there is no point in discussing the changes in the forest cover. The main goal in the next pages will not be to suggest numbers on the extent and change of the forest cover in this period but will primarily be to identify the impacts that the Ottoman-Hungarian frontier and the lasting military struggles might have had on changes in the vegetation. Similar to the previous main chapters, the focus will be on Transdanubia, but a more extensive outlook will be provided on the hinterlands on both sides of the frontier. As the area was home to numerous military campaigns, the impact of the war on the forests should be evident in this area – probably more so than for the Great Hungarian Plain, about which even less data is available.

4.3 Forms of Wood Consumption in Early Modern Hungary – Paved Road to Deforestation?

The change in the forest coverage certainly has connections with shifts in the wood consumption of the population in a given area. As indicated in the previous chapters, scorched-earth tactics were used only seldom and only locally. Therefore, one may presume that by far the most forest clearance and wood cutting took place in the early modern period to provide timber and other building materials as well as to provide fire- and fuelwood for the local population and the different industries. To understand the

33 Probably the only exception is Eszter Magyar's work on the woodland management of the Lower-Hungarian mining towns (in present-day Eastern-Slovakia). Eszter Magyar, *A feudalizmus kori erdőgazdálkodás az alsó-magyarországi bányavárosokban, 1255–1747* [Feudal-age forest management in the mining towns of Lower Hungary, 1255–1747] (Értekezések a Történeti Tudományok Köréből [New Series], 101) (Budapest: Akadémiai, 1983).

shifts in forest cover, we first need to discuss the changes and the scale of households' demands for wood, industrial needs as well the consumption of military purposes. First, the firewood consumption of the population will be examined, followed by an overview of some aspects of the wood consumption for military architecture and other war-related industries. Without at least a partial estimate of the pressure that these needs put on the woodlands of the central part of the Carpathian Basin, it is impossible to conclude what role the Ottoman presence and the attendant wars played in deforestation.

4.3.1 Firewood Consumption of Households in Early Modern Transdanubia – A Very Rough Estimate

The historical firewood consumption of households in the Carpathian Basin has never been, to my knowledge, thoroughly analyzed by scholarly literature. Probably this is due to the fact that at least three basic items of data would be needed to calculate the annual firewood consumption and its impact on the forest cover of the early modern Carpathian Basin: the exact population, the per capita firewood demand, and the annual yield of woodlands; and none of these is easy to estimate.[34]

Of the three above-mentioned items, research has – unsurprisingly – primarily addressed the demographic situation in the Ottoman period in the Carpathian Basin. Despite long periods of war and several major epidemics that struck Hungary, Transylvania, and the Ottoman Empire at the time, none of the estimates suggests a decrease in the population of the Carpathian Basin during the entire early modern period nor specifically during the presence of the Ottomans.[35] Therefore, solely based on the population growth, it would probably be safe to say that firewood consumption in the

34 For similar calculations, see Richard. W. Unger, "Thresholds for Market Integration in the Low Countries and England in the Fifteenth Century," in *Money, Markets and Trade in Late Medieval Europe Essays in Honour of John H.A. Munro*, eds. Lawrin Armstrong, Ivana Elbl, and Martin M. Elbl (Leiden and Boston: Brill, 2007), 369–370, and Paolo Malanima, "The Energy Basis for Early Modern Growth, 1650–1820," in *Early Modern Capitalism Economic and Social Change in Europe, 1400–1800*, ed. Maarten Prak (London and New York: Routledge, 2001), 53 and 61.

35 For the late medieval period, see László Solymosi, "Az Ernuszt-féle számadáskönyv és a középkor végi népességszám," [The account book of Ernuszt and the late medieval population] *Történelmi Szemle* 28 (1985): 414–436, András Kubinyi, "A magyar királyság népessége a XV. század végén," [The population of the Kingdom of Hungary at the end of the 15th century] in *Magyarország történeti demográfiája*, 93–110 and for the early modern period, see Dávid, "Magyarország népessége," and H. Németh, "Háború és népesség." For plague epidemics, see András Vadas, "A Batthyány uradalomrendszer és a 17. század első felének pestisjárványai"

Carpathian Basin was probably higher in the late seventeenth century than in the early sixteenth.[36] Despite the limited knowledge of the climatic processes of the Little Ice Age in the Carpathian Basin, there is every reason to believe that the early modern period was slightly colder than the High Middle Ages (i.e., the tenth through the fourteenth century) or the period that began in the mid-nineteenth century, which also suggests a higher demand for firewood.[37]

Even so, the impact of two factors – population and climate – on forest cover in the Carpathian Basin was probably close to zero. As has been demonstrated at different sites, firewood consumption was based on the use of coppicing (cutting trees regularly to ground level to promote growth) and pollarding (cutting off the top of a tree and its lower branches to encourage new growth), which seldom led to deforestation.[38] Nevertheless, it is important to note that major areas of the forests had to be managed to provide coppice for the inhabitants of the Carpathian Basin in the late medieval and the early modern periods. It is also important that firewood was seldom gathered from distant forests, and thus the rich woodlands of the marginal areas in the Carpathians may have played only a minor role in the firewood supply of the population of Hungary.

As far as I know, no calculation has been made of the extent of coppice forests required for the medieval or early modern Carpathian Basin or one of its sub-regions, but similar estimates have been calculated for medieval London, Paris, and Moravia. These numbers can be applied to the Carpathian

[The Batthyány Estate Complex and the Plagues in the First Half of the 17th Century], *Századok* 156 (2022): 25–46.

36 Firewood demand raised considerably elsewhere in Europe as well in the early modern times. For England, see Alfred W. Crosby, *Children of the Sun. A History of Humanity's Unappeasable Appetite for Energy* (New York: W. W. Norton, 2006), 69; for France: Chantal Camenisch et al., "The 1430s: A Cold Period of Extraordinary Internal Climate Variability during the Early Spörer Minimum with Social and Economic Impacts in North-Western and Central Europe." *Climate of the Past* 12 (2016): 2116.

37 András Vadas and Lajos Rácz, "Climatic Changes in the Carpathian Basin during the Middle Ages. The State of Research," *Global Environment* 12 (2013): 198–227.

38 See Oliver Rackham, *Trees and Woodland in the British Landscape* (London: J.M. Dent, 1976), 8. On coppicing: Julian Evans, "Coppice Forestry-An Overview," in *Ecology and Management of Coppice Woodlands*, ed. G.P. Buckley (New York: Chapman and Hall, 1992), 18–27 and on pollarding, see Péter Szabó, "'There is hope for a tree': Pollarding in Hungary," *Medium Aevum Quotidianum* 44 (2001): 41–60, Szabó, *Woodland and Forests*, 77–83. See also numerous publications connected to the project called: Long-Term Woodland Dynamics in Central Europe (Longwood). Online: http://longwood.cz/?page_id=109 (last accessed: 18 April 2017). See also the example of England: Paul Warde, *Energy Consumption in England & Wales 1560–2000* (Energy Consumption, 2) ([n.p.]: Consiglio Nazionale delle Ricerche Istituto di Studi sulle Società del Mediterraneo, 2007), 33.

Basin or, in this case, Transdanubia, to provide at least a rough estimate.[39] In London in 1300, the estimated population was no more than 80,000, and the town's households consumed c. 141,000 tons of wood each year. The annual yield of a managed coppice wood per hectare varied from place to place; there are references to yields of as much as five tons per hectare, but this may have seldom been the case. The average yield of a hectare of coppice wood in the Middle Ages and the early modern period may have been between two and three tons.[40] Based on the estimates on yields from England, in the case of London, this would mean 500 to 700 square kilometers of coppice to sustain the city's firewood needs.[41]

Estimating the population of the Carpathian Basin in the Ottoman period is a challenge, especially in areas like Transdanubia, which was partly in Habsburg and partly in Ottoman hands. To reach a rough estimate, I have calculated using the per capita firewood consumption of medieval London. Two population counts will be used: the estimated population based on the account books from 1494–1495 and the numbers that have been inferred based on the first official census in the Kingdom of Hungary ordered by King Joseph II in 1784/85. Bridging the three-hundred-year gap between the two is a major challenge in the historical demography in Hungary. According to the latest estimates, the population of the Kingdom of Hungary (including Transylvania and Slavonia) around 1500 may have been around 3.3 million at most, while in the 1780s it was 9.5 million, almost three times as much as in the late medieval period. If there had been a significant decrease in population during the Ottoman-war period, as has been suggested in earlier handbooks, this multiplication of the population would not be realistic. This does not mean that no areas were depopulated due to the Ottoman wars,[42] but in all probability, even the sixteenth century brought a slow increase in the population of the Carpathian Basin and even in Transdanubia, which

39 James Galloway, Derek Keene, and Margaret Murphy, "Fuelling the City: Production and Distribution of Firewood and Fuel in London's Region, 1290–1400," *Economic History Review* 49 (1996): 455–458, and Yvonne-Hélène Le Maresquier-Kesteloot, "L'approvisionnement de Paris en bois (XIVe–XVe siècles)," *Franco-British Studies* 20 (1995): 69–83. See also: Roland Bechmann, *Trees and Man: The Forest in the Middle Ages*, translated by Katharyn Duncan (New York: Random House, 1990), 141–142, and Warde, *Ecology, Economy, and State Formation*, 226–242.

40 Péter Szabó, Jana Müllerová, Silvie Suchánková, and Martin Kotacka, "Intensive Woodland Management in the Middle Ages: Spatial Modelling Based on Archival Data," *Journal of Historical Geography* 48 (2015): 8–9.

41 In the case of counting with the highest-yielding woods, 300 square kilometers could have been enough. Galloway, Keene, and Murphy, "Fuelling the City," 458–465. Hoffmann, "Footprint Metaphor," 305 accepts their numbers.

42 See e.g., the case of Tolna or Somogy Counties.

was heavily affected by military campaigns as the previous chapters have showed. This growth, however, was far below the numbers in Western Europe or other countries of the region such as Poland or Bohemia.[43] The territory of the Transdanubian counties was c. 45,000 square kilometers, and their population may have been around 700,000 in 1494/95. Applying the per capita consumption of medieval London and the yields in tons from the secondary literature adds up to a requirement for 4,000 to 6,000 square kilometers (c. ten percent of the territory of the Transdanubian counties) of coppice woodland to meet the needs of households in Transdanubia around 1500. Of course, one must keep in mind that not all the firewood was supplied from managed coppices. Although little evidence is preserved, collecting the smaller branches and boughs was a widespread practice for which tenant peasants usually did not have to pay. Based on several instructive documents issued for the leading administrators of royal domains, people were allowed to not only collect the firewood of the fallen trees in forests but also fell trees for building their own dwelling houses in some cases. Of course, many similar documents also refer to the importance of protecting the forests given their importance for related incomes such as providing acorns to keep herds of pigs.[44]

The overall consumption of firewood in London or Paris probably differed from the per capita consumption in the Carpathian Basin partly because of the different climatic conditions, the house sizes, their building materials, and numerous other factors. Although the industrial consumption of wood in the Hungarian countryside was fairly low in general compared to that of the highly concentrated craft activities in the English and French cities, the above numbers on London only indicate the consumption of households, which were comparable with that of the Hungarian households to some extent.

The estimates from Moravia may be closer to the per capita consumption in the Carpathian Basin because of the rather similar climatic and economic conditions. A recent study gave a rough estimate that was calculated using a similar method to the one used here. According to Péter Szabó and

43 Dávid, "Magyarország népessége," 151–152. See also: H. Németh, "Háború és népesség."

44 For examples, see e.g., *XVI. századi uradalmi utasítások. Utasítások a kamarai uradalmak prefektusai, udvarbírái és ellenőrei részére*, 2 vols. [16th-century estate instructions. Instructions to the administrators of the estates of the Hungarian Chamber], eds. István Kenyeres and Péter Kis (Fons Könyvek, 2) (Budapest: Szentpétery Imre Történettudományi Alapítvány, 2002), vol. I. 354 (cap. 25), I. 411 (cap. 16) and II. 620 (cap. 32). For instructions that forbid this practice, see ibid., vol. I. 212 (cap. 15), 226 (cap. 28), 337 (cap. 13), and vol. II. 574 (cap. 10). This practice is also mentioned in: Anon., "Regéczi uradalmi erdőrendtartás."

his colleagues, the c. 900,000 people who lived in Moravia in 1400 each consumed 1.5 cubic meters of firewood a year – not including the industrial production – which adds up to an annual consumption of 1,350,000 cubic meters. Based on their estimate, the firewood consumption of the population required c. 450,000 hectares of coppice, which is roughly 20 percent of the overall territory of Moravia and probably the majority of the forested areas there.[45] Accepting the 1.5 cubic meters per capita consumption and the 3 cubic meters per hectare yield, less than eight percent of the surface of the counties (c. 3500 square kilometers) in Transdanubia could have supplied the households with firewood in around 1500, significantly less than in the case of the calculation based on the London example.

As noted above, based on the present state of research, the extent of woodlands in Transdanubia was probably around 20 percent in the late medieval period. This seemingly low percentage of woodland could nonetheless easily have sustained the population and was probably still sufficient at the time of the census in the 1780s, when the population of the same area was above 1.6 million (see Fig. 4.2 and Appendix 2), although by that time, most of the forests had probably been used intensively for coppicing.[46] The 1780s provide a good research perspective for understanding the proportion of coppicing in the forests of Transdanubia, as this was the period not only of the first census but also of the First Military Survey. As mentioned above, no attempts have been made to reconstruct the exact forest cover for Transdanubia at the time. Dénes Bartha and Sándor Oroszi, however, have published an estimate of the forest cover in the 1780s for the whole of the modern territory of Hungary: 29.7 percent of the total area. The scholars neglect to mention the method they used, but more importantly for our purposes, this number seems to be greatly exaggerated. At the time, the forest cover on the Great Hungarian Plain, which covers almost 50 percent of present-day Hungary, was less than five percent (c. 3.5 percent). An estimate of 29.7 percent of forest cover for the whole of present-day Hungary would mean that more than 50 percent of Transdanubia was covered with woodland, which is highly improbable.[47]

45 Szabó et al., "Intensive Woodland Management."

46 Tamás Faragó and Péter Őri, *Az 1784–1787 évi népszámlálás II. Az Alföld, a Délvidék és a Dunántúl népességi adatai* [The census of 1784–1787, II. The population data of the Great Hungarian Plain, the Southern Frontier and Transdanubia] (Budapest: [n. p.], 2008), 44–45. Available online: http://real.mtak.hu/2144/1/46348_ZJ1.pdf (last accessed: 7 July 2017).

47 Dénes Bartha and Sándor Oroszi, "Magyar erdők" [Hungarian forests], in *Pannon Enciklopédia. Magyarország növényvilága* [Pannonian encyclopedia. Flora of Hungary], ed. Magda Járainé Komlódi (Budapest: Dunakanyar 2000 Könyvkiadó, 2000), 221–231, Dénes Bartha, "Történeti

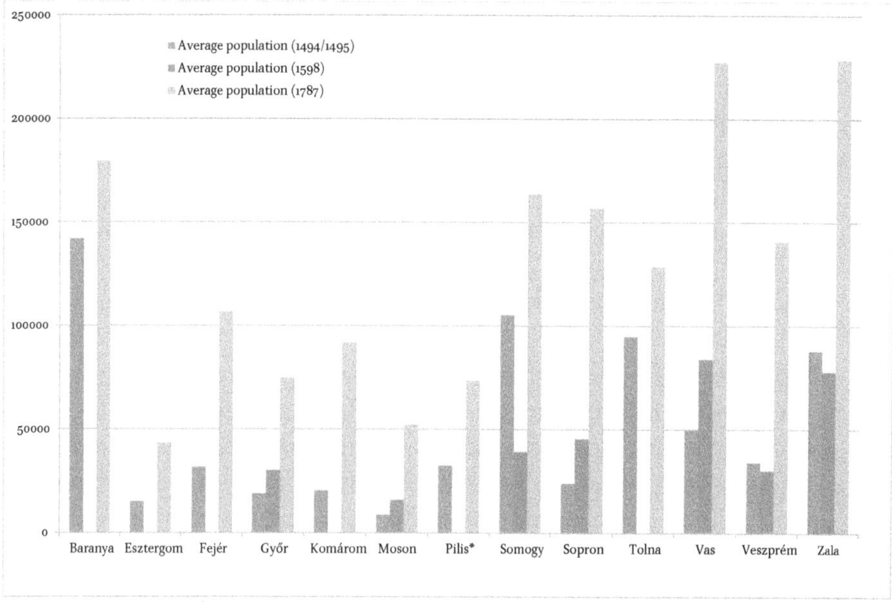

Figure 4.2 The population fluctuation of the counties of Transdanubia from the late medieval period to the late eighteenth century[48]

The above numbers on the possible territories used as coppice forests show that despite the growing population in Transdanubia, it may not have been a major problem supplying the population with firewood in the sixteenth and seventeenth centuries nor in the eighteenth. The numbers suggest that, at least during the Ottoman period, the forests in this area could have easily sufficed for the population's needs. Of course, with the growing population, more and more forests were cleared to create new plowlands as well as to re-inhabit former settlements, but in all likelihood, this process was accelerated in Transdanubia only from the period after Rákóczi's War of Independence (1703–1711), with the reorganization of the country's economy.[49] Nonetheless, the presence of permanent militaries in the central

erdőhasználatok Magyarországon" [Historical forest exploitation in Hungary], *Magyar Tudomány* 48 (2003): 1566–1577. Cf. Bíró and Molnár, "Az Alföld erdei," and Szabó, *Woodland and Forest*, 47–55.

48 For the data and the sources, see Appendix 2 as well as the explanation of the asterisk at Pilis County.

49 See most recently, with an excellent summary of the existing literature: Gábor Máté, "Landscape Reconstruction of the Southern Transdanubian Puszta (1683–1735) Based on 18th-Century Border Litigations," *Acta Ethnographica Hungarica* 62 (2017): 105–134. See also the classic study by Lajos Takács, *Egy irtásfalu földművelése* [The agriculture of a clearance village] (Budapest: Akadémiai, 1987), esp. 36–53.

part of the Basin along both sides of the frontier meant the introduction of a population that certainly was a new and significant consumer of firewood and also a major provisioner of food and drinks. The primary goal of this chapter is to explore and present the overall wood consumption of the area in the early modern period; however, other factors of consumption must also be discussed such as the needs of food and other industries, especially those that are related in some form to the persistent military conflicts.

4.3.2 Industries and Wood Consumption – Some Considerations

Before the beginning of the systematic use of coal in different industries, wood was the most important source of industrial energy followed by water and wind, the latter two being negligible in importance in energy production compared to fuelwood.[50] Several industries were significant consumers, sometimes even comparable in scale to the wood consumption of households. Three of them will be discussed along these lines: industries that satisfied primary needs such as food and drink (probably mass consumers); war-related industries such as gunpowder and firearm production; and finally, the building industry. It is important to note before going any further that it is almost impossible to provide numbers on the approximate industrial consumption in a certain region. Nonetheless, the available data suggest the scale of the resource use of the above-mentioned industries, which as I will try to argue can lead us to conclusions regarding changes in the forest cover in consequence of the Ottomans' presence and the attendant military struggles.

4.3.2.1 Food and Drink

Bread and beer accounted for a significant part of the calorie intake of pre-modern people, and the Kingdom of Hungary was no exception.[51] There are too many uncertainties – including the questionable population

50 Human and animal power is not discussed along these lines. Cf. Warde, *Energy Consumption*, passim, John Langdon, *Horses, Oxen and Technological Innovation. The Use of Draught Animals in English Farming from 1066–1500* (Cambridge: Cambridge University Press, 1986), 20 esp. note 51, and idem, "The Use of Animal Power from 1200 to 1800," in *Economia e energia, secc. XIII–XVIII*, ed. Simonetta Cavaciocchi (Florence: Le Monnier, 2003), 213–221.

51 See: Árpád Nógrády, "A középkor végi Magyarország mindennapi kenyere (Beregszász lakosságának gabonavásárlásai 1530-ban és a Jagelló-kori malomvámok)" [The everyday bread of late medieval Hungary. The grain provisioning of the population of Berehove in the 1530s and the mill customs in the Jagiellonian period], *A Nyíregyházi Jósa András Múzeum Évkönyve* 42 (2000): 155–169.

estimates in early modern Transdanubia, the very different data available on the average calorie intake of a historic population,[52] and the lack of precise data on the diets of different social layers[53] – to calculate the total food consumption and its distribution between the different sources of the calorie intake in the sixteenth and seventeenth century, which would be the basis for reconstructing the firewood consumption of the related industries. Nonetheless, two industries within the food industry must have been dominant in the demand for firewood, that is, baking and brewing. In London, the energy requirements of these two industries added as much as 25 percent to the annual per capita firewood consumption of the city dwellers.[54] Baking bread was less commercialized in early modern Hungary and the areas under Ottoman control than in London around 1300 or 1400, but this does not mean that related firewood consumption was more efficient. It was rather the other way round: most likely the better baking ovens in the bakeries of London consumed less energy per kilogram of bread than the rather small ovens used by families in Transdanubia. The baking industry in Transdanubia may have thus contributed to wood consumption in much the same way that it did in London. With brewing, it was probably somewhat different. Wine consumption, typical of Hungary, certainly required less wood than that of beer, although the production of barrels consumed approximately the same amount of wood in the case of wine and beer.[55] It is important to add, however, that it was in the early modern period that the Kingdom of Hungary also became a significant consumer of beer. Scattered data suggest that the garrisons on the Hungarian side

52 For another extreme, see the diet of priors in the Norwich cathedral priory in the Middle Ages: Philip Slavin, *Bread and Ale for the Brethren: The Provisioning of Norwich Cathedral Priory, 1260–1536* (Studies in Regional and Local History, 11) (Hatfield: University of Hertfordshire Press, 2012), 169–172.

53 On the consumption aristocracy: Borbála Benda, *Étkezesi szokások a magyar főúri udvarokban a kora újkorban* [Dietary practices in Hungarian aristocratic courts in the early modern period] (Archivum Comitatus Castriferrei, 6) (Szombathely: Magyar Nemzeti Levéltár Vas Megyei Levéltára, 2014).

54 Galloway, Keene, and Murphy, "Fuelling the City," 456, and 469–470. See also: Bruce M.S. Campbell, James A. Galloway, Derek Keene, and Margaret Murphy, *A Medieval Capital and Its Grain Supply: Agrarian Production and Distribution in the London Region, c. 1300* (Historical Geography Research Series, 40) ([n. p.]: Historical Geography Research Group, Institute of British Geographers, 1993), 31–36.

55 On the problem, see Renáta Skorka, "Pozsony a bécsi közvetítőkereskedelem árnyékában" [Bratislava in the shadow of intermediary trade], in *Tiszteletkör. Történeti tanulmányok Draskóczy István egyetemi tanár 60. születésnapjára* [Lap of honor. Studies in honor of the 60th birthday of István Draskóczy], eds. Gábor Mikó, Bence Péterfi, and András Vadas (Budapest: ELTE Eötvös Kiadó, 2012), 306 and 308. I am thankful to Renáta Skorka for drawing my attention to the data.

contributed to a growth in the consumption of this product. As garrisons only formed about five percent of the population of Transdanubia at the time, they may not have been such a significant source of consumption overall,[56] nonetheless, their bread and beer consumption can be directly associated with the war period. Brewing according to all probabilities was less widespread than in London or in the Low Countries, where it contributed significantly to the rise in the price of firewood at the time.[57] Unfortunately, there is no comparative data that can be used to estimate the firewood consumption of baking and brewing. But based on the fairly detailed documentation from London, it is likely that it added hundreds of square kilometers to the Transdanubian woodland used for coppicing in the early modern period. This use, of course, grew in line with the population in the eighteenth century,

4.3.2.2 *War Materiel*

The early modern period brought a change in the wood consumption of the military.[58] Changing fortification techniques and an increasing number of buildings raised the requirements for wood, as did the production of war materiel. The two most important uses were gunpowder production, founding cannons, and other weapons. Both activities were important in the context of the Ottoman period in Transdanubia as well as in other parts of the Carpathian Basin.

The sources studied in the context of the areas occupied by the Ottomans provided more comprehensive data than what is available for the Kingdom of Hungary. The literature on the production of gunpowder – particularly saltpeter (potassium nitrate), its main ingredient – indicates that the production centers in Ottoman Hungary in the mid-seventeenth century required around 1300 metric tons of wood every year.[59] As noted above, a hectare of

56 On beer provisioning of borderline castles, see Szabolcs Marton, "Gondolatok középkori katonaságunk szeszesital-ellátásával kapcsolatban" [Notes on the alcohol consumption of the garrisons in medieval Hungary], *Hadtörténelmi Közlemények* 120, no. 2 (2007): 577–589. See also: Gergely Csiffáry, "Végvárak és hadiipari létesítmények" [Borderline fortifications and military industrial complexes], in *Végvár és ellátás*, 127, and Sándor Takáts, *Művelődéstörténeti tanulmányok a XVI–XVII. századból* [Studies in 16th–17th-century intellectual history] (Budapest: Gondolat, 1961), 97–98.

57 Richard W. Unger, *Beer in the Middle Ages and the Renaissance* (Philadelphia: University of Pennsylvania Press, 2004), 137–142.

58 Gábor Ágoston, *Guns for the Sultan: Technology, Industry, and Military Power in the Ottoman Empire* (Cambridge: Cambridge University Press, 2004). See also: White, *Climate of Rebellion*, esp. 20–39.

59 Saltpeter is the main constituent of gunpowder.

well-managed coppice could yield between two and three tons every year on average. This suggests that the two most important production centers used approximately 500 hectares (five square kilometers) of coppice wood to meet the needs of the Ottoman gunpowder industry in the vilayets covering the former territory of the Kingdom of Hungary. Although this is not a negligible amount of firewood, it is not significant compared to the coppice wood used by households for heating.[60]

The sources for the Kingdom of Hungary are more scattered in this respect. The aforementioned Géza Pálffy, when studying the military complex of Košice in the last decades of the sixteenth century, concluded that the scale of the military-related industries in the Kingdom of Hungary did not require significant quantities of wood, which can be confirmed based on the different estimates in the literature on the consumption of the military complexes on both sides. The data gathered by Pálffy demonstrated that the wood was supplied from sometimes surprisingly big distances because of the lack of woodlands in the surroundings of Košice (which is a rather forested area in present-day Slovakia), and the growing special needs of the saltpeter industry; specific species of wood needed to be supplied. To give but one example, data on the supply record show that charcoal was burned from lime (linden) trees especially for saltpeter production, a tree seldom found in the area around Košice.[61]

Even though gunpowder (and saltpeter) were important elements of the military industry at the time, the quantities of wood required may have been insignificant compared to those of other industries. However, this in no way means that the military industry was not a significant factor in wood consumption. The increased demand for iron used in warfare, partly due to the spread of cannons and other military materiel, was probably more important in the consumption of wood than gunpowder production. It has been demonstrated in the context of Mediterranean forests that producing one ton of iron required the annual yield of twelve hectares of coppice wood (c. 30 tons). Considering the significant increase in the need for iron in Europe,[62] it is no surprise that fuelwood was seen in many parts of Europe as

60 Cf. Ágoston, "Where Environmental and Frontier Studies Meet," 74–75.

61 Pálffy, "A főkapitányi hadiipari műhely," 196 note 81. Data from: Béla Iványi, "A tüzérség története Magyarországon kezdettől 1711-ig, III." [History of artillery in Hungary from the beginnings to 1711], *Hadtörténelmi Közlemények* 27, no. 3 (1926): 263–264.

62 Richard Hoffmann, *An Environmental History of Medieval Europe* (Cambridge: Cambridge University Press, 2014), 215. A slightly different result can be found in David A. Tillman, *Wood as an Energy Resource* (New York: Academic Press, 1978), 3. According to Tillman, one ton of iron ore requires 14 tons of charcoal. One ton of charcoal is produced from c. 4 tons of wood, which

a limited resource, especially in forested areas where iron ore was mined.[63] This certainly had an impact on areas where iron-smelting and gunpowder production were major sources of income. It has been argued recently, however, that this may not have affected the forest cover in the Ottoman Empire to a large extent because of conscious woodland management,[64] and the situation may have been somewhat similar in the Kingdom of Hungary.[65] This nonetheless suggests that more and more woodlands had to be managed as coppice in the Carpathian Basin not only for the population's growing demand but also for smelting.

In areas of iron smelting, it was not only the smelting itself that was a major consumer of wood but also the mines themselves.[66] In the surroundings of mining towns, measures dating back to the Middle Ages protected the woodlands.[67] A special piece of evidence of the intensive exploitation of woodlands can be identified on a panel painting from the mining area in Upper Hungary (present-day Slovakia) from 1513. The painting in Rožňava depicts Saint Anne, the Virgin, and Child in the milieu of early-sixteenth-century Hungary. The mountains, painted by the unknown master, show the mining activities around Rožňava itself. The mountains, which were certainly wooded until the beginning of mining activities, had lost their

would add up to 56 tons of wood for one ton of iron: Joachim Radkau, *Wood: A History*. Transl. Patrick Camiller (Cambridge: Polity, 2012), 119.

63 Hoffmann, *An Environmental History*, 202 and 219, Bechmann, *Trees and Man*, 149, and Radkau, *Wood*, 119.

64 White, *Climate of Rebellion*, 289, based on: Purcell Nicholas and Horden Peregrine, *The Corrupting Sea: A Study of Mediterranean History* (Oxford: Blackwell, 2000), 184. See also: Alan Mikhail, "Anatolian Timber and Egyptian Grain: Things That Made the Ottoman Empire," in *Early Modern Things: Objects and Their Histories, 1500–1800*, ed. Paula Findlen (New York: Routledge, 2013), 274–294, Mikhail, *Nature and Empire*, 124–169. See further: Bechmann, *Trees and Man*, 151–154.

65 Vadas, "For the Benefit of Generations."

66 Beatrix F. Romhányi, Zsolt Pinke, and József Laszlovszky, "Environmental Impacts of Medieval Uses of Natural Resources in the Carpathian Basin," *Hungarian Historical Review* 9 (2020), 241–283.

67 Tagányi, *Magyar erdészeti oklevéltár*, vol. I, 20, and 25 (nos. 69 and 87). For the charters themselves, see *Codex diplomaticus Hungariae ecclesiasticus ac civilis*, 11 vols. in 44 parts, ed. Georgius Fejér (Buda: Typis Typogr. Regia Universitatis Ungaricae, 1829–1844), vol. IX/1, 497–503 (no. 277), and *Zsigmondkori oklevéltár*, 14 vols. [1387–1427] [Sigismundian cartulary] [A Magyar Országos Levéltár kiadványai II. Forráskiadványok 1, 3–4, 22, 25, 27, 32, 37, 39, 41, 43, 49, 52, 55, 59], eds. Elemér Mályusz et al. (Budapest: Akadémiai and Magyar Országos Levéltár, 1951–2020), vol. XIII, 326 (no. 927). See also for the latter: Gusztáv Wenzel, *Magyarország mezőgazdaságának története* [History of Hungarian agriculture] (Budapest: Magyar Tudományos Akadémia, 1887), 320. See also on hydrological impacts of floods: Kiss, *Floods and Long-Term Water-Level Changes*.

trees, probably mostly because they were used as props in mines.[68] Even if similar processes operated in some mining areas, this certainly was not representative of large areas of the Carpathian Basin.[69]

Similar to the problem of calculating firewood consumption in early modern Transdanubia, identifying the wood consumption for gun founding or manufacturing other war materiel is rather speculative without systematic analysis of related source material. Small firearms were mostly made of iron, while the larger caliber cannons used in the sixteenth and seventeenth centuries were made of either bronze or iron.[70] In the fifteenth century, iron predominated, but in the sixteenth and seventeenth centuries, bronze was used almost exclusively in cannon founding in Hungary. Although I found no relevant archival data on the amount of wood needed to produce one kilogram of copper, tin, or lead, the firewood demand probably did not differ significantly from that of iron.[71] The size and weight of cannons and bombards in medieval and early modern Europe varied significantly, but the largest examples weighed 15 to 18 tons.[72] The smelting of these bombards may have required 400 to 600 tons of firewood each. Most firearms were significantly smaller in size, however, and would have seldom required more than a few hectares of managed coppice wood. Based on the surviving inventories, the number of firearms kept in the fortifications strongly varied, but none of these documents suggests that their production required significant amounts of woodland even in the production centers. Of course, not only firearms consumed firewood; so did cannonballs, which unlike cannons were most often locally produced.[73]

Just as in the case of the saltpeter industry, the unavailability of certain species of wood required for charcoal presented a problem, which deserves

68 István Batta, "Középkori bányászatunk és kohászatunk a Metercián" [Mining and metallurgy in medieval Hungary as represented on the Metercia], *Bányászati és Kohászati Lapok. Kohászat* 121 (1988): 277–285.

69 See the case of the Lower Hungarian mining towns: Magyar, *A feudalizmus kori erdőgazdálkodás*.

70 György Domokos, "A kassai királyi hadszertár fegyverzete és felszerelése a XVI–XVII. századi inventáriumok tükrében" [Weapons and munitions of the royal armory of Košice based on inventories in the 16th and 17th centuries], *Hadtörténelmi Közlemények* 110, no. 4 (1997): 687–691.

71 Probably it was less because of the lower melting point of copper than that of iron.

72 Ágoston, *Guns for the Sultan*, 64–67.

73 György Domokos, "Vasfelhasználás és vasmegmunkálás a várakban a 16–17. században" [Iron usage and iron manufacturing in castles during the 16th–17th centuries], in *A vasművesség évezredei a Kárpát-medencében* [Thousands of years of iron crafts in the Carpathian Basin] (Anyagi kultúrák a Kárpát-medencében, 3), eds. Zoltán Nagy and János Szulovszky (Szombathely: Vas Megyei Múzeumok Igazgatósága, 2009), 177–188.

more attention. Both historical and scientific data suggest that beech was used in bronze production and pine in cannon founding, which may have led to shortages in these specific species of wood in some areas.[74] Pálffy also draws attention to the fact that the industrial complexes in Hungary were far from being self-supplying. A major proportion of the cannons and muskets were provisioned from other areas of the Habsburg Empire, the Holy Roman Empire, and elsewhere.

To sum up, limited data are available on the overall output of the military complexes in early modern Hungary, and the data is also incomplete for the Ottoman Empire, which makes it difficult to estimate the overall consumption of war-related industries. Nonetheless, it is probably not an overstatement to say that their production was only a minor factor in the wood consumption of Transdanubia and for that matter the whole of the Carpathian Basin.

4.3.2.3 Building Industries

Timber was certainly the most important building material in pre-modern Europe. The dwelling houses in medieval villages used significant amounts of timber, as did noble residences and the houses of burghers, and dozens of infrastructural elements. Instead of a general account of the use of timber in medieval society,[75] I will discuss two elements of the built environment that have been addressed in the existing literature in relation to timber consumption: mills and bridges.

As discussed in Chapter 3, water mills were important buildings not only because they supplied the population with flour and functioned as a significant source of income for their owners but also because they consumed a large amount of timber. By the mid-sixteenth century, grain was ground in water mills where the hydrogeographic conditions allowed it. Wind mills in the Carpathian Basin were seldom used until modern times. Instead, besides the continuous use of rotary hand mills, dry mills were used on the Great Hungarian Plain and elsewhere in the early modern period.[76] Nonetheless,

74 Mihály Détshy, "A sárospataki ágyúöntőház története" [History of the cannon foundry of Sárospatak], *Technikatörténeti Szemle* 5 (1968–1970): 79, István Ringer, Péter Barkóczy, and Árpád Kovács, "A sárospataki ágyúöntő műhely régészeti kutatása és a régészeti leletanyag metallurgiai vizsgálata" [Archaeological research of the Sárospatak gun foundry and the metallurgical study and the metallurgical study of the finds], *Archeometriai Műhely* no. 4 (2011): 354, 356, and 360, Ídris Bostan, "A szultáni ágyúöntő műhelyben (Tophâne-i Âmire) folyó tevékenység a 16. század elején" [Cannon founding at the Sultans Foundry (Tophâne-i Âmire) at the beginning of the 16th century] *Aetas* 18, no. 2 (2003): 17–18.

75 For similar works, see Radkau, *Wood*, and Behcman, *Trees and Man*.

76 See Katalin Szende, "Mills and Towns: Textual evidence and cartographic conjectures from Hungarian towns in the preindustrial period," in *Extra muros Vorstädtische Räume in*

probably at least 90 percent of the grain was ground in water mills by the late medieval period in Transdanubia, and the situation probably did not change during the Ottoman occupation.[77] Enumerating the number of water mills that operated in Transdanubia in the early modern period would be a complex task – although not impossible – based on several studies covering smaller regions or in one case a whole county. This would greatly contribute to the debate about several problems.[78] Having at least a rough estimate of the population in the late fifteenth century would make it easier to assess how many families relied on one water wheel in the Middle Ages. This may be possible if the sources allowed an almost complete count of the water mills. The more extensive the area covered by the different studies, the more reliable the estimate would be. In the context of the early modern period, the most important collection available is about Tolna County, one of the traditionally rich and densely populated counties in medieval Transdanubia. Using terriers, Ottoman tax records (*defterler*), historical maps, and other sources, the archaeologist András K. Németh identified approximately 200 mill wheels (!) that operated in the sixteenth century throughout the county.[79] Some of the mills mentioned in the seventeenth century or some referred in medieval charters were lost later, and some surely served different purposes than grinding grain at the time, but the c. 200 mill wheels identified by Németh may be close to the actual number existing in the second half of the sixteenth century. The population of Tolna County in the second half of the century may have been fewer than at the time of the tax conscription of 1494/95 when it was c. 95,000. Because it lies in one of the most important corridors that Ottomans utilized in their military campaigns – the Danube valley and its proximity – from the south

Spätmittelalter und Früher Neuzeit. Espaces suburbains au bas Moyen Âge et à l'époque moderne (Städteforschung, 91), eds. Guy Thewes and Martin Uhrmacher (Cologne: Böhlau, 2019), 505–508, and András Vadas, "Technologies on the Road between West and East: The Spread of Water Mills and the Christianization of East Central Europe," in *The Medieval Networks in East Central Europe: Commerce, Contacts, Communication*, eds. Balázs Nagy, Felicitas Schmieder, and András Vadas (New York: Routledge, 2019), passim.

77 See, e.g., Gyula Káldy-Nagy, *A budai szandzsák 1546–1590. évi összeírásai. Demográfiai és gazdaságtörténeti adatok* [The conscriptions of the sanjak of Buda, 1546–1590. Demographical and economic historical data] (Pest megye múltjából, 6) (Budapest: Pest Megyei Levéltár, 1985), passim or Géza Dávid, *A simontornyai szandzsák a 16. században* [The sanjak of Simontornya in the 16th century] (Budapest: Akadémiai, 1982).

78 E.g., for Körmend and its surroundings, see Vadas, *Körmend és a vizek*, but most importantly, see András K. Németh's study on the water mills in Tolna County: idem, "Vizek és vízgazdálkodás a középkori Tolna megyében I. Vízimalmok" [Waters and water management in medieval Tolna County, I. Water mills], *A Wosinsky Mór Múzeum Évkönyve* 35 (2013): 121–151.

79 For the meaning of the term, see: Vadas, "Some Remarks."

towards the center of medieval Hungary, the population loss in Tolna County was probably the most significant of all medieval counties. It is likely to have reached a minimum of around 50,000 people in 1571.[80] With 200 mill wheels noted above, this means that one mill wheel served approximately 250 people, which is very close to Western European parallels as well as the existing estimates for medieval Hungary.[81]

Based on the example of Tolna County, one can give at least a very rough estimate of the number of mill wheels in operation in Transdanubia in the Ottoman period based on the population numbers discussed above. Using the population numbers in Transdanubia around the year 1500 and in the 1780s leads to estimates of c. 2800 and 6400 water wheels, respectively. This is far more than suggested by the existing literature, which has estimated no more than 5,000 to 6,000 mills for the whole country in the late medieval period. One factor that might affect this estimate is that there were likely major regional differences in the prevalence of water mills even in the Late Middle Ages and the early modern period.[82] However, the numbers are probably not overestimating the operating wheels, as there were hundreds of industrial mills spread out over the country.

From the point of view of this chapter, however, the interesting question is how much timber was required to maintain the mills and the associated dams in Transdanubia. The buildings that housed mills were certainly less prone to destruction by water than their dams. Based on the archival material used in Chapter 3, it is clear that mill dams were rebuilt on more or less an annual basis. In the case of the mills of Körmend, Csákány, and Szecsőd discussed above, the usual annual provisioning was 40 to 100 fir stakes for the dam and some additional material for the water wheels themselves, for the mill building, and in some cases for the mill race.[83] Most of these

80 Ferenc Szakály, "Tolna megye negyven esztendeje a mohácsi csata után, 1526–1566" [Forty years of Tolna County after the battle of Mohács], *Tanulmányok Tolna megye történetéből* 2 (1969): 27–29. Cf. the results on this county with the more significant losses in the counties south of the River Dráva: Pál Engel, "A török dúlások hatása a népességre: Valkó megye példája" [The impact of the Ottoman plundering on the population of Valkó County], *Századok* 134 (2000): 267–321.

81 László Makkai, "Östliches Erbe und westliche Leihe in der ungarischen Landwirtschaft der frühfeudalen Zeit," *Agrártörténeti szemle* 45 [Supplementum] (1974): 45. See also: Water Endrei, "A magyar malomipar, 1550–1800" [Hungarian milling industry], in *Műszaki innovációk sorsa Magyarországon: malomipar, vaskohászat, textilipar* [Fate of technological innovations in Hungary: milling industry, iron smelting and textile industry], ed. idem (Budapest: Akadémiai, 1995), 48–54.

82 Károlyi, *A vízhasznosítás, vízépítés*. See also Ferenczi, "Water Management in Medieval Hungary," in *The Economy of Medieval Hungary*, 249, and Vadas, "Who Stole the Water," passim.

83 E.g., MNL OL P 1314 no. 24 094 (17 October 1622), MNL OL P 1314 no. 16 076 (22 September 1641) MNL OL P 1314 no. 48 109 (14 March 1646), or MNL OL P 1314 no. 33 770 (19 February 1650).

required stakes of significant sizes and not only smaller branches from fast-growing coppices to attach the stakes.

Pine was the best material to use for the stakes standing in water, but this was not always accessible, so oak and other species of trees were also used as building materials. The trees used for stakes were usually 40- to 50-year-old pines or oaks.[84] The cut trees probably produced an average of two stakes each. Most of the mills had more than one wheel served by the same dam, and of course not all the mills needed dams. Therefore, my calculations are based on approximately 1500 dams for mills in Transdanubia in the Late Middle Ages, and 50 stakes for each mill. In the case of an annual rebuilding, this adds up to as much as 75,000 logs each year for the dams. The wheels themselves were damaged less frequently but required pieces of timber that had to be very specific in length and diameter.[85] Still, the quantity used for them is probably negligible compared to those used for the dams. The mill buildings themselves also had to be reconstructed from time to time, which of course also required some amount of timber. After the stakes were built into dams, perhaps the two most important sources of consumption were smaller branches of trees that were used for solidifying dams and the timber that was used to form mill races. This sometimes required as many as 1,000 cartloads of branches, mostly from coppices.[86] Totaling everything, there is every reason to believe that water mills throughout Transdanubia required at least 80,000 stakes, that is 40,000 trees annually along with a few thousand other trees used for building mill houses (pine, oak, and beach mostly) as well as short-cycle coppices for the smaller branches. Given that 40,000 trees required about 100 hectares of forest each year[87] and the growing cycles may have been 40 years on average, up to 4,000 hectares or 40 square kilometers of woodland would be needed just for the mill wheels and the mill houses. Although this may not have put significant environmental pressure on the woodlands in the region, mills probably still used significantly more forest resources than the saltpeter industry and gun founding.

The data available for bridges is even less comprehensive than what we have for mills. Little research has been carried out on bridges, although by the

84 Older specimens were identified at the earth and wood fortification at Szentgotthárd by the Rába. It is important, as their construction dated very likely to the 1550s: Gábor Ilon, András Grynaeus, and Andrea Torma, "A szentgotthárdi török kori palánk kutatásáról" [On the archaeological research of the Ottoman palisade of Szentgotthárd], *Savaria* 31 (2007): 307–328.

85 Bechmann, *Trees and Man*, 166.

86 MNL OL P 1314 no. 48 109 (14 March 1646). A cartload of firewood (!) was probably around 1.7 cubic meters. Cf. Szabó, *Woodland and Forests*, 67.

87 I thank Péter Szabó for providing me with this number.

sixteenth century, thousands of bridges were built of varying sizes because they eased communication in the Carpathian Basin. Up until modern times, the bridges were usually made of wood, and the Transdanubian bridges were no exception.[88] The few built of stone were usually referred to as stone bridges, leaving little doubt that this was not the normal material used in such constructions.[89] Perhaps the best-known early modern bridge in the area was the one at Osijek; this was probably the longest bridge in the whole of the Carpathian Basin at the time.[90] Lacking in related archaeological findings, little research has been conducted on this bridge, but another Ottoman-period wooden bridge that stood in the central part of the Great Hungarian Plain at Szolnok on the Tisza River has been investigated in detail. Parts of the bridge were discovered during the extremely low water level of the river in the summer of 2003.[91] The structure and the materials used for this bridge give us useful insight into the wood consumption of similar bridges built by the Ottomans. The bridge at Szolnok may have been 190 meters long and spanned both the Tisza and the Zagyva Rivers. The material used for the bridge and even for the stakes that stood in the water was the less resistant oak. Even though oak decomposes faster in water than pine does, it was probably more accessible on the Great Hungarian Plain. According to the estimates, the construction of the bridge required c. 370 cubic meters of oak wood. Based on the average growth rate of the 40- to 50-year-old oaks – which were used not only in mill dams but also at

88 See Szilágyi, "Árpád Period Communication Networks," 150–155, eadem, *On the Road*, 186–193, and Oana Toda, "Economic and Material Aspects of the Late Medieval Bridges from Transylvania: The Written Sources," *Banatica* 27 (2017): 361–397.

89 There is data on the reconstruction of the bridges leading to the castle of Kanizsa during its Ottoman occupation using stone: Stein, *Guarding the Frontier*, 52.

90 On the length and the running of the bridge, see most recently: István Pánya, "Az eszéki Dráva-híd" [The bridge over the Dráva at Osijek], *Várak, kastélyok, templomok. Évkönyv, 2018* [Castles, Palaces, Churches. Yearbook, 2018], 166–170.

91 Róbert Kertész et al., "Tisza-hidak a török hódoltság korából radiokarbon és dendrokronológiai vizsgálatok tükrében" [Radiocarbon and dendrochronological investigations of Tisza bridges built during the Ottoman period], in *Az erdő és a fa régészete és néprajza (kézművesipar-történeti megközelítésben) / Archaeology and Ethnography of Forest and Wood (in Approximation of Handicraft History)* (Az anyagi kultúra a Kárpát-medencében, 2), ed. János Gömöri (Sopron: MTA VEAB Iparrégészeti, Archaeometriai Munkabizottság és Kézművesipar-történeti Munkabizottság, 2007), 145–178. On bridges over the Tisza River, see also: György Domokos, "Egy itáliai várfundáló mester Magyarországon a XVI. század második felében. Ottavio Baldigara élete és tevékenysége" [An Italian castle-builder in Hungary in the second half of the 16th century. The life and the activity of Ottavio Baldigara], *Hadtörténelmi közlemények* 111, no. 4 (1998): 767–856. See also an extended version of the works: idem, *Ottavio Baldigara: egy itáliai várfundáló mester Magyarországon* [An Italian castle-builder in Hungary, Ottavio Baldigara] (Budapest: Balassi, 2000).

earth and wood fortifications, as we shall see, and were probably the most frequently required wood – the above-mentioned amount of oak wood would have needed about 1.5 to 2 hectares of forest.[92] Even if the bridge had to be maintained regularly – as is evident based on accounts related to another bridge, the one at Bratislava in the late medieval period[93] – and some of the timber had to be replaced, it is unlikely to have required more than two or three hectares of managed oak forest. The width of the bridge at Osijek may have been similar to the one at Szolnok, but its length certainly was not. Although the different accounts vary considerably (from c. 2 to 8 kilometers), it was certainly many times longer than the bridge at Szolnok and may have been one of the largest such constructions in contemporary Central Europe.[94] Even if it was 40 times as long as the one at Szolnok, which seems the most likely (7.5 kilometers), and also slightly wider as suggested by the account of the famous Ottoman traveler Evliya Çelebi,[95] it may not have required more than 30 times as much wood as the one at Szolnok. It still would not have required more than a maximum of about 100 hectares, that is, one single square kilometer of forest. The bridge was burned down in the winter campaign of 1664 but was rebuilt shortly thereafter (probably in a different form), which again allows the assumption that it may not have been a major problem to find timber.[96]

Of course, this was "just one" bridge, but it probably consumed timber on a greater scale than any other bridge in Transdanubia. Most of the bridges that spanned the smaller rivers such as the Rába, the Zala, or the Rábca did not use more than one or two percent of the consumption of the bridge at Osijek. This means that probably no more than a few square kilometers of forest had to be managed to provide timber for the bridges in Transdanubia.

92 For the growth rate of oak, see Dezső Radó, "Bel- és külterületi fasorok EU-módszer szerinti értékelése" [Assessment of esplanades in settlements and in outskirts based on an EU scheme], *Lélegzet* no. 7–8 [Supplement] (1999): 12, Béla Keresztesi, "Az akác erdőművelési tulajdonságai és erdőgazdasági jelentősége a Magyar Alföldön" [The forestry characteristics of acacia and its role in forest management in the Great Hungarian Plain], *Az Erdő* 3 (1954): 188, Norbert Frank, "A kocsányos tölgy (Quercus robur L.) erdőművelési tulajdonságai" [The forestry characteristics of pedunculate oak (Quercus robur L.)], *Erdészeti Lapok* 150 (2015): 314–315, and Kovács and Sümegi, "Palánkvárak, fák, erdők," 117.

93 Tivadar Ortvay, *Pozsony város története*, vol. II/2 [The history of the town of Bratislava] (Bratislava: Stampfel Károly, 1898), 398–409.

94 For the different accounts of the bridge, see József Molnár, Török emlékek: Eszék–Dárdai híd a XVII. században" [Ottoman monuments: Osijek–Bridge of Dárda in the 17th century], *Művészettörténeti Értesítő* 7 (1958): 259–261 and Pánya, "Az eszéki Dráva-híd."

95 *Evlia Cselebi török világutazó magyarországi utazásai, 1660–1664* [The travels of Evliya Çelebi in Hungary, 1660–1664], ed. Imre Karácson (Budapest: Magyar Tudományos Akadémia, 1904), 189.

96 Pálffy, "Scorched-Earth Tactics," 193–194.

Nonetheless, their maintenance may have been just as frequently required as the mills' dams, which makes it clear that locally, the wood consumption of bridges may have affected the availability of timber for construction.[97]

Many other problems that had some impact on wood consumption could have been examined here, such as shipbuilding and the construction of dwelling houses.[98] The former may not have had much of an impact in Transdanubia due to the small size of the ships that were used by the Ottomans and the Hungarians.[99] Domestic buildings probably had a much more significant influence than any other consumers I have discussed in the context of timber consumption, but I simply have no means of assessing the forest consumption of the thousands of domestic structures of different sizes and materials. Nonetheless, I hope that this short survey on the scale of wood consumption of water mills as well as that of bridges has pointed to the fact that some of the consumers less frequently discussed in this context may have been more important in the use of forest resources than some of the above-mentioned ones. Existing scholarship, however, associates the Ottoman-Hungarian wars with large-scale woodland loss due to a further factor, namely the construction of hundreds of defense fortifications located in the central part of the Carpathian Basin, which is explained below.

4.4 The Biggest "New" Consumer? The Wood Consumption of the Military Defense in the Sixteenth and Seventeenth Centuries

Almost all the literature that discusses the environmental impact of the Ottoman wars on the forest cover refers to the construction of earth and wood fortifications as the most important factor putting pressure on forest resources in the early modern Carpathian Basin.[100] So far, however, no studies have produced an estimate of the consumption of timber to build and maintain the fortification system on a regional scale, although the consumption for some individual fortifications has been estimated recently.[101] The

97 Cf. Rohr, "Measuring the Frequency and Intensity of Floods."

98 On dwellings, see Bechmann, *Trees and Man*, 160–166.

99 Ágoston, "Where Environmental and Frontier Studies Meet," 58–60, and Végh, "A balatoni 'hadiflotta'."

100 Rácz, "From Steppe," Ágoston and Oborni, *A tizenhetedik század*, 88–92, Ágoston, "Where Environmental and Frontier Studies Meet," and most recently, Sárosi, *Deserting Villages*, 35, 41, and esp. 55.

101 Most importantly: Kovács and Sümegi, "Palánkvárak, fák, erdők."

main goal of this subchapter is to analyze how much timber was used for the building and maintenance of the more than 300 castles, fortifications, and fortified watchtowers that are likely to have stood in Transdanubia in the Ottoman period. The findings of this analysis will be used to argue for or against the importance of the Ottoman-Hungarian wars in the landscape changes in this respect. Just as in the case of the firewood, the needs of the population, or the timber consumption of water mills, the calculations to follow will be rather indicative and will only allow for cautious conclusions on the scale of resource use. Still, I believe these calculations will serve as a starting point for further, more, and more precise studies that address the same issue.

Different materials were used to build castles in the Carpathian Basin in the Middle Ages and the early modern period.[102] In the period of early statehood, castles were predominantly of earth and wood, but from the thirteenth century onwards the number of stone fortifications started to increase, although stone never became the sole building material, and stone castles themselves of course also used significant amounts of wood in their structures.[103] In the lowland areas of the Great Hungarian Plain and to a lesser extent in the Little Hungarian Plain, stone was a scarce and thus more expensive resource; therefore brick was frequently used as a building material as well. The firing of these bricks also required major quantities of firewood, but this was a completely different field of consumption than the stakes used in wooden fortifications.[104]

The long-lasting border defense line that successfully withstood the Ottoman advance until the 1520s consisted primarily of stone-built castles. After the fall of the linchpin fortification of Belgrade in 1521, as discussed in Chapter 1, this line of fortifications had to be replaced rapidly with new strongholds in the inner areas of the Carpathian Basin. In these fortification attempts, the use of timber prevailed. This does not mean that no stone or brick was used for fortifications, but the overwhelming majority of the newly erected fortifications used wood as their main material. The amount of timber used in the different fortification works varied significantly. An entirely different number of stakes had to be used at a small village church

102 See in detail: Erik Fügedi, *Castle and Society in Medieval Hungary* (Studia Historica Academiae Scientiarum Hungaricae, 187), translated by János M. Bak (Budapest: Akadémiai, 1986).

103 Bechmann, *Trees and Man*, 156–187.

104 For the consumption of fuelwood in the construction of Vienna's brick fortifications in the sixteenth century, see Christoph Sonnlechner, Severin Hohensinner, and Gertrud Haidvogl, "Floods, Fights and a Fluid River: The Viennese Danube in the Sixteenth Century," *Water History* 5 (2013): 188–189.

where the locals built a surrounding fence to provide at least some protection in case of Ottoman raids than for wooden fortifications with thick earth packing, wood walls, and bastions.

Relatively little attention has been paid to earth and wood fortifications until recently because they were believed to be quite simple in their architectures[105] and because only a few of them had been properly excavated by archaeologists.[106] The study of the historical sources and results of archaeological excavations have proved that many of the fortifications built in the sixteenth and seventeenth centuries, primarily of wood and earth, represent the highest technical knowledge of the period and should not be regarded as insignificant architectural remains. This has changed the perception of these buildings, and more and more studies are being dedicated to them. One of the crucial questions regarding the fortifications from the point of view of the analysis here is the structure of the walls. When trying to estimate the amount of timber used to build and maintain the different fortifications, one of the basic features is the number of stakes used for a meter of a wall. As the buildings behind the fences in many cases were not built in this period, or if so then there is limited knowledge, I chose to focus on the walls themselves to understand at least the scale of their wood consumption.

Before giving a brief overview of the fortification types, it is important to note that there was no major difference between the building techniques used by the Ottomans and the Hungarians.[107] The scholarly literature usually refers to three major types of fortifications; the first is the so-called *latorkert*. This may have been the simplest solution to providing a building with at least some protection (see Fig. 4.3). It was a palisade created with

105 See the influential work of Vidor Pataki, "A XVI. századi várépítés Magyarországon" [Castle-building in 16th-century Hungary], *A Bécsi Magyar Történeti Intézet Évkönyve* 1 (1931): 98–133. A slightly different view is reflected in: Sándor Takáts, "A magyar vár" [Hungarian castles], *Századok* 41 (1907): 726–741, 815–837, and idem, *Rajzok a török világból*, vol. II, 1–132. For a more recent discussion of wooden fortification, see most importantly the works of György Domokos as well as Gergely Tolnai, *Palánkvárak Magyarországon* [Earth and wood fortifications in Hungary] (Budapest: Martin Opitz, 2011).

106 For a short overview, see Maxim Mordovin, "Szécsény városának kora újkori palánkerődítése (A szécsényi Pintér-háznál feltárt maradványok alapján)" [The early modern plank fortification of Szécsény (in the light of the excavations at the Pintér house)], in *Várak nyomában*, 149–150, and idem, "The Post-Medieval Fortifications of Earth and Timber in Hungary," in *Études de castellologie médiévale. Château et frontière* (Château Gaillard, 26), eds. Peter Ettel, Anne-Marie Flambard Héricher, and Kieran O'Conor (Caen: Presses Universitaires de Caen, 2014), 273–282.

107 See: Takáts, *Rajzok a török világból*, vol. II, 1–132 (for the Hungarian side) and Stein, *Guarding the Frontier*, 53 (for the Ottoman side).

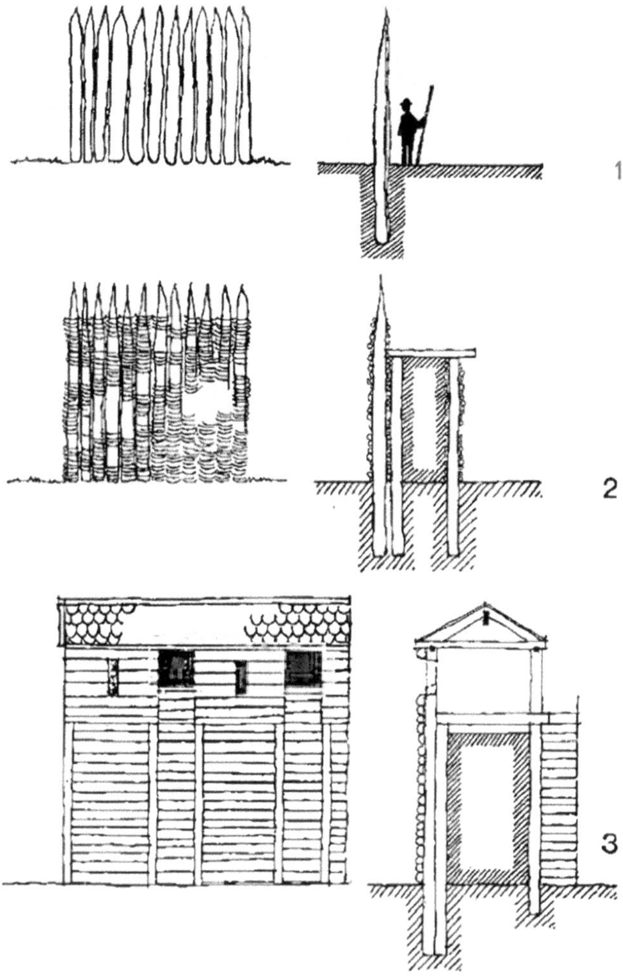

Figure 4.3 The main types of earth and wood fortifications used in the Carpathian Basin[108]

stakes driven in next to each other at small intervals and had a wattling to connect these stakes. This did not require much technological knowledge from the builders and was used frequently to provide some protection to watchtowers (the *górés* discussed above), manor houses, late medieval palace

108 László Gerő, "A török elleni harcokban átépített, vagy újonnan épített bástyás várak kialakulása" [The new fortifications with bastions that were built or rebuilt during the period of the Ottoman wars], in *Várépítészetünk* [Castle building in Hungary], ed. idem (Budapest: Műszaki Kiadó, 1976), 337.

buildings, or churches. It was also used as an exterior defense for more elaborate earth and wood fortifications. The amount of timber needed for these fences was relatively significant compared to their limited defensive role, as the stakes were driven in at narrow intervals, which did not provide strong protection.[109]

More elaborate types of wooden fortifications had stakes driven in two parallel lines. The two lines were formed by stakes at somewhat greater intervals than in the case of the simple palisades. To strengthen the structure, sticks were added to the stakes. The best type of wood for holding the stakes together was probably willow, but as it was relatively difficult to access, it seems that all kinds of sticks were used to strengthen the structure of the walls.[110] Earth was filled in between the two lines of stakes, which gave considerable thickness and strength to the walls and provided much better protection than simple wooden palisades. The earth filling may have been huge in quantity, as in some cases the two lines of logs were relatively distant – sometimes well more than a meter – from each other. In some cases, as has recently been demonstrated with the example of the fortifications of the town of Szécsény in the Northern Hungarian Mountains, apart from the two rows of stakes, a thick embankment was erected which also was strengthened with beams.[111] The third type of fortifications that has been identified in the scholarship is similar to the previous one, but the stakes in these fortifications were attached not only by sticks but also by iron bands, making the fence even harder to besiege.[112]

There were hundreds of fortifications in Transdanubia in the Ottoman period, which certainly required major quantities of logs, branches, and other materials. Understanding at least the scale of this consumption is certainly challenging and raises numerous methodological problems. However, as I argue in the following section, it is probably not impossible.

109 For an overview of the types of fortification, see Takáts, "A magyar vár," and idem, *Palánkvárak*.

110 Imre Szántó, "A végvári rendszer kiépítése Magyarországon" [Formation of the borderline defense in Hungary], *Acta Universitatis Szegediensis: Acta Historica* 38 (1971): 6–7. See also: Lajos Bende, "A törökkori magyar végvárakról" [On Ottoman-period Hungarian borderline fortifications], *Hadtörténelmi Közlemények* 18, no. 3 (1971): 514.

111 Mordovin, "The Post-Medieval Fortifications," 280. Similarly thick walls were identified by archaeological excavations at Bajcsa and other sites as well.

112 See, e.g., the building works of Kanizsa using iron bonds in 1558: Takáts, *Rajzok a török világból*, vol. II, 21.

4.4.1 Estimating the Wood Consumption of Earth and Wood Fortifications: Possibilities and Limitations

As a starting point in giving at least an estimate of the consumption of Ottoman-period fortifications in Transdanubia, a database of all the defensive structures had to be built. At least a few parameters must be defined to calculate the timber needed for a specific fortification: first, its size and shape, and second, its structure. It is to be noted here that the following calculations do not include the timber consumption related to buildings inside the different fortifications. These buildings inside the walls were also significant consumers, however I chose to exclude these buildings from the research, as their timber need well exceeds the uncertainties that are related to the walls themselves. Moreover, the most important consumers were probably the fortification walls themselves, and they had to be maintained more regularly than timber built into buildings. It should also be noted that in several cases there was a palisade that encircled a large area, and inside this perimeter, there was an earth and wood defensive wall to protect the inner stone- or brick-built castle, or a town in the area. In the case of the castle of Kanizsa mentioned above, for instance, at least three parallel walls protected the stronghold.[113]

Significantly fewer stakes were needed for a castle that stood for a decade compared to one that was in use for more than 150 years of the Ottoman presence in Transdanubia. Besides the actual building work of the fortifications, they also had to be reconstructed regularly because of the natural deterioration of the materials as well as the damage caused by sieges, fires – which were rather frequent – and so on. It is also challenging to estimate how long the stakes could last on average, how often the sticks had to be repositioned, and the frequency of minor earth filling works that seem to have been carried out continuously at each fortification.

Some of these problems are easier to deal with than others. There is little chance of knowing what type of fortifications stood in different places. In some cases, written evidence gives detailed insight into the structure and the size of fortifications, for instance at the better studied Körmend, Kanizsa, Győr, Pápa, or Bajcsavár, and about a few dozen others; however, there were many more small watchtowers and minor fortifications about which little is known apart from their approximate location. Similarly, the length of time that the fortifications were in operation is difficult to ascertain. In some cases, there are precise data on the beginnings of the

113 Takáts, *Rajzok a török világból*, vol. II, 21.

fortification in a certain place, on extensions during the wars, etc. And in other cases, either because of the lack of research into a fortification, the loss of sources, or simply because of its short life, there are only a few or perhaps one single reference to them. This is especially true for small fortifications which may have been similar to *górés* or small watchtowers surrounded by a simple wooden fence.

These points set several barriers to a reliable calculation of the timber consumption of the fortifications in Transdanubia. However, the goal here – just as in the case of the gunpowder industry, mill and bridge construction – is to achieve a rough understanding of the scale of the need for timber to understand how important the Ottoman period and the military activities may have been in the change of forest cover in Transdanubia. Therefore, I chose to make methodological decisions and restrictions to arrive at some conclusions. I had to accept that no matter how promising, it was impossible to collect all the archaeological and historical data on each of the fortifications as part of the present research. To conclude, I had to find a way to make some generalizations. I had to make three methodological assessments: first, I had to estimate the average perimeter of fortifications at different levels of importance; second, I had to decide whether the subject of the examination is a single palisade or a more complex earth and wood fortification wall; and third, I had to estimate how long the timber of the fortifications lasted and how often the stakes in fortifications had to be replaced by new ones.

Thanks to the meticulous research of the past few decades, one can come to a relatively precise estimate of the size and perimeter of several fortifications and also the sizes of the garrison stationed there by the Ottoman and the Habsburg-Hungarian military administrations.[114] In the case of literature that referred to the size of the fortifications, my task would of course be relatively easy; however, in the majority of the cases only the size of the garrisons stationed in different years is available. To estimate the number of stakes used for those fortifications whose sizes are unknown, I have used the average of the size of fortifications with similar garrisons. I assume that the number of men in a garrison correlated to some extent with the size of the fortification, but of course, this is not always straightforward. Another difficulty that I had to face, which will be discussed below, was the fact that

114 For the Hungarian fortifications, see most importantly: Géza Pálffy, "A magyarországi és délvidéki végvárrendszer 1576. és 1582. évi jegyzékei" [The Registers of the Hungarian and Croatian-Slavonian Border Fortresses of 1576 and 1582], *Hadtörténelmi Közlemények* 108, no. 1 (1995): 114–185, idem, "A török elleni," and Hegyi, *A török hódoltság*.

the same fortification had garrisons of very different sizes in the different periods of the Ottoman wars, even though according to the present state of the research in this field, many of these fortifications themselves did not change much in size. Despite these not negligible shortcomings, estimating the scale of the perimeter of these fortifications still seems to be the best way to make the calculation.

As noted above, I also had to generalize the wood consumption of the three different basic types of fortifications. One meter of a wall in three different earth and wood fortifications required different amounts of wood. I have estimated both the average consumption of a fence as well as that of the fortification that had stakes palisaded in two rows. The former is relatively easy. The post holes at different fortifications suggest that the stakes prevailing in these constructions were 20 to 30 centimeters in diameter. Estimating with smaller stakes of 20 centimeters in diameter and relatively limited gaps (c. 10-centimeter) between them, this meant that a meter of these walls consumed c. 3 stakes.

How many stakes the more complex fortifications consumed is seemingly less evident. The stakes used here were pretty much the same in size, but their intervals varied more, of course. Recent research has yielded important results in identifying the number of stakes used for building these earth and wood fortifications. Based on the examples of the fortifications of Barcs, Bajcsavár, and Szécsény, the interval between two logs may have varied between 40 and 60 centimeters.[115] Estimating similarly with the smaller, 20-centimeter logs set at 40-centimeter intervals to identify the greatest possible consumption, each meter of earth and wood fortification needed 3.2 stakes for the more complex fortifications that had the stakes poled in two parallel lines (1.6 stakes for each lines of stakes). Because of the fairly similar results in the case of the number of stakes needed for both fences and earth and wood fortifications, I chose to calculate using three stakes per meter for all kinds of structures. Some fortifications had inner and outer walls, or at least an outer palisade in addition to earth and wooden wall, but this only modifies the calculation in a few individual cases.

As noted above, the size of the garrisons stationed at the fortifications varied significantly in different years on both sides of the frontier. Where data are available, I have used numbers based on the sixteenth-century garrisons from the earliest sources published by a Hungarian Ottomanist, Klára Hegyi, on the Ottoman side of the frontier, and on the Hungarian

115 Kovács and Sümegi, "Palánkvárak, fák, erdők," and Mordovin, "The Post-Medieval Fortifications," 278.

side from the sources published mostly by historians such as Géza Pálffy and József Kelenik and archaeologists such as László Vándor. I have tried to avoid including numbers that might be distorted due to various factors – for instance, the large garrisons during the Fifteen Years' War at the turn of the sixteenth century as well as the slightly lower mid-seventeenth century numbers. I used the earliest possible data for what is missing for the sixteenth century. To classify the garrisons of the fortifications, I identified three categories. The smallest fortifications had fewer than 50 men. The size of fortifications with a small but continuous presence of troops could, of course, differ significantly in wood consumption from a mere watchtower. Based on the size of well-studied fortifications (Csány, Főnyed, Szentgyörgy, Szenyér, and Tótfalu[116]), however, and estimating two stakes from a single tree,[117] c. 300 trees might have been enough to build the fortifications of a castle or watchtower of this scale. Castles housing between 50 and 199 troops must have been larger based on the fortifications held by the Ottomans (Vál, Paks, Újpalánk, and Dombó[118]) and the Hungarians (Babócsa, Lenti, Fonyód,

116 On Csány: Pálffy, "A magyarországi és délvidéki végvárrendszer," 170, Károly Sági, "A zalacsányi török kori várak" [Ottoman-period castle of Zalacsány], in *A Göcseji Múzeum jubileumi emlékkönyve 1950–1960* [Jubilee volume of the Göcsej Museum] (A Göcseji Múzeum Közleményei, 8), ed. Imre Szentmihályi (Zalaegerszeg: Göcseji Múzeum, 1960) 131–135, on Főnyed: Kálmán Magyar and Gyula Nováki, *Somogy megye várai a középkortól a kuruc korig* [The castles of Somogy County from the Middle Ages until the *kuruc* period] (Kaposvár: Somogy Megyei Múzeumok Igazgatósága, 2005), 48, on Szenyér: László Vándor, "A zalai végvárrendszer a 16–17. században" [The frontier defense of Zala County in the 16th–17th centuries], in *Zala megye ezer éve. Tanulmánykötet a magyar államalapítás millenniumának tiszteletére* [Millennium of Zala County. Studies in honor of the millennium of the Hungarian state foundations], ed. idem (Zalaegerszeg: Zala Megyei Múzeumok Igazgatósága, 1996) 95, on Szentgyörgy: Gábor Szatlóczki, "Szentgyörgyvár a török időkben" [Szentgyörgyvár in the Ottoman period], in *Szentgyörgyvár története* [History of Szentgyörgyvár] (Zalai Kismonográfiák, 7), ed. Róbert Müller (Zalaegerszeg: [n. p.], 2002), 33–48, and on Tótfalu: Ferenc Végh, *Egerszeg végvár és város a 17. században* [The borderline fortification of Egerszeg and the town in the 17th century] (Zalaegerszegi füzetek, 10) (Zalaegerszeg: Millecentenáriumi Közalapítány: 2010), 156, and Baráth, "A Rába mint védelmi vonal," 45.

117 Based on: Kovács and Sümegi, "Palánkvárak, fák erdők," 115.

118 On Vál: Hegyi, *A török hódoltság*, vol. II, 614–626, and György Terei et al., *Fejér megye várai az őskortól a kuruc korig* [The castles of Fejér County from Prehistory to the *kuruc* period] (Magyarország várainak topográfiája, 3) (Budapest: Castrum Bene Egyesület and Cívertan Bt., 2011), 101–102, and 222, on Paks: Hegyi, *A török hódoltság*, vol. II, 1153–1161 and Zsuzsa Miklós, *Tolna megye várai* [Castles of Tolna County] (Budapest: Históriaantik Könyvesház Kiadó, 2007), 293–294, on Újpalánk: Hegyi, *A török hódoltság*, vol. II, 1162–1166, and Attila Gaál, "Turkish Palisades on the Tolna-County Stretch of the Buda-to-Eszék Road," in *Archeology of the Ottoman Period in Hungary* (Opuscula Hungarica, 3), eds. Ibolya Gerelyes and Gyöngyi Kovács (Budapest: Magyar Nemzeti Múzeum, 2004), 105–108, and on Dombó: Hegyi, *A török hódoltság*, vol. II, 1216–1224, Miklós, *Tolna megye várai*, 171–180, and Géza Szabó and Viktor Csányi, "Werbőczy két Tolna

and Csákány[119]). The average number of trees required for the wall of these fortifications may have been c. 600 trees. It has to be noted, however, that the numbers are not always clear. Just to note but one problem, some of the fortifications in the largest category – forts with more than 200 troops – were not larger or only slightly larger than those with sometimes significantly smaller garrisons. The category of the largest garrisons (200 and above) is more problematic than the smaller ones. They often housed not only troops but a significant civil population as well. This makes the estimates less accurate, but because by far the most fortifications in Transdanubia fell into the second and third categories, i.e., housed fewer than 200 troops, this problem does not seriously affect the overall calculations. Based on the size of some of the fortifications that belong to this first category – Berzence, and Paks[120] on the Ottoman side, and Körmend, Egerszeg, Kiskomár, and Bajcsavár[121] on the Hungarian (and Habsburg) side – I chose to calculate that c. 1,000 trees were required for the construction of their walls.

Wherever precise data are available for individual settlements, I used them, not the standardized sizes. In some cases, the two are very different; one example is the town of Keszthely, which had a relatively small garrison – around 100 people – but since the fortification surrounded the town's houses, it required significant amounts of wood. The area that the earth and wood

megyei vára: Dombó és Döbrököz az újabb régészeti megfigyelések tükrében" [Two castles of Werbőczy in Tolna County. Dombó and Döbrököz in the light of archaeological observations], *A Wosinsky Mór Múzeum Évkönyve* 34 (2012): 179–212.

119 On Babócsa: Pálffy, *A császárváros védelmében*, passim, Endre Marosi, *XVI. századi váraink* [16th-century castles in Hungary] (Budapest and Miskolc: Hungária-Európa Alapítvány and Borsod-Abaúj-Zemplén Megyei Levéltár, 1991), 9, and Magyar and Nováki, *Somogy megye várai*, 23–24, and later under Ottoman authority: Hegyi, *A török hódoltság*, vol. II, 1312–1319 and vol. III, 1578–1583, on Lenti: Marosi, *XVI. századi váraink*, 16, and Végh, *Egerszeg végvár*, 156, on Fonyód: Magyar, and Nováki, *Somogy megye várai*, 46–47, and on Csákány: Vadas, "Vízgazdálkodás és háborús védekezés," 220–221, and Baráth, "A Rába mint védelmi vonal," 52.

120 On Berzence: Hegyi, *A török hódoltság*, vol. II, 1320–1326 and vol. III, 1583–1589, and Magyar, and Nováki, *Somogy megye várai*, 38–39., and Paks, see note 449.

121 On Körmend, see Chapter 4. See also: József Kelenik, "A nemzetiségi megoszlás, a veszteségek és a fluktuáció mértéke. Tizennégy Kanizsa elleni végvár helyőrségében (1633–1640)" [Ethnic composition, the losses, and the scale of fluctuation. On the example of 14 garrisons against Kanizsa], in *Végvárak és régiók*, 108, on Egerszeg: Végh, *Egerszeg végvár*, 30, Kiskomár: Pálffy "A magyarországi és délvidéki végvárrendszer," 149, and Róbert József Szvitek, "Kiskomárom végvár építéstörténete" [The building history of the borderline fortification of Kiskomárom], *Castrum Bene* 14 (2011): 45, and on Bajcsavár: Leopold Toifl, "Bajcsavár története a stájer levéltári források alapján" [The history of Bajcsavár in the light of Styrians archival data], in *Weitschawar/Bajcsa-Vár. Egy stájer erődítmény Magyarországon a 16. század második felében* [Weitschawar/Bajcsa-Vár. A Styrian fortification in Hungary in the second half of the 16th century], ed. Gyöngyi Kovács (Zalaegerszeg: Zala Megyei Múzeumok Igazgatósága, 2002), 28.

fortification surrounded was c. 400 by 500 meters.[122] The structure in this case was more elaborate than just a wooden palisade; it had two parallel lines of stakes with earth filling between them. Building these fortification walls probably required c. 5,400 stakes, which equal 2,700 fallen trees. There are also some contrasting examples when a large military population was ordered to defend rather small fortifications. This was possible, as many of the garrisons did not actually live inside the encircled areas except during wars. A good example of this is the well-studied fortification of Bajcsavár. The fortification's size and shape are well-documented by archaeological research. The rather peculiar, pentagonal fortification – partly brick, partly earth and wood – required less than 800 trees or 1,600 stakes but housed as many as 550 men.[123]

The next problem is to find information on how often stakes had to be replaced in fortifications. There are numerous sources on reconstructions, for instance, in the case of the above-discussed fortifications at Körmend and Csákány. The documents on reconstructions in some very fortunate cases even refer to the number of stakes that were built in to replace the decomposing ones, but the data in the sources are usually not precise enough to give a clear picture of the speed of the decomposition process. The data available are controversial. In the case of Körmend and Csákány, the reconstruction works probably were not very frequent, but when the fortifications were rebuilt and its stakes replaced, the sources refer to major quantities of material built in.[124] At Bajcsa, the picture is somewhat different. Here, probably due to the unfavorable site selection – namely the unstable foundations in a marshland – continuous rebuilding activities are suggested in the sources, with sometimes large quantities of stakes.[125]

The need for frequent maintenance was well known to contemporaries. A source from present-day eastern Hungary describes how frequently the wooden parts of fortifications had to be replaced. In the 1670s, the nobility of Szabolcs County, which lay in the contact zone of the Ottomans, the Kingdom of Hungary, and the Principality of Transylvania, wrote a supplication to be read out at the noble gathering held at Banská Bystrica in 1670:

122 Végh, *Birodalmak határán*, 56, and 124–125. See also: Pálffy, *A császárváros védelmében*, passim.

123 Kovács, *Weitschawar/Bajcsa-Vár* and Kovács and Sümegi, "Palánkvárak, fák, erdők."

124 In one case, as many as 2,600 new logs are mentioned: MNL OL P 1314 no. 19 237 (13 March 1641). For Körmend, apart from the present book, see Vadas, *Körmend és a vizek*, and Baráth, "A Rába mint védelmi vonal," 35.

125 Toifl, "Bajcsavár története," 28–34. See also: Kovács and Sümegi, "Palánkvárak, fák, erdők," 119–120.

> as the castle of Kálló [Nagykálló] is built of wood we humbly ask your majesty [Emperor Leopold I] to provide some financial assistance so that the castle could be rebuilt of brick, as the wooden bastions and planks last short, their wood as well as the branches rot fast.[126]

This source demonstrates that the materials had to be renewed regularly, which is also echoed in other sources. In Kecskemét, a town in the central part of present-day Hungary, the local judge or his representative had to check the ditches and the earth and wood fortification four times every year by literally perambulating the town.[127] This is probably a good indication of how fast the material of wood fortifications could deteriorate. This does not mean, however, that almost the entire material of the fortifications was replaced every once in a while. In the cases of Körmend and Csákány, sources also refer to works when only a few stakes were replaced.[128] The sources suggest that such works were regularly necessary – which mostly did not involve replacing stakes, however, but rather were carried out using new branches and earth filling to strengthen the braid. Major rebuilding works at these fortifications probably took place only a few times during the Ottoman period. Similar conclusions were drawn by archaeological observations and written sources at one of the most important fortifications in the frontier zone east of the Danube, the aforementioned Szécsény. Here the dendrochronological examination of the excavated wooden materials shows that some of the stakes were never replaced in the fortification during the Ottoman wars; there were, however, some smaller maintenance works done during that period. One rebuilding activity, however, is worth noting here. The castle changed hands three times during the Ottoman period: it was in Turkish hands for a short period during the Fifteen Years' War and then from 1663 to 1683.[129] After the second occupation, the walls had to be rebuilt because the fortifications had been set on fire by the Hungarians

126 *Kalló vára, mivel csak fából vagyon építve őfelsége kegyelmesen provideáljon felőle és rendelje arra való contributiót, mellyel hozáfogván építéséhez, téglából építtessék meg, mivelhogy az fából épített bástyák és palánkok kevés ideig tartanak, hamar elrothadván mind fájok és sövények* – MNL SzSzBML IV. A. 1. Fasc. 88. no. 29/1670. Quoted in: Gyula Koroknay, "Kálló véghely két ostroma (1670, 1672)" [The two sieges of borderline fortification of Kálló], *Hadtörténelmi Közlemények* 103, no. 3 (1990): 72.

127 *Kecskeméti szabályrendeletek 1659–1849* [Kecskemét town statutes], ed. Tibor Iványosi-Szabó (Kecskemét: Bács-Kiskun Megyei Levéltár, 1991), 75–76 (no. 83). See also: Sárosi, *Deserting Villages*, 178.

128 See, e.g., MNL OL P 1314 no. 24 356. (7 March 1641). The source is also quoted by: Baráth, "A Rába mint védelmi vonal," 45 note 120.

129 Mordovin, "Szécsény városának kora újkori palánkerődítése."

before leaving Szécsény. The available data are slightly controversial because some sources suggest that the rebuilding process only included smaller works, but an Ottoman source implies something else. According to an order sent to the *beylerbey* (district governor) of Eger on the materials needed for the reconstruction of Szécsény in the period when it was occupied by the Ottomans, the quantities are almost unheard of and are interesting even if the fortification itself is not in the geographical focus of the present work.[130] The Ottoman janissaries requested the beylerbeyi send 10,000 stakes for the inner and outer tower, a further 5,000 for the four sides, and 4,000 deck planks for its foundations. Furthermore, for the rebuilding of a bridge that led to the castle, they required 1,000 special planks, along with 4,000 further bridge deck planks. Although the quantities were probably intentionally exaggerated, they are huge. This is difficult to reconcile with the data from Csákány or Körmend. Because of the controversial data, I decided to calculate using a 10-year cycle, which means that within this period each stake was replaced by a new one. Some stakes probably lasted well more than 10 years, but those affected by water had to be replaced almost annually.[131] This calculation probably overestimates the speed of the replacing of the stakes in the earth and wood fortifications, but as I aim to understand the possible scale, it is better to count with a relatively higher frequency than to underestimate the amounts required.

The next step in the calculation process is to find a way to see how long the different fortifications were in operation in the Ottoman period. Some of them changed hands between the Hungarians or the Ottomans a couple of times and were used by both parties. Some had short occupations and were probably in operation for not more than a few years or a decade. Where I found references to the length of their operation, I used that; in every other case, I estimated that the fortification was in existence throughout the entire period discussed here. This leads to numbers higher than the reality, but it enables us to calculate the upper boundary of timber consumption. If a fortification fell to the Ottomans and was then included on the list of Ottoman fortifications, I did not include it in the list of both sides.

Based on this methodology, the approximate demand for timber in the different fortifications can be reconstructed with a large margin of error. The aim, however, as was noted above, is to identify the scale of the use of forest

130 Lajos Fekete, *A hódoltság török levéltári forrásai nyomában* [In pursuit of the archival sources of the Ottoman period] (Budapest Oriental Reprints, Ser. A 6.), ed. Géza Dávid (Budapest: MTAK, 1993), 290 [90] (no. 252).

131 See, e.g., the problems at Bajcsavár: Toifl, "Bajcsavár története," 28–34.

resources rather than the precise number of trees or square kilometers of woodland that had to be managed for this purpose. The calculation offered here is by no means precise enough to argue for or against a change in the extent of woods in Transdanubia in consequence of the Ottoman period, but it can be a tool in determining whether building a significant number of new fortifications would have consumed large quantities of woodland resources or whether other areas of consumption were more important and can be instrumental in assessing the impact of the wars on the forest resources.

The database I compiled at the time of writing the book includes c. 220 fortifications on the Hungarian side and 95 on the Ottoman (for their lists, see Appendix 3). Because of the nature of the Ottoman sources, the list based on them may be more complete than the list of the Hungarian fortifications. There were dozens of smaller watchtowers and fortified churches on the Hungarian side, but the lists of fortifications I used did not include them. Neither did I systematically include fortified – walled – towns in Transdanubia. Some of the towns in Transdanubia had stone walls dating back to the Middle Ages. These were in some cases also augmented by new outworks made of wood as well as the foundations of stone fortifications, which also used significant amounts of wood.

According to my calculations, the roughly 320 fortifications in Transdanubia, most of which can be associated with the Ottoman wars, needed on average c. 3,600 (for the Ottoman side) and 6,300 trees (for the Hungarian/Habsburg side), that is, 7,200 and 12,600 stakes in each year during the 150 years between 1541 and 1690. This adds up to c. 534,000 (Ottoman) and 938,000 (Hungarian/Habsburg) trees altogether. It is important to emphasize two things. First, as noted, these numbers represent only the amount consumed for building the earth and wooden fortifications themselves; it does not include the amount built into the castle buildings and fortified palaces themselves nor the other different buildings that were part of the castles such as stables and storage houses. Second, fortification work was not distributed evenly throughout the Ottoman presence in Transdanubia. The 1540s and especially the 1550s certainly brought a major rise in demand for wood, and wood was probably consumed less consciously in these periods than from the 1560s and 1570s onwards, when a more refined defense strategy came into effect. In these years, probably many more than 10,000 trees were felled yearly to build fortifications in Transdanubia, while in the 1670s, for instance, there may have been significantly less demand for timber.[132]

132 Pálffy, "A török elleni," 192–199, and idem, "The Origins."

The most important question has yet to be answered: how much area was required to provide 10,000 trees annually? There is limited information on the yields of medieval forests, but there are relatively many studies on nineteenth- and twentieth-century forest management that I could use – with some limitations, of course. For forests used for similar purposes – i.e., producing stakes of similar sizes –trees in modern times are planted c. 4 meters apart. If we estimate less intensive management, with a tree every five meters, a hectare could have provided c. 400 trees. This means that on an annual basis, c. 25 hectares of wood had to be clear cut to provide timber for the walls of Transdanubian fortifications. During the 150 years discussed in this chapter, the same wood plot could have gone through approximately three cycles (assuming a 50-year cutting cycle), which means that c. 1,250 hectares – or 12.5 square kilometers – of woodland had to be managed in order to build and maintain the fences and outworks of the fortifications in the region.

There is no question that this number is remarkably low. Even if my results are greatly underestimated – and because of several factors listed above they may be – leading to the conclusion that in reality, ten times more woodland had to be managed to produce the trees necessary for the earth and wood fortifications (which I believe is unlikely), that would still mean a relatively limited area for a territory like that of the entire Transdanubia. This would probably be less than or comparable to the needs of the above-mentioned water mills, for instance. This is surprising because as discussed in the introduction to this book, scholars in many cases refer to the newly built fortification systems as the key reason for the supposed deforestation during the Ottoman period both on a regional scale and at individual sites.[133] Transdanubia in the Ottoman period was probably the area with the densest network of fortifications. While there were many fortifications on the northern edges of the Great Hungarian Plain at the hilly area in present-day northern Hungary and southern Slovakia, larger areas, such as the lowland itself at the center of the Carpathians, had very few fortified places. Therefore, if the new fortification works had a relatively limited impact on the forest resources at Transdanubia, it is doubtful that they could have seriously impacted the woodlands in the Great Hungarian Plain or elsewhere in the frontier zone between the Ottomans and Hungarians.

Although a systematic collection of the written evidence on the problem has yet to be made, some general observations can be summarized here,

133 On the suggested deforestations related to castle building in the Dráva valley, see: Kovács and Sümegi, "Palánkvárak, fák, erdők."

which strengthens my assessment of wood consumption. While the source material discussed in Chapter 3 makes hundreds of references to the problem of water, only a few more than a dozen sources refer in some way to the problems of forest resources. Although there is scattered evidence of supplies of wood and timber having been brought from far away, wood was mostly brought to fortifications from within a reasonable distance on both the Hungarian and Ottoman sides in the frontier zone.[134] Both the Ottoman and Hungarian sources frequently mention the need for rebuilding the palisades and fortifications, but there is seldom a reference in these sources to a lack of timber.[135] Although there is limited data, timber does not seem to have been an expensive material at the time. I will only include one single example, which is not even from Transdanubia but is certainly telling of the timber prices in the region. At the building work of the Ottoman fortification of Gradişka on the Sava River somewhat to the south of the region discussed here, almost four times as much money was spent on the ropes (!) that held the stakes together than on the timber itself.[136]

This does not mean that Transdanubia in the early modern period did not periodically experience a local shortage of timber, but this was probably seldom the case or was connected to a specific species of wood, or a timber suited a certain use such as pine for stakes built into the dams of water mills, or willow and other species used by the different industries.

4.5 Forests in the Hinterlands – What Were They Used For? – A Brief Outlook

Although this chapter focuses on the forests of Transdanubia, it is important to take at least a brief look at the hinterlands on both sides of the frontiers to understand what factors may have been important in the changes of the forests and landscapes there. For the Hungarian side of the frontier, I will briefly refer to the most important form of forest resource consumption

134 On Ottoman provisioning, see Pál Fodor, "Török várerődítési munkák Magyarországon a XVI–XVII. században" [Ottoman fortification works in 16th–17th-century Hungary], *Hadtörténelmi Közlemények* 26, no. 3 (1979): 382–383 and idem, "Bauarbeiten der Türken an den Burgen in Ungarn im 16–17. Jahrhundert," *Acta Orientalia Academiae Scientiarum Hungaricae* 35 (1981): 55–88.

135 See, e.g., the numerous references to the reconstruction of earth and wood fortifications in Ottoman sources: Fekete, *A hódoltság török levéltári forrásai*, 228 (no. 38), 231 (no. 50), 261 (no. 154), 282 (no. 226), 290 (no. 252), 296 (nos. 274 and 275), 386 (no. 159), 398 (no. 194) 404 (no. 211).

136 Stein, *Guarding the Frontier*, 53. See also: Fodor, "Török várerődítési munkák," 382–383.

– mining – followed by a brief overview of the impact of large-scale cattle herding in the central part of the Great Hungarian Plain, which for the period discussed in this book was almost fully controlled by the Ottomans.

The area of the valley of the Hron was one of the most industrialized landscapes in pre-modern Hungary.[137] This was not the case until the second half of the medieval period. The scarce Slavic and Hungarian population that lived in the area did not engage in the exploitation of the ore deposits up to the mid-twelfth century. It only received an impetus from the late twelfth and early thirteenth century onwards, when mostly German-speaking settlers, receiving special privileges, moved to the area and started systematic mining activity.[138] The most important ores mined were gold and silver, soon followed by copper mining, still in the first half of the thirteenth century. Other ores, including iron, were also mined in the vicinity of the Hron River valley, but their importance was much smaller than that of the precious metals and copper.[139] The output of gold and silver peaked in the Late Middle Ages, but copper, along with other metals including tin and lead as well as iron, remained important in the sixteenth century and after.[140] By this time, many of the settlements in the area grew to be relatively significant towns, and their inhabitants enjoyed a wide variety of freedoms.[141] They not only

137 Pavol Maliniak, "Mlynárstvo na strednom Pohroní v stredoveku a na začiatku novoveku" [Milling in the Central Hron Regon in the Middle Ages and at the beginning of the modern age], in *Vydavateľstvo: Z histórie technicko-hospodárskeho využitia vodných tokov na strednom Pohroní* [From the history of technological and economic use of water courses in Central Hron Region], eds. J. Žilák and P. Hrонček (Banská Bystrica: Centrum vedy a výskumu FHV UMB, 2011), 35.
138 Erik Fügedi, "Das mittelalterliche Königreich Ungarn als Gastland," in *Die deutsche Ostsiedelung des Mittelalters als Problem der europäischen Geschichte* (Vorträge und Forschungen, 18), ed. Walter Schlesinger (Sigmaringen: Thorbecke, 1974), 471–507, and more recently: Katalin Szende, "Iure Theutonico? German Settlers and Legal Frameworks for Immigration to Hungary in an East-Central European Perspective," *Journal of Medieval History* 45 (2019): 360–379.
139 In general, see Zoltán Batizi, "Mining in Medieval Hungary," in *The Economy of Medieval Hungary*, 166–181 (with further literature). For mining towns in Slovakia with special regard to Banská Bystrica in the center of the Hron valley: Martin Stefánik, "Die Anfänge der slowakischen Bergstädte. Das Beispiel Neusohl," in *Stadt und Bergbau* (Städteforschung A/64), eds. Karl Heinrich Kaufhold and Wilfried Reininghaus (Cologne: Böhlau, 2004), 295–312.
140 Gusztáv Heckenast, *A magyarországi vaskohászat története a feudalizmus korában* [History of iron smelting in Hungary in the age of feudalism] (Budapest: Akadémiai, 1991). See for the new mining activities in the sixteenth century: Petra Mátyás-Rausch, "A liptói nemesércbányászat a 16. század második felében" [Precious metal mining in Liptó County in the second half of the 16th century], *Urbs. Magyar Várostörténeti Közlemények* 12 (2018): 107–123 and eadem, "Az ismeretlen földjén – Thorda Zsigmond jelentése a liptói bányászatról (1560)" [On the land of the unknown – The report of Zsigmond Thorda on mining in Liptó County], *Lymbus – Magyarságtudományi Forrásközlemények* 17 (2019): 29–84.
141 Szende, "Iure Theutonico?."

became important centers of mining but had significant crafts activities as well,[142] and some of them became centers of the mining administration in Hungary, attracting more administrative staff (and thus people) to the settlements in the Late Middle Ages and the early modern period.[143]

The increasingly deep and long mine shafts that were built to bring ores to the ground required large amounts of timber, which was a crucial problem going back to the Middle Ages. This problem did not, of course, go away when the Habsburgs took over power in Hungary.[144] In response to the potential scarcity, or rather to gain full control of timber, the Lower Austrian Chamber had the resource areas surveyed in the second half of the 1540s in much the same way as the Rába River was surveyed a few years before. The survey was led by a trusted administrator of King Ferdinand I, Wolfgang Hohenwarter.[145] He suggested that the Hron and its tributaries could be used not only for milling but for drifting wood to the mining centers more systematically than the way in which it had been done before.[146] He proposed that Banská Bystrica, which became the most important copper mining area by the mid-sixteenth century, could be supplied by introducing dams, sluices, and shoots to the Hron and one of its sources branches, the Čierny Hron. Hohenwarter's plans seem to have been very much welcomed by the Lower Austrian Chamber and King Ferdinand I, as construction of the proposed infrastructural elements began already a

142 For the town types in a comparative perspective, see Katalin Szende and Magdolna Szilágyi, "Town Typology in the Context of Historic Towns Atlases: A Target or a Tool?," in *Political Functions of Urban Spaces and Town Types through the Ages. Making Use of the Historic Towns Atlases in Europe*, eds. Roman Czaja et al. (Cracow, Toruń, and Vienna: Böhlau, 2019), 267–302, esp. 273–276.

143 Oszkár Paulinyi, "A Garam-vidéki bányavárosok lakosságának lélekszáma a XVI. század derekán" [Population of the mining towns along the River Hron in the mid-16th century], *Történelmi Szemle* 1 (1958): 351–378.

144 Boglárka Weisz, "Az alsó-magyarországi bányavárosok kiváltságai a Zsigmond-korban," [The privileges of the Lower Hungarian mining towns in the age of Sigismund] *Urbs. Magyar Várostörténeti Közlemények* 12 (2018): 21–48, and Magyar, *A feudalizmus kori erdőgazdálkodás*. See also: Miroslav Lacko, "Frühneuzeitlicher Bergbau und Umwelt in Mitteleuropa. Probleme und Perspektiven der Forschung," in *Bergbau und Umwelt. 15. Internationaler Montanhistorischer Kongress: Sterzing/Hall in Tirol/Schwaz*, eds. Wolfgang Ingenhaeff and Johann Bair (Wattens: Berenkamp Buch- und Kunstverlag, 2017), 191–211.

145 For his titles, see e.g., MNL OL E 21 (Benignae resolutiones) 1564. no.3 (after 10 June 1564); MNL OL P 632 (Szentiványi család levéltára. A család által rendezett iratok [Archive of the Szentiványi Family. Documents arranged by the family]) IV. nos. 143, 146 and 147 (17 April 1560, 7 May 1560 and 10 May 1560), etc.

146 Drifting wood was widespread going back to the Middle Ages, but usually not from major distances. F. Romhányi, Pinke, and Laszlovszky, "Environmental Impacts of the Medieval Exploitation."

year after Hohenwarter submitted his proposal. It seems that in the second half of the 1540s, access to woodlands for burning charcoal was crucial for the Habsburg administration. This was inseparable from the ongoing surveying of the metal deposits, the processing of which required large quantities of wood. Probably the most important outcome of this pursuit of conscious management of this important income[147] is the *Constitutio Maximiliana seu norma silvas camerales propagandi et colendi* of Emperor Maximilian II issued in 1565.[148] This rather long patent not only provides detailed instructions to local land stewards about handling forests at royal domains but also gives a summary of the detailed surveys of the forests in Upper Hungary. It may not be an overinterpretation of the situation to say that this intense care for these strategic resources was a direct consequence of the constant military threat in the central basin areas.[149]

Earlier it was suggested that due to the loss of parts of the lowland areas to the Ottomans and the internal struggles, the output of the mines decreased in the 1540s and the 1550s in Upper Hungary.[150] This view has relatively recently been challenged by the discovery of new account books on iron production at the manor of the Muránsky Hrad already mentioned above.[151] It seems that not even the double taxation and the rather odd political situation at Muránsky Hrad affected the iron output of the mines and the iron-smelting complex. This can be explained by the fact that there was a need for metal not only for the usual purposes but also for the military at that time. Because of this growing interest in the exploitation of metal deposits in the sixteenth century, the kings and local stewards are likely to have increasingly focused on avoiding the overexploitation of woodlands not only in the lowland areas, as discussed in Chapters 2 and 3, but also in

147 For the importance of the mining incomes in the budget of Hungary at the time, see István Kenyeres, "A királyi Magyarország bevételei és kiadásai a 16. században," [The incomes and expenses of Hungary in the 16th century] *Levéltári Közlemények* 74, no. 1–2 (2003): 59–103, esp. 93–94, and Pálffy, *The Kingdom of Hungary and the Habsburg Monarchy*, 129–134.

148 ÖSta FHKA Sonderbestände, Sammlungen und Selekte Patente 3.49 (Waldordnung für Ungarn) (15 May 1565). For the edition: Tagányi, *Magyar erdészeti*, vol. 1. 96–167 (no. 143). See also Jenő Zivuska, *A besztercebányai m. kir. erdőigazgatóság okiratainak tartalomjegyzéke* [Summary of the documents of the Hungarian Royal Forestry of Banská Bystrica] (Banská Bystrica: Magyar Királyi Erdőigazgatóság 1906). The document was produced in German and was then translated into Latin. For the Latin translation: Gábor Almási, *The Uses of Humanism: Johannes Sambucus (1531–1584), Andreas Dudith (1533–1589), and the Republic of Letters in East Central Europe* (Brill's Studies in Intellectual History, 185) (Leiden and Boston: Brill, 2010), 159.

149 Vadas, "For the Benefit of Generations."

150 Heckenast, *A magyarországi vaskohászat története*, 89–184.

151 Sarusi Kiss, "Vasgyártás és vasgazdálkodás."

the mountainous areas. The already existing mining activities as well as the mines that the Habsburg administration were thinking of opening in the future made it vital that the remaining woodlands be managed carefully and consciously. It is safe to assume, even based on the few sources referred to here, that woodland in the mining area of the Hron valley was not scarce even in the vicinity of the mining towns in the sixteenth century.[152] In the context of the continuous political clashes and recurrent military campaigns in the 1540s to the mid-1560s initiated by the Ottomans against Hungary, it may have been more relevant than ever to have a solid and sustainable basis for gold, silver, iron, copper or other metal production in the decades to come, as metal was one of the largest assets of the Habsburgs.

For the Ottomans, the most important factor in managing and consuming forests was not mining, as the lowlands on this side of the frontier lacked such resources, but rather agriculture, which was largely based on cattle herding. Maintaining the balance between herding animals in wooded areas and allowing forests and wood pastures (which may have been important elements of the landscape in the past millennia in the Great Hungarian Plain[153]) to regenerate was a difficult endeavor. Although cattle is less damaging to pastures than goats, sheep, or even horses, the number of herded animals was probably a factor in the loss of the otherwise also low proportion of woodlands (discussed above in Chapter 4.1).[154] Even if it may not seem so at first sight, the area of the Great Hungarian Plain is also a mosaic landscape made up of different soils originating from the different hydrographies and different vegetations that existed before human activities started to heavily influence the landscape. The areas that are covered with loess are amongst the richest plowlands in the region and were therefore never abandoned or allowed to become a wasteland (*puszta*), not even during the Ottoman occupation. Although the cultivation of the land in the early Árpádian period was probably also significant, the less fertile clay and sand soil areas likely became exhausted from the later stages of the Middle Ages on and prone to salinization not suited to plowing, or at least it proved to be much less profitable than other activities in the same areas.[155] While plowing on clay and sand soil was surely an unsuccessful endeavor in the long run,

152 Tagányi, *Magyar erdészeti oklevéltár*, vol. 1.

153 Bíró and Molnár, "Az Alföld erdei."

154 Bartha, "Történeti erdőhasználatok."

155 Marianna Bálint, "Az Árpád-kori településhálózat rekonstrukciója a Dorozsma-Majsai Homokhát területén" [Reconstruction of the Arpadian Age Settlement Network in the Dorozsma-Majsa Sandy Ridge] (PhD diss., Eötvös Loránd University, 2006), 1. See also Romhányi, Pinke, and Laszlovszky, "Environmental Impacts of the Medieval Exploitation."

these wastelands did provide significant incomes in other forms. From the second half of the Middle Ages, most of these areas became *puszta*. The dating of this process is rather unclear, but many of the settlements were either abandoned due to the Mongol invasion or as part of the economic reorganization that, as recent research has pointed out, likely started before the cataclysmic Mongol devastation.[156]

The reorganization of the local economies along with the transformation of the settlement pattern have been shown using different case studies, mostly building on the examples of the largest market towns that came into existence in the second half of the Middle Ages, such as Debrecen, Kecskemét, Kőrös (today's Nagykőrös), or Cegléd.[157] The most complex approach so far has been to study the economic and environmental transformation of the borders of Kecskemét in the late medieval and the early modern periods. Edit Sárosi, an archaeologist, dedicated a monograph to the changes of the settlement network in the surroundings of the market town. According to Sárosi, from the foundation of the Hungarian state and the intensification of arable farming, the forest-steppe started to shrink and gave way to the *puszta*, a process that she mostly attributes to the forest clearance during the Ottoman period. Although there is no question that by the early eighteenth century the region lacked woodlands, as was nicely described by contemporaries such as Matthias Bel, the extent to which the woodlands were victims of the Ottoman occupation in this area is as unclear as it is in Transdanubia.[158]

156 József Laszlovszky et al., "Contextualizing the Mongol Invasion of Hungary in 1241–42: Short and Long-Term Perspectives," *Hungarian Historical Review* 7 (2018): 419–450, and Beatrix F. Romhányi, "Szempontok a Kárpát-medence térszervezésének változásaihoz (5–14. század)" [Changes in the early medieval settlement structure of the Carpathian Basin (5th–14th centuries)], in *Hatalmi központok az Avar Kaganátusban – Power Centres of the Avar Khaganate*, eds. Csilla Balogh, József Szentpéteri, and Erika Wicker (Kecskemét: Katona József Múzeum and MTA BTK MŐT, 2019), 399–420.
157 András Kubinyi, *Városfejlődés és vásárhálózat a középkori Alföldön és az Alföld szélén* [Urbanization and the Network of Markets in the Great Plain Region in the Medieval Period] (Dél-alföldi évszázadok, 14) (Szeged: Csongrád Megyei Levéltár, 2000), and Sárosi, *Deserting Villages*.
158 "It completely lacks forests; its grass and feces are used for burning. Here and there one finds insignificant brush, but these suited neither firewood nor building." Mátyás Bél, "A kunok és jászok avagy filiszteusok kerületei" [District of the Cumans, and Jazyians or Phylistines], eds. Bálint Illyés and Rudolf Szőts in *Bács-Kiskun megye múltjából, I.* [From the past of Bács-Kiskun County, vol. 1], ed. Tibor Iványosi-Szabó (Kecskemét: Bács-Kiskun Megyei Levéltár, 1975), 31. The translation is based on the following manuscript: OSZK K. Fol. Lat. 3370. Districtus Cumanorum, et Jazygum, seu Philistaeorum. The best manuscript is at Esztergomi Főszékesegyházi Könyvtár [Esztergom Archbishopric Library]. Cf. Gergely Tóth, "Bél Mátyás 'Notitia Hungariae novae...'

Two aspects that concern the forest resources of the town must be raised here briefly. First, there is limited information about the medieval forest cover in the area, but except for a relatively extensive forest used by the three largest market towns in the area – Kőrös, Kecskemét, and Cegléd – there are only a few references to forests around the town.[159] This of course could easily be attributed to the source situation; it is not necessarily proof that the area was treeless in the medieval period. However, not even paleoenvironmental (pollen) data point to the existence of extensive woodlands in the Middle Ages nor in the early modern period. However – and this is one of the two points that I hope to emphasize here – the market towns in the area, including Kecskemét, paid amongst other things firewood as tax to the Ottoman administration.[160] This was not atypical; many of the market towns throughout the plains area paid this tax to the pashas of Buda or to the heads of the local administrations who kept claiming a lack of firewood there.[161] In the case of Kecskemét, the sources show that an annual 100 cartload of firewood was sent to Buda. It is not the quantity that made the tax costly but the expenses of transporting the firewood to the center of Ottoman Hungary, which was about 100 kilometers from Kecskemét. This means that apart from the firewood for the not insignificant local population, this additional amount of firewood was also necessary. In this area, however, not only coppice forests and high forests were used for providing the population with heat but also grass, reek, and animal feces.[162] These were probably relatively

című művének keletkezéstörténete és kéziratainak ismertetése" 2 vols. [The History of the Formation of the 'Notitia Hungariae novae...' of Matthias Bel and its manuscripts], (PhD diss., Eötvös Loránd Tudományegyetem, 2007), vol. 2. 119–121.

159 László Bártfai Szabó, *Pest megye történetének okleveles emlékei, 1002–1599* [The history of Pest County in documentary sources, 1002–1599] (Budapest: [author's edition], 1938), 84–86 (no. 394). See also: Kaán, *Alföldi kérdések*, 119–120, Sárosi, *Deserting Villages*, 51–55. See also for the broader landscape: Tóber and Kiss, "Landscape History of the Medieval Sand Ridge Area."

160 János Hornyik, *Kecskemét város története, oklevéltárral*, vol. 2 [History of the town of Kecskemét, with a cartulary] (Kecskemét: [author's edition], 1864), 351–352 (no. 76), 355 (no. 85), 363 (no. 106), 367 (no. 113), 369 (no. 118 [only 60 cartloads of wood]), 369–370 (no. 119), 424 (no. 211), or 438 (no. 233). The last document tells of a complete lack in wood in the court of the vizier at Buda.

161 E.g., Gyula Káldy-Nagy, *A csanádi szandzsák 1567. és 1579. évi összeírása* [Conscription of the Sanjak of Cenad, 1567, and 1579] (Dél-alföldi évszázadok, 15), passim (Szeged: Csongrád Megyei Levéltár, 2000) and idem, *A szegedi szandzsák települései, lakosai és török birtokosai 1570-ben* [The settlements, population, and Ottoman owners of the Sanjak of Szeged in 1570] (Dél-alföldi évszázadok, 24) (Szeged: Csongrád Megyei Levéltár, 2008), passim.

162 E.g., *Le voyage d'outremer de Bertrandon de la Broquiere, premier écuyer tranchant et conseiller de Philippe le Bon, duc de Bourgogne* (Recueil de voyages et de documents pour servir à l'histoire

abundant throughout the less fertile areas of the Great Hungarian Plain. Nonetheless, the above note indicating that the pashas lacked firewood at Buda and had to be partly supplied from Kecskemét is of course more a ruse used by the scribe who sent warrants for the firewood than an actual problem. The nearby forests on the hills around Buda have never stopped providing firewood for the population of Buda and Pest up to modern times. Still, it seems that it was not such a significant burden for Kecskemét to provide some amount of firewood.

The second point to be made here concerns the regeneration of forests in these areas, a point that has only been touched upon but may be of great importance. As was noted in section 4.1 and as will be revisited in the concluding section of this chapter, in the context of Tolna County, in areas where significant population loss can be presumed, or at least in areas that became uninhabited and where previous plowlands were not in use anymore, which probably happened in many areas in the low hilly areas throughout the Basin, forests could regenerate within a few decades if the environmental conditions were favorable. The *puszta* areas, many of which had once been forested or had at least small patches of forested areas previously, were potentially capable of regenerating. However, these areas were not abandoned, as testified by hundreds of lease contracts, but were used extensively for animals to fodder. Under these circumstances, these areas where forests had probably long disappeared due to human activities had virtually no chance to regenerate.[163] It is highly unlikely that in the mid-sixteenth century one would have found extensive woodlands in these areas that then fell victim to timber mining by the Ottomans. It is more likely that by then, there was not much left to mine.

These two points aimed to provide at least some indication as to why determining the impact of war on forest cover in different parts of the Carpathian Basin is a complex task. One has to consider not only factors that can be directly associated with the war and the military activities but also ones that are only "side products" such as the more systematic exploitation of natural resources – partly for military consumption perhaps, partly

de la géographie depuis le XIIIe jusqu'à la fin du XVIe siècle, 12), ed. Charles Schefer (Paris: Ernest Leroux, 1892), 232, or Bél, "A kunok és jászok," 13, 31, and 37.

163 For the leasing of puszta, see Szakály, *Magyar adóztatás*, 406–448. For the state of forest cover, see Bíró and Molnár. "Az Alföld erdei," as well as the maps of the region from the period of the aftermath of Ottoman presence. E.g., Samuel Mikoviny, *Mappa Ichnographica Comitatuum Pest Pilis et Solth* (1731) HIM B IX a 649, or *Mappa unitorum comitatum Pesthiensis Pilisiensis et Solthens* ([Posonii], 1740). OSZK T 1086.

to provide more finances for wars – or an acceleration of the economic transformation.[164] All such factors affected the availability of forests.

4.6 Conclusions

This chapter examined how the early modern period, more specifically the period of the wars (c. 1540–1690) between the Ottomans and the Hungarians, brought changes in the use of forest resources in the Transdanubia. With few but notable exceptions, scholars attribute a wave of deforestation to resource overuse in the Carpathian Basin during the presence of the Ottomans and the long period of war. My goal has been to reconsider some of the forms of wood and timber consumption to present a more nuanced view than what currently exists in the literature. To offer a relatively detailed survey, I chose not to discuss the whole of the Carpathian Basin but focused only on Transdanubia, which, as I have argued, may be considered indicative of many regions in the Carpathian Basin. The different sections in this chapter considered the firewood needs of the population in the early modern period and the demand for firewood from some industries. The sources allow a relatively accurate population estimate, which was instrumental in identifying the firewood demand in the area, enabling me to conclude that by the late eighteenth century most of the available woodlands were probably managed as coppices with a relatively short cut cycle to supply the need for firewood in Transdanubia, similarly to what research has revealed for the neighboring Moravia. This is certainly true not only for Transdanubia but even more so for the Great Hungarian Plains, which in the past millennium was certainly more treeless. Except for regions with very low population densities, it is unlikely that the population's firewood needs could have been satisfied from local coppice forests there. There is, of course, a significant local variance in the relatively high pressure on the forests from the late eighteenth century in Transdanubia, and probably already from the Middle Ages in the Great Hungarian Plain. As I demonstrated in the previous section, despite the low overall proportion of forests in the surroundings of Kecskemét, the town was very likely to have been self-sustaining, but in other settlements in the central part of the Great Hungarian Plain, such as Debrecen, this may have been slightly more critical.

This chapter has considered not only the firewood needs of the populations but also a significant area of woodland that was used for fuelwood

164 See the special issue of *Századok* as referred to in note 51.

production to sustain the populations' other needs. There is little information on the firewood consumption of the food industry, but based on Western European parallels, particularly of London, it is likely that baking and brewing used significant amounts of firewood. There is slightly more information available on military industries, but in the absence of comprehensive research, there are limitations to reconstructing its firewood needs. The principal military industries were gun, cannon, and gunpowder production. Based on limited references, there is no ground for suggesting that either of them consumed major quantities of firewood. As research has identified, however, the firewood demand of some production phases of military materiel required particular species of wood, which may have affected their local availability.

Besides food and military industries, this chapter's main goal was to analyze the wood consumption of the building industries. Three areas were considered: water mills, bridges, and most importantly, earth and wood fortifications. Despite the numerous local histories written on settlements in Transdanubia, there is no comprehensive survey of the bridges in this region in pre-modern times. Without that, it would be rather speculative to quantify the amount of wood used for bridges. Instead, I tried to estimate the timber demand of the largest wooden bridge in Transdanubia at the time, a structure that bridged the marshes along the Dráva River at Osijek. Based on that bridge, it is safe to assume that the stakes built into the bridges in Transdanubia did not require more than a few dozen hectares of forests to be clear-cut annually.

The timber demand of water mills may have been more significant than that of bridges and may even have been comparable to the consumption of wooden fortifications. This is rather surprising in the light of existing scholarly opinions. This does not mean that water mills, and most importantly their dams, consumed hundreds of square kilometers of forests in Transdanubia or elsewhere in the Carpathian Basin. However, because of the large number of these buildings and the frequent rebuilding of their structures, usually built almost exclusively of wood, they simply required more woodlands than one would expect. Conversely, the timber consumption of fortifications, despite their large number in Transdanubia, may have been more limited than previously thought. The most important lesson learned here is that none of the building industries that used stakes of full-grown (c. 40- to 50-year-old) oak, beech, pine, and other species of trees were major factors in the deforestation of Transdanubia. Based on the amount of wood consumed by these industries, the amount of woodland in Transdanubia could very well have increased in the Ottoman period – and probably did indeed do so.

The impact of the Ottoman wars on forest resources may also be assessed by looking at the significant numbers of troops stationed at garrisons on both sides of the frontier. The impact of the production of the meat and bread they ate and the beer they drank may have been more important than the impact of the fortifications themselves. Even though a mercenary's footprint in the early modern period was more significant than that of a tenant peasant,[165] one should not think that hundred-year-old oaks were ever felled along the frontiers to provide heat for the garrisons. Nonetheless, the growing importance of animal herding and growing food locally changed demand, which may have made the regeneration of forests after felling timber a less successful strategy than before. The growing demand for food or beer may not have been specific to the military population; the demands of the nobility and urban populations also impacted forest resources.

This does not mean that the loss of forests did not concern contemporary people. Numerous statutes dating from the mid-sixteenth century onwards mention the importance of protecting the forests. Hungarian historians have frequently quoted these statutes as proof of the lack of forest resources, but I believe that these statutes and mandates are to be interpreted quite differently. Considering the discussions in Chapters 2 to 4, I argue that these sources simply reveal more conscious woodland management than what historians of East Central European woodlands have until recently attributed to the pre-modern period. The mandates issued by the Habsburg emperors point to an increasingly sophisticated forest management system not only in the Kingdom of Hungary but also in other parts of the Habsburg Empire. Furthermore, this is not specific to Central Europe: research recently has highlighted similar developments in several early modern states. In the Venetian Republic, the beginnings of conscious forest management date back to the medieval period, but in the sixteenth and seventeenth centuries, the French Kingdom, the Holy Roman Empire, and the Habsburg-ruled Spanish Empire went through similar changes.[166] Even in northern Finland, this was the period when conscious management was initiated despite a relative

165 For a comparison with a non-producing population, see the footprint of Benedictine monks in the Middle Ages: Richard C. Hoffmann, "Footprint Metaphor and Metabolic Realities Environmental Impacts of Medieval European Cities," in *Natures Past. The Environment and Human History*, ed. Paolo Squatriti (Ann Arbor, MI: The University of Michigan Press, 2007), 295.
166 Paul Warde, *The Invention of Sustainability: Nature and Destiny, c. 1500–1870* (Cambridge: Cambridge University Press, 2018), *Conservation's Roots: Managing for Sustainability in Preindustrial Europe, 1100–1800.* (Environment in History: International Perspectives, 19), eds. Abigail P. Dowling and Richard Keyser (New York: Berghahn Books, 2020).

abundance of forest resources.[167] Finally, and it is crucial for this book, recent studies have suggested the introduction of conscious forest management in the Ottoman Empire as well. It is therefore highly unlikely that any of the actors wanted to consume one of their most valuable resources. Moreover, because of the wars, it was probably ever more important to preserve them.

167 Jakob Starlander, "Conflict and Negotiation: Management of Forest Commons in Seventeenth-Century Northern Finland," *Scandinavian Economic History Review* 69 (2021): 177–194.

5 Conclusions

Abstract
In this concluding chapter, the problem of war-associated environmental change in pre-modern Hungary is revisited. It is argued that the Ottoman-Hungarian wars had important impacts on the landscapes of Transdanubia and elsewhere in the Carpathian Basin, but the most important landscape transformations were hydrological changes and not the loss of forests, as has been argued.

Keywords: Environmental history, Kingdom of Hungary, Ottoman Empire, landscape change, forest history, water history

The present book has aimed to deepen our understanding of what kind of environmental legacies a pre-modern war left and how one can demonstrate the environmental transformations that occurred in areas that were probably more fundamentally changed in the centuries to come after. The case studies tried to demonstrate how the changing political situation and the lasting military conflict in the Carpathian Basin in the sixteenth and seventeenth centuries affected environmental conditions. Two factors were considered in detail: first, the war's impact on water management and its resultant fluvial landscapes and local farming possibilities, and second, how forest cover and the exploitation of this resource were affected by the war. This chapter has provided some general remarks on the specific context and on the possibilities for studying the environmental legacy of pre-modern warfare in general.

From the very beginning of the military activities in the Carpathian Basin in the early modern period, the Habsburg-Hungarian armies systematically utilized the features of the landscape in their military strategy. Although the documentary sources for the Ottomans are less abundant or rather less exploited, for as far as we know, their military elite seems to have made major efforts to understand the local environmental conditions. Hence, by the time they took possession of the central part of the Carpathian Basin,

Vadas, A., *The Environmental Legacy of War on the Hungarian-Ottoman Frontier, c. 1540–1690.* Taylor & Francis Group, 2023
DOI: 10.5117/9789463727938_CH05

they already had considerable knowledge of the geographies of the area. However, at this point, there is no data available from the Ottoman Empire suggesting the existence on the Ottoman side of a similarly complex defense system based on the landscape, or at least not in this part of the Empire.[1]

The defense of the Kingdom of Hungary, as the previous chapters have demonstrated, was designed with the systematic use of landscape features, most significantly the marshes and rivers in the lowlands. The area that I chose for a more in-depth study was the valley of the Rába River, which was regarded as one of the most important elements in the defense system from the mid-sixteenth century onwards. In that period, the river was less like a border than an element in the buffer zone of the Ottoman-Hungarian frontier zone. From the turn of the sixteenth century, however, from the time marked most importantly by the fall of Kanizsa in 1600, the river became the most important landscape element of the defense in the central part of Transdanubia. This new function had a major impact on the river and its environment: it transformed from a river used by local economies to become part of a semi-militarized landscape in which the economic benefits seem to have been secondary to the military aspects. The river surveys (discussed in Chapter 2) demonstrate the different ways the river was meant to be transformed for the sake of border protection. The gallery forests – which were important resources for the local economies for feeding pigs, providing firewood, and occasionally supplying the settlements with timber for structures – were now used for frontier protection. These forests became reserves of trees to be felled into the river to flood areas where the crossing would otherwise have been easy; they were also transformed into places for watchtowers (*górés*).

The militarization of the river led to environmental problems: the numerous backwaters along the river significantly impacted the flood regime and resulted in siltation. Private letters were written by the local land stewards and military personnel from the settlements of Körmend and Csákány (see Chapter 3) – important sources previously unused in environmental history research – reveal that major floods became very frequent. High or elevated flood frequencies can be attributed to numerous factors: to name but a few, changes in the vegetation of the catchment area, climatic change, land-use transformation, and geomorphological changes. None of these, however, fully explains what the written sources reported, that is, that from the first third of the seventeenth century the lowlands along the

1 This, however, is not true for the whole of the Ottoman Empire. Cf. Husain, "Changes in the Euphrates River."

Rába River experienced disastrous flood events on an annual basis. The detailed accounts of the floods allow us to identify not only the frequency of the flooding but also the floods' impacts. In the 1640s and 1650s, they not only affected the meadows and the gallery forests but also damaged the houses in parts of the town of Körmend that had previously been protected from floods. So far, the only explanation for the elevated flood frequencies is the influence of the changing exploitation of the river.

The problem of the changing local environmental conditions is not merely a modern scholarly construction; some of its elements were identified by contemporaries. The acts passed at the diets of Hunary from the mid-seventeenth century onwards kept referring to the problem of the siltation of the Rába and the need to remove the accumulated sand from the bed of the river to foster the downstream flow of the river. One of the first Habsburg decrees on this matter refers to the river's obstructed flow (*in facto obstructionis fluvii Rábae*).[2] Royal decrees on several occasions in the seventeenth century repeated the need to clean the riverbed of the Rába to protect the settlements both from the Ottomans and the floods by the river. However, the decrees kept revisiting the problem, which testifies to the fact that although the riverbed was cleaned in some sections, this work was never completed in the seventeenth century. The inhabitants of the villages and towns by the Rába appeared to have faced recurring economic losses, which most probably were involuntary sacrifices so that the hinterland of the Kingdom of Hungary and other parts of the Habsburg realm could be protected more effectively. It was not in the interest of the local landlords such as the Batthyány family to have their lands flooded, but neither was it in the interest of any of the actors on the Hungarian side to provide the Ottomans with easy access to the left bank of the Rába River.

Even though the scholarly literature seldom refers to rivers having been transformed in this way as a consequence of the wars against the Ottomans, I have argued that this aspect of the environmental legacy of pre-modern warfare is certainly worth considering. One last point that I touched upon earlier and should be emphasized again here is that the environment was transformed in this area not by the Ottomans but by the Hungarians with the support of the Habsburg military leadership. The traditional view of

2 *Magyar törvénytár. 1608–1657. évi törvényczikkek* [Corpus Juris Hungarici. Laws of Hungary], ed. Dezső Márkus (Budapest: Franklin, 1900). For the decrees on the need for the clearance of the riverbed of the Rába, see ibid., 214–215 (1622 no. 42), 246–247 (1625 no. 16), 288–291 (1630 no. 14), 338–339 (1635 no. 64), 600–603 (1655 no. 30). For decrees of county gatherings in Western Transdanubia from the same time, see Dominkovits, "Folyóvizek és a XVII. századi vármegyei."

the Ottomans devastating the country and transforming fertile land into wastelands therefore has to be reconsidered in this respect.

The other main environmental issue that was discussed in the book is the change in the extent of forests in the Transdanubian area during the Ottoman period. The loss of woodlands is one of the most widespread allegations made against the Ottoman presence in the Carpathian Basin. This assumption is heavily based on the same traditional anti-Ottoman narrative mentioned in the last paragraph. Recent literature has shed light on the question and showed that practices on the two sides of the frontier did not differ significantly. The Ottomans neither used scorched-earth tactics systematically nor exploited forest resources without considering the long-term impact of their actions. Scholarly literature, however, still has very different views of changes in the extent of woodlands during the war period in general. Until recently, the major deforestation in the central part of the Carpathian Basin was attributed to the c. 150 years of Ottoman presence there. Most scholars have listed the construction of earth and wood fortifications and the gunpowder industry as the most important factors in deforestation.

By providing a cautious estimate, I have argued that these factors may not have contributed significantly to the change in the extent of woodlands in the Transdanubia. Many factors in this context, however, deserve further study – including, most importantly, the firewood needs of the population of the Basin as well as the firewood consumption of the military garrisons in particular. The other main lesson from this part of the book is that even if the extent of woodland did not decrease as a result of the direct demand for the war, forest resources may have become more and more scarce due to the increasing need for firewood because, despite the decrease in population in some counties in the sixteenth century, the 150 years of Ottoman presence brought an increase in the population of the lowlands. The firewood needs of these people may have led to the management of most of the woodlands in Transdanubia and the Great Hungarian Plains as coppice forests.

While the extent of woodlands may not have decreased, their functions certainly did. In areas where the population loss was rather significant, such as Somogy and Tolna Counties in southern Transdanubia, there is little doubt that the extent of wooded areas grew. In areas where the population became more concentrated, however, forests probably became carefully protected areas to provide the locals with firewood. The conclusions I came to by no means indicate that the forest cover did not decrease in some micro-regions or even in areas on the scale of a county, but this may not have been connected directly to military activity, but rather to changes in farming practices – most importantly, the cattle-herding briefly discussed above in this context. The spread of cattle herding in the Great Hungarian Plain certainly contributed to the loss of woodlands, as the stocks kept in the parklands did not allow forests to regenerate. Although this process may have impacted much of the areas in Transdanubia, one other factor also needs to be mentioned in this context. Some sources suggest that forests that were extensively used in animal herding in the late medieval period lost their value because of the threat of raiding Ottomans. It is no surprise that in such raids, the easiest plunders were the animals that could be easily driven away. This occurred in many settlements by the frontier, and it significantly decreased the value of forests and probably indirectly contributed to their better conservation in some areas.[3]

The case study provided in this book has revealed that the wars had important impacts on the landscapes of Transdanubia and elsewhere in the Carpathian Basin. Although this has been demonstrated to some extent in previous studies, the ways in which the environment was affected may have been considerably different than what was previously suggested. Nonetheless, it is clear that the period of war between the Ottomans and the Hungarians changed the environments and the ecosystems in many parts of the Carpathian Basin.

5.1 Epilogue – What Came After?

Perhaps one of the most important aspects that has been touched upon but not examined here in detail is what came after the Ottomans had gradually been expelled from Transdanubia and eventually from the Carpathian Basin. What were the environmental problems the local societies had to face, and what actions did different actors take?

The Rába River lost much of its importance as a frontier after the treaty of Vasvár, which in 1664 put an end to a short but major series of military campaigns between the Habsburgs and the Ottomans.[4] The signing of the

3 On the driving away of animals, see: Péter Illik, "Török kártételek a nyugat-dunántúli hódoltsági peremvidéken a 17. század első felében" [Ottoman raids in Transdanubia. Plundering on the fringes of the Ottoman Empire's Transdanubian parts in the first half of the 17th century] (PhD diss., PPKE BTK, 2009), 36 table 5, 55 table 14, 45 table 15, 63, etc. On the loss of the value of these forests, see Lajos Juhász, "A Vas megyei Farkaserdő a XVII–XVIII. században" [Farkaserdő is Vas County in the 17th–18th century], *Századok* 71 [Supplement] (1937): 553–575 [37–69].

4 On the treaty and the Ottoman military campaign that preceded it, see: *Die Schlacht von Mogersdorf/St. Gotthard und der Friede von Eisenburg/Vasvár*, eds. Karin Sperl, Martin Scheutz, and Arno Strohmeyer (Eisenstadt: Amt der Burgenländischen Landesregierung, 2016).

peace treaty created stability for 20 years. During this period, the intensity of border protection decreased significantly in parallel with the number of raids by the Ottomans, although this did not mean that there was no plundering of villages on the left bank of the Rába after the 1660s. The Rába River remained key in the Habsburg defense against the Ottomans until the late 1680s. This is well illustrated by the fact that when the Habsburgs confiscated the estates of the Nádasdy family in the 1670s on account of treason, the provost of Csorna (a settlement close to the left bank of the Rába in Sopron County) sent an official complaint to the Hungarian Chamber drawing attention to the insufficient protection of the Rábaköz region against both floods and Ottoman raids.[5] In the decade that followed, however, the Rába lost its importance in the defense of the Kingdom of Hungary and Vienna. From the mid-1680s, the Ottomans were gradually expelled from the area along the river. From that time on, the river once again had only one role: to serve the local economies. Many of the obstacles built into the river as well as almost all the earth and wood fortifications and watchtowers built in the preceding 150 years became useless. From the 1690s, plans were made to reorganize the water system of the Rába River. One of the first steps was to conduct a new survey of the mills and dams along the river (already mentioned in Chapter 2). This was probably not the only survey that was made, although the only other one discovered so far dates back to 1699 (and was published by Kálmán Thaly in the nineteenth century). This document is a detailed description of the lower section of the Rába ordered by King Leopold I and carried out under the supervision of the Nádasdy family, which by then had been partly rehabilitated.[6] The survey ordered by Ferenc Nádasdy only gives an account of the lower section of the Rába, but it is a uniquely detailed source describing the problems the region had to face after the Ottomans had been expelled. As the document explains, the numerous dams by the river were a constant source of flood danger for the

5 Letter of Mihály Czuppon provisor: MNL OL E 41 Fasc. 58. 1674 no. 122. (21 April 1674) See also: László Benczédi, "A Habsburg-abszolutizmus indítékai és megvalósulása az 1670-es évek Magyarországán" [The motivations and execution of Habsburg absolutism in Hungary in the 1670s], *Történelmi Szemle* 21 (1978): 548 note 27.

6 The survey was edited in low quality in Thaly, "Az Rába rectificatiójáról." Thaly in many cases neglected to refer to where he found the different archival materials he published. According to him, the copy he edited was kept in the Archives of the Csesznek Branch of the Esterházy Family kept in Bratislava. The document was lost for a while but has recently been rediscovered. See: SNA Esterházy 6 [Miscellanea]. The list of the records in the fond is available online: http://mnl.gov.hu/download/file/fid/40827 (last accessed: 11 July 2017).
For the mandate of Leopold I, 1 or 17 July 1699. MNL OL A 14 (Insinuata Consilii Bellici) 3. csom. For the micro-fiche copy of the archival unit, see MNL OL X 5136.

villages on the flood plain. In order to solve the problem of floods without harming the economic interest of the villages by the river, it was decided that an old, channelized bed of the Marcal River would be connected with the Rába. This would allow some of the water to be diverted into this channel and then into the Marcal when there were floods on the Rába.[7] The work was probably completed soon after the decision had been made, but it did not solve the problem in the long run and had immediate impacts on the local environment. The village called Malomsok, which lay closest to the diversion of the Rába into the old–new channel (*Ásvány*), gradually had to move to a less flood-prone area, as from that time onwards it lay between the two branches of the Rába, that is, its old bed and the channel leading to the Marcal. In the late eighteenth century, the location of the completely abandoned old settlement (*Ómalomsok*) is clearly visible on the sheet of the First Military Survey. Besides this immediate problem, this solution was also unsatisfactory in the long term, as the Rába riverbed continued to silt up gradually in the eighteenth century (see Fig. 5.1).

By the middle of the eighteenth century, it was clear that the situation required a reaction. Several regulatory projects were carried out both in the valley of the Rába and the Marcal rivers from the 1760s, but the flood protection works were halted until the second half of the nineteenth century.[8] This was not unique to this region, as the eighteenth century was a period of major water-control projects and the beginning of river regulation on a larger scale all over the Kingdom of Hungary.[9] However, in many sections of the rivers, these works were not completed before the end of the nineteenth century.

The changes in the exploitation of forest resources in the eighteenth century can be less clearly seen than the changes in the water management

7 *Unde deventum est ad molendinum Episcopatus Iaurinensis ad Arcem Keszö spectans, in comitatu Castriferrei habitum, cuius caput supra aquam consistentem, repertum est elevatum digitis 12 aquagium rotas vertens super caput altitudinis digitorum 25. Habetur super illud molendinum certum quoddam antiquum fossatum vulgo Asvány dictum, quo olim aquae superfluae ex dicto alveo in fluvium Marczal, illinc non procul praeterlabens se effuderat. Quod denuo aperiendum et in fluvium Marczal derivandum determinatum est* – Thaly, "Az Rába rectificatiójáról,"160.

8 Klára Dóka, "A Rába szabályozása, 1762–1895" [The regulation of the River Rába], *Technikatörténeti Szemle* 11 (1979): 85–101, 86 and Dóka, "A Rába-szabályozás." On the nineteenth-century regulations, see Márton Simonkay, "Csorbuló önrendelkezés vagy állami siker? Érdekek és vélemények a Rába-szabályozás kapcsán (1886–1893)" [Confined autonomy or state success? Interests and opinions regarding the regulation of the River Rába], in *Víz és társadalom*, 343–376.

9 Zsigmond Károlyi, *A vízhasznosítás, vízépítés és vízgazdálkodás története Magyarországon.* (Vázlat) [History of water use, water construction, and water management in Hungary (A sketch)] (Budapest: Tankönyvkiadó, 1960). See also: *A magyar vízszabályozás története* [History of water regulation in Hungary], ed. Dénes Ihrig (Budapest: Országos Vízügyi Hivatal, 1973), 82–101. (The relevant part is the work of Zoltán Károlyi.)

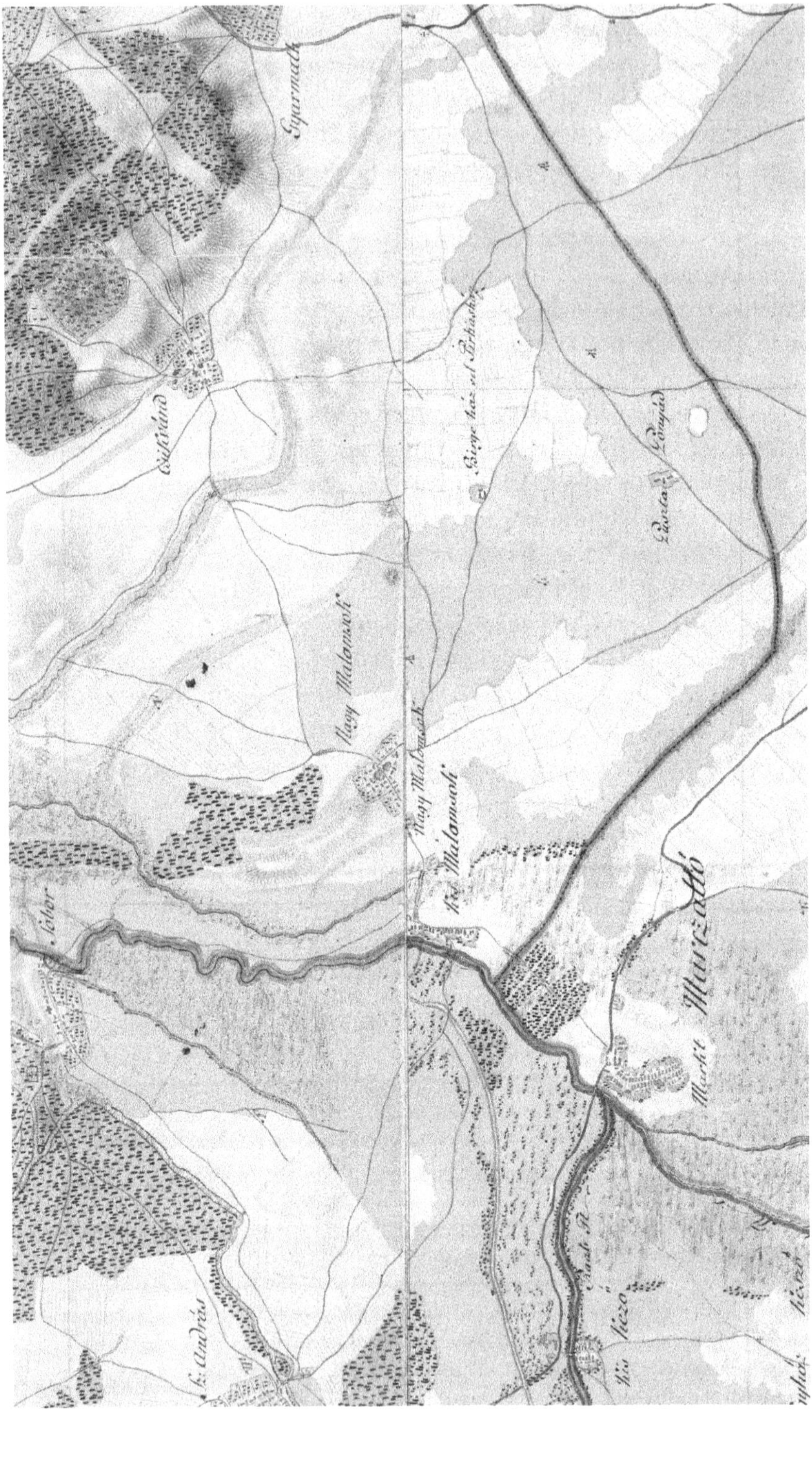

Figure 5.1 Malomsok and the branches of the Rába on the First Military Survey

system in this region. With the expulsion of the Ottomans from Transdanubia and the treaty of Karlowitz (1699), which put a formal end to the war in the region, the numerous earth and wood fortifications in the region became more and more obsolete. There was still a need for a defense system against the Ottomans, but it was built well to the south of Transdanubia. Most of the fortifications were not maintained regularly after the 1660s and were in poor shape by the time the Ottomans left the area. The fortifications' moats started to silt up, and at best they were used for fishing and other purposes, the fortification walls themselves having completely lost their protective roles. Without regular maintenance, they quickly decayed, and many of the fortifications were demolished by Habsburg forces. This demolition process took place mostly at the turn of the century. As has recently been demonstrated,[10] the intention with this process was on the one hand to provide building material for civic buildings and on the other hand to get rid of the "free troops," the mercenaries whose role was sharply restricted by that time. Many of them did not leave the fortifications after the wars had ended, which resulted in many conflicts with both local communities and the Habsburg military administration.[11]

As I argued, however, the war and the construction of the fortifications themselves were probably less important in the exploitation of forest resources than the need to supply the population with firewood and wood for baking and brewing. Consequently, it would be better to associate the changes in the use of forests with the increasing population numbers than the regaining of the territory of the former Kingdom of Hungary. The rapid population growth certainly influenced the change in the usage of forests. The woodlands were affected not only by the growing population but also by the construction of the dwellings of the peasantry moving to the areas that lost parts of their population. Nonetheless, the eighteenth-century surveys frequently refer more to the abundance of forests in the counties of Transdanubia and their untamed state rather than their shortage.[12] Péter

10 András Oross, *A Magyar Királyság törökellenes határvédelmi rendszerének felszámolása és átszervezése* [The dissolution and reorganization of the anti-Ottoman border defense system of the Kingdom of Hungary] (Fons Könyvek, 4) (Budapest: Szentpétery Imre Történettudományi Alapítvány, 2013).

11 Oross, *A Magyar Királyság törökellenes határvédelmi rendszerének felszámolása*, 113–178.

12 See most importantly the immense work of Matthias Bel, *Notitia Hungariae novae historico geographica*, 6 vols., eds. Tóth Gregorius et al. (Budapest: Magyar Országos Levéltár and Magyar Tudományos Akadémia Történettudományi Intézete, 2011–2020). See also: Klára T. Mérey, "Az erdőgazdálkodás Somogy megyében (1700–1879)" [Forest management in Somogy County (1700–1879)], *Agrártörténeti Szemle* 5 (1963): 133–152. In general, see also: Imre Wellmann,

Szabó's claim that while the forest coverage during the Ottoman period may have grown in percentage, the eighteenth century probably witnessed a decrease in the extent of forests is very likely to be valid.

* * *

Let me recapitulate the questions addressed in the introduction to this book. I proposed to discuss whether it is valid to speak of the environmental legacy of pre-modern warfare. I have tried to show that the answer is a definite yes. How the environment was affected by this particular war was not as the existing literature has portrayed it. I was unable to demonstrate a clear impact of the Little Ice Age on the environmental conditions nor that of the cessation of the maintenance of the flood protection systems in the Great Hungarian Plain or the deforestation of the Carpathian Basin. It is clear, however, that other kinds of environmental pressures can be directly associated with the war period and the frontier zone that lay between the Ottoman Empire and the Kingdom of Hungary. The second related question was whether this war left an environmental legacy and if so, what that legacy was. I have argued that the ways in which water was manipulated for the sake of frontier defense probably changed local landscapes and significantly affected agricultural practices. Whether or not there was a change in the vegetation or changes in the forest cover in the frontier zone, however, is much more difficult to determine. Perhaps it is best to say that the war changed the ways in which the forests were used and that the areas covered in wood probably expanded in the Basin area.

The case study here was intended to show that there are sources and methods that can help us understand how the environmental conditions in this region were affected by war in the pre-modern period. These results are certainly not directly applicable to the environments of the Hundred Years' War or the Thirty Years' War, nor to every area of the frontiers of the Ottoman Empire and the Habsburg realm. This does not mean that there is no ground for discussing pre-modern war and militarized environments – on the contrary, many other environments have yet to be understood.

"Népesség és mezőgazdaság a XVII. és a XVIII. század fordulóján" [Population and agriculture at the turning of the 17th century], *Történelmi Szemle* 18, no. 4 (1978): 701–730, esp. 713.

Appendices

Appendix 1. The writers, function, and temporal coverage of the private letters studied in Chapter 4

Name of letter writer[1]	His or her function[2]	Archival reference (number of letters)[3]	Dates
György Babos Szabó [and Márton Tekenyős]	Innkeeper, master of the tailor guild, and foot soldier at Körmend	No. 1177–1208. (31)	1 May 1648–10 October 1660
Magdolna Batthyány	Sister of Ádám Batthyány, pledgee of the abbey of Szentgotthárd	No. 5071–5141., 8595. (71)	24 April 1635–3 January 1658
Bálint Becsföldy	Bridge toll collector at Csákány	No. 5333–5334. (2)	1617
Benedek Bér	Tipstaff	No. 5979–5981. (3)	8 May 1623–29 March 1624
István Blaskovics	Captain to Ferenc Batthyány (II)	No. 1475–1480. (6) No. 6822–6849. (28)	3 July 1608–24 May 1618
Benedek Borz	Voivode of Körmend	No. 7428–7443. (15) RHM-Ht 90.1.434 (1)	6 June 1614–10 January 1644
László Bozay	Provisor of Körmend	No. 7485–7619. (136)	15 October 1615–4 May 1638

1 The names appear in many forms in the letters. I use the standardized form of their name, the one that is used by the MNL OL database as well: http://adatbazisokonline.hu/adatbazis/batthyany-missilisek (last accessed: 24 July 2021).

2 On the function of the different letter writers, in most cases the letters themselves are the main sources. Apart from them, terriers and house conscriptions were the most important sources I used. See, e.g., MNL OL P 1322 IV. 52. fol. 10–15; IV. 68. f. 3–59. I also made some additions based on payment lists from Körmend. See MNL OL P 1322 1. nos. 46–54.

3 Unless otherwise indicated, they are all from MNL OL P 1314.

Name of letter writer[1]	His or her function[2]	Archival reference (number of letters)[3]	Dates
Balázs Czegléd	Merchant (?)	No. 9352–9357., No. 50 406. (7)	3 April 1651–23 June 1654
Miklós Czellecz(i)	Literate and accountant	No. 9362–9409. (48)	12 December 1643–31 October 1657
György Csabay	Voivode of Szengrót	No. 8139–8161. (23)	30 March 1648–21 October 1658
András Csákányi	Literate at Csákány	No. 8197–8203. (7)	7 October 1610–28 February 1624
Inhabitants of Csákány	-	No. 8208–8209. (2)	4 August 1614–2 July 1623
László Csáky (the Elder)	Chief justice of Hungary, pledgee of the abbey of Szentgotthárd	No. 8317–8517. (201)	28 November 1631–24 June 1657
Zsigmond Csáky	Son of László Csáky	No. 8628–8643., 8657–8658. (18)	15 January 1651–7 June 1658
János Csanádi	Lieutenant at Körmend	No. 8667–8724. (57) No. 108 285–108 297. (12)	18 August 1649–2 November 1658
Bernát Csány	Captain of Csákány, brother of György Csány	No. 8732–8920. (189)	9 July 1650–14 October 1658
György Csány	Vice-captain of the cavaliers and the footmen at Körmend	No. 8950–8998. (48)	27 March 1634–20 April 1658
Bálint Erdődy	?	No. 11 468–11 475. (8)	4 March 1620–25 April 1625
György Erdődy	Imperial and royal chamberlain, then master of the treasury	No. 11 505–11 684.; 108 400. (180)	12 October 1636–17 January 1660
Erzsébet Erdődy Györgyné Batthyány	Wife of György Erdődy	No. 3310–3484. (174)	14 December 1640–20 May 1663
György Falusy	Provisor of Körmend, literate	No. 13 030–13 316. (284) No. 108 447. (1)	30 April 1622–19 October 1660
György Falusy Jr.	Provisor of Körmend	No. 13 317–13 328. (12)	21 April 1640–18 June 1655
András Fejér	Soldier (?)	No. 13 580–13 589. (9)	1 August 1642 1– 7 April 1655
Gáspár Francsics	Chief-captain of Körmend	No. 14 936–15337/a. (402) No. 108 518–108 530. (13) RHM-Ht 90.1.456–467 (11)	1622–25 April 1653

Name of letter writer[1]	His or her function[2]	Archival reference (number of letters)[3]	Dates
András Fülöp	Lieutenant at Csákány	No. 15 530–15 557. (28)	27 March 1649–6 June 1658
János Geilberger Keresztelő (Johan Baptist Geilberger)	German captain of Körmend	No. 15 735–15 775. (40)	20 September 1653–13 May 1658
Ferenc Gencsy	Provisor of Csákány	No. 15 813–15 944. (132)	2 October 1636–20 June 1658
Mátyás Gerdákovics	Provisor of Körmend	No. 15 971–16 291. (284) No. 54 750–54 787. (38) RHM-Ht 90.1.823–90.1.841., 90.1.986. (20)	7 June 1638–26 June 1654
István Gosztonyi	Accountant of Gasztony	No. 16 556–16 586. (31)	25 July 1613–31 March 1622
János Gosztonyi	Accountant of Gasztony	No. 16 589–16 597. (9)	26 October 1655–13 December 1658
Bálint Györkény	Fisherman at Csákány	No. 108 610–108 612. (3)	21 February 1648–28 February 1648
István Hagymássy	Captain of Szentgrót	No. 17 971–18 082., 54 498., 54 796. (114)	28 April 1635–24 August 1651
János Hagymássy	?	No. 18 090–18 147.,108 643–108 655. (71)	2 September 1637–23 February 1646
Kristóf Hagymássy	?	No. 18 149–18 150., 108 656–108 674. (21)	18 August 1605–31 October 1624
Miklós Hagymássy	?	No. 18 151–18 157. (7)	13 November 1620–14 October 1621
Gáspár Hall (Gaspar Hall)	German captain of Körmend	No. 18 194–18 236. (37)	22 February 1653–30 August 1656
Ádám Hertelendi	Accountant of Gasztony	No. 18 687–18 764. (78)	10 December 1635–23 July 1656
András Hidasy	Vice-captain of Körmend	No. 19 069–19 314. (246) No. 54 800–54 801. (2) RHM-Ht 90.1.166–90.1.1.167. (2)	19 January 1620–17 July 1643.
György Hidasy	Lieutenant at Körmend	No. 19 319–19 338. (19).	24 March 1644–30 June 1645
István Hidasy	Flag-bearer soldier	No. 19 339–19 343. (5)	14 November 1656–15 July 1658

Name of letter writer[1]	His or her function[2]	Archival reference (number of letters)[3]	Dates
János Hidasy	Voivode of Körmend	No. 19 344–19 353. (16)	18 February 1612–20 (?) May 1656
Benedek Hollósy	?	No. 19 539–19 585. (47)	4 April 1633–1 July 1657
Sándor Hollósy	?	No. 19 595., 108 720–108 730., RHM-Ht 90.1.468–90.1.470. (14)	24 February 1604–16 September 1616
Bálint Horváth	Chief-captain of Körmend	No. 19 727–19 909. (230) No. 108 753–108 754. (2) RHM-Ht 90.1.481–90.1.484., 90.1.492.–90.1.494. (7)	16 July 1605–3 December 1633
Dániel Jobbágy	Judge at the court of Rechnitz	No. 21 934–21 935., 21 968. (3)	2 August 1649–6 June 1650
László Joó	Brother of the landlord of Körmend [János Joó]	No. 22 811–22 823. (12)	22 October 1603–27 July 1615
Balázs Kaczor	Voivode of Egervár	No. 22 930–22 965. (36)	20 April 1636–1 January 1647
Ferenc Káldy	Vice-count of Vas County and chief captain of Körmend	No. 23 009–23 569. (560)	23 April 1615–13 October 1660
Péter Káldy	?	No. 23 588–23 703. (116)	21 January 1643–24 April 1660
Imre Kállóczfalvay	Voivode of Egervár	No. 23 710–23 720. (11)	23 June 1637–10 May 1656
Pál Kardos	Lieutenant of Körmend	No. 24 041–24 048. (8)	1 January 1656–18 July 1683 (?)
Benedek Károl	Provisor of Csákány	No. 24 018–24 167. (149)	1618–1625
Ferenc Károly	Innkeeper at Csákány	No. 24 181–24 190. (10)	1648–1654
János Keczer	Captain of Csákány, vice-commandant of the Transdanubian forces	No. 24 352–24 688. (338)	1641–1654
Erzsébet Keczer Jánosné Csányi	Widow of János Keczer	No. 24 689–24 691. (3)	1654
György Kelemen	Voivode of Körmend	No. 25 386–25 462. (76)	7 March 1646–26 November 1658
István Keserű	Captain of Csákány	No. 26 388–26 442., 109 064. (55)	1595–1616

Name of letter writer[1]	His or her function[2]	Archival reference (number of letters)[3]	Dates
Balázs Kisfaludy	Captain of Csákány	No. 26 712–26 783. (71)	1650–1658
György Kondoray	Accountant of the castle of Körmend and Zala County	No. 27 145–27 185. (39)	8 August 1605–7 June 1624
The citizens of the castle of Körmend	–	No. 27 458–27 471. (14)	4 November 1607–15 February 1638
The judges and jurors of the town of Körmend	–	No. 47 472–47 476. (7)	9 June 1624–19 May 1652
The Cavaliers and the footmen of Körmend	–	No. 27 487–27 525. (38)	4 January 1609–2 January 1699
Citizens of Körmend	–	No. 27 537–27 539. (4)	Without date (before 1659)
Mátyás Krajnai	Provisor of László Csáky	No. 27 786–27 795. (10)	1647–1657
Mihály Krisanics	Tipstaff at Körmend	No. 27 818–27 825. (7)	14 January 1635–17 January 1639
Markó Kuprisics	Soldier (?)	No. 27 899–27 956. (58)	21 August 1594–4 November 1650
János Lantsák	Innkeeper at Csákány	No. 28 400–28 401. (2)	13 March 1653–21 May 1658
Ferenc Lippich	Accountant at Murska Sobota–Rakičan (?)	No. 29 408–29 480. (73)	8 June 1643–31 March 1676
Márton Lippich	?	No. 29 358–29 407. (49)	4 August 1638–18 October 1655
István Lutár	Accountant at Murska Sobota–Rakičan (?)	No. 30 100–30 117. (18)	4 July 1650–20 January 1658
Mihály Lutár	Soldier (?)	No. 39 217–39 229. (13)	13 January 1652–26 November 1658
Mihály	Literate, accountant	No. 31 193–31 207. (15[?])	2 December 1620–25 February 1643
János Nagy Magyari	Accountant of the abbey of Szentgotthárd	No. 33 345–33 385. (41)	3 April 1624–3 April 1646
Mihály Nagy	?	No. 33 392–33 403. (12)	8 January 1612–1 May 1614

Name of letter writer[1]	**His or her function**[2]	**Archival reference (number of letters)**[3]	**Dates**
Ferenc Nádasdy (III.)	*Comes* of Vas County, later chief justice of Hungary	No. 32 016–32 656. (641)	15 August 1636–20 December 1660
István Nemsem	Provisor of Körmend	No. 33 636–33 948. (328) No. 54 873–54 889. (17)	12 June 1640–10 November 1654
Boldizsár Niczky	?	No. 34 065–34 066., 109 449–109 458. (12)	9 October 1645–3 August 1658
György (Gersei) Pethő	Noble at the court of László Csáky, customs officer of Sopron	No. 36 595–36 616. (21)	5 August 1642–20 October 1658
Pál Pethő	Provisor of Szentgotthárd	No. 37 067–37 142. (76)	25 March 1605–18 January 1627
Ádám Poppel	Cavalier captain	No. 37 890–37 903. (15)	16 March 1607–11 April 1626
Éva Poppel	Wife of Ferenc Batthyány Ferenc	No. 4597–4638., 37 905–38 110., 48 726., 54 906–54 908., 109 588–109 592. (257)	19 October 1604–6 November 1640
István Potyondi	Captain of Csákány	No. 38 251–38 297. (46)	18 January 1621–23 April 1639
György Pulyai	Provisor of Szentgrót	No. 38 752–38 790., 38 792–38 793. (41)	2 March 1653–26 December 1658
János Rácz	Accountant	No. 38 977–38 994. (18)	23 April 1613–1647
Menyhért Ráttky	Vice-commandant of the Transdanubian forces, captain of Körmend	No. 39 427–34 963. (36)	4 December 1604–27 March 1625
Balázs Rettkes	Burgher of Körmend, lawyer of the Batthyány family	No. 39 757–39 772. (16)	21 October 1649–5 October 1660
István Ságodi	Captain of Körmend	No. 40 420–40 575. (156)	12 March 1652–23 December 1658
Imre Saly	Soldier (?), provisor	No. 40 802–40 890. (88)	30 April 1612–1627
Simon	Literate	No. 43 283–43 287. (5)	2 May 1605–3 June 1605

Name of letter writer[1]	His or her function[2]	Archival reference (number of letters)[3]	Dates
András Somogy	Provisor of Körmend	No. 43 693–43 773. (82)	15 June 1600–18 August 1623
István Svastics	Captain of Körmend	No. 44 987–45 175. (188)	16 September 1635–20 December (?)1656
János Szarka	Court judge of Szentgotthárd	No. 46 030–46 081., No. 110 151–110 173. (72)	23 January 162–18 April 1621
István Szecsődi	Captain of Csákány	No. 46 553–46 671. (118)	20 May 1643–25 April 1658
Gáspár Szentgyörgyi	Provisor of Turnišče	No. 47 045–47 135. (90)	26 January 1612–16 June 1625
Mihály Szokoly	Accountant of the castle of Körmend, literate	No. 48 089–48 133. (44)	14 January 1639–20 March 1655
Mihály Tarródy	?	No. 109 921–109 923. (3)	11 February 1636–23 August 1657
Balázs Temlin	Provisor of Rakičan	No. 48 027., 48 722–48 890., 54 980–54 983. 100 926. (173)	10 June 1640–17 June 1649
György Wragovics	?	No. 51 706–51 716. (11)	22 April 1641–18 February 1658

Appendix 2. Change in population of the counties of Transdanubia from the late medieval period (1494/1495) to the census of Joseph II (1787)[4]

County	Territory (km²)	Mean estimated population (1494/1495)	Mean estimated population (1598)	Mean estimated population (1787)
Baranya	5717.60	141,887.5	-	179,515
Esztergom	1166.06	15,035	-	43,027
Fejér	4132.62	31,580	-	106,521
Győr	1513.31	18,687.5	30,000	74,811
Komárom	2598.02	20,187.5	-	91,680
Moson	1611.02	8680	15,750	52,168
Pilis*	981.91	32,312.5	-	73,414
Somogy	6339.28	105,110	39,000	163,560
Sopron	2407.57	23,607.5	45,500	156,981
Tolna	4502.97	94,742.5	-	128,591
Vas	4482.99	49,675	84,000	227,174
Veszprém	4080.91	33,817.5	30,000	140,789
Zala	5881.64	87,937.5	78,000	228,415
Total	45415.90	705,300		1,666,646

* After the expelling of the Ottomans, Pilis County became part of Pest-Pilis-Solt County. The numbers in the last column refer to the former area of Pilis County.

4 Source of the data: for the late medieval period, Kubinyi, "A magyar királyság népessége," for the late medieval period, Szakály, "Tolna megye negyven esztendeje," (for 1598), and Aranka Szaszkóné Sin, ed. *Magyarország történeti helységnévtára 1773–1808 – Pest-Pilis-Solt megye és a Kiskunság* [Historical settlement names of Hungary, 1773–1808 – Pest-Pilis-Solt County and the Kiskunság] (Budapest: Központi Statisztikai Hivatal Könyvtár és Dokumentációs szolgálat, 1988).

Appendix 3. Fortifications in the Transdanubia in the Ottoman period (c. 1540–1690)

Appendix 3a. Hungarian/Habsburg fortifications in Transdanubia (c. 1540–1690)

Name of fortification	Material	Number of garrisons (at the earliest period)	Category	Functioning	Related literature
Abda	Timber (?)	25	3	1560–1686	Pálffy 1999. 169 and 171–172.
Alsórajk	Timber	39	3	1550–1600	Pálffy 1995. 150, Vándor 1996. 91, and Koppány 2014. 222–223.
Árpás	Timber (?)		3	1680–1690	Tolnai 2011. 43.
Ásványráró	Timber (?)		3	1680–1690	Tolnai 2011. 43.
Asszonyfa	Stone and timber	10	3	1550–1686	Pálffy 1999. 114, 128, 215–216, and 219 and Koppány 2014. 125–126.
Babócsa	Brick and timber (?)	50	2	1541–1600	Magyar–Nováki 2005. 23–24 and Pálffy 1999. 101–107.
Bajcsavár	Brick and timber	550	1	1578–1612	Toifl 2002. 28 and Kovács–Sümegi 2011.
Bakháza–Pá-cod	Timber (?)	41	3	1540–1560	Pálffy 1999. 49.
Bakonybél	Stone and timber (?)		3	1540–1550	Tolnai 2011. 47.
Balatonfőkajár	Stone and timber (?)		3	1570–	Pálffy 1995b. 74.
Balatonhídvég	Timber (?)	50	2	1550–1660	Tolnai 2011. 75.
Balaton-mogyoród	Timber (?)	20	3	1650–1690	Tolnai 2011. 93.
Balatonszabadi	Timber		2	1590–1690	Tolnai 2011. 148.
Balatonszárszó	Timber		2	1560–1690	Tolnai 2011. 148–149.
Balatonszent-györgy	Stone and timber (?)	44	3	1560–1690	Tolnai 2011. 123.
Balf	Stone and timber (?)		3	1650–1690	Tolnai 2011. 149.
Belezna	Brick and timber (?)		3		Tolnai 2011. 52.
Berhida	Stone and timber (?)		3		Tolnai 2011. 149.

Name of fortification	Material	Number of garrisons (at the earliest period)	Category	Functioning	Related literature
Berzence	Stone and timber (?)		3	1541–1560	Pálffy 1999. 98, 114, and 117.
Bodonhely	Stone and timber (?)	45	3	1550–1683	Pálffy 1999. passim and Ráth 1865. 275–280.
Boncodfölde	Stone and timber		3	1575–1600	Takáts 1915. II. 55 and Koppány 2014. 131–132.
Bozsok	Stone and timber		3	1580–1690	Koppány 2014. 134–139.
Bozsok (2)	Stone and timber		3	1610–1690	Tolnai 2011. 163 and Koppány 2014. 141–142.
Börcs	Timber (?)	5	3	1588–??	Pálffy 1999. 171 and Tolnai 2011. 55.
Bük	Stone and timber (?)		3	1600–1690	Koppány 2014. 140–142.
Celldömölk–Dömölk	Stone and timber		3	1560–1690	Pálffy 1999. 216, Ivicsics 1993. 298. and Koppány 2014. 155–156.
Celldömölk–Izsákfa	Stone and timber (?)		3	1541–1670	Pálffy 1999. 216 and Koppány 2014. 173–174.
Chernelháza-damonya–Csernelháza	Stone and timber (?)	33	3	1550–1690	Koppány 2014. 149–150.
Csákány	Stone and timber	100	2	1541–1600	Pálffy 1995. 149.
Csákány-doroszló	Timber	100	2	1590–1686	Vadas 2014. 220–221 and Koppány 2014. 144–146.
Csány	Stone and timber	10	3	1540–1610	Sági 1960, Pálffy 1995. 179, Tolnai 2011. 57–58. and Koppány 2014. 146–147.
Csatár	Stone and timber (?)		3	1540–1600	Vándor 1996. 94, Tolnai 2011. 59, and Koppány 2014. 147–148.
Csepreg	Stone and timber (?)		3	1540–1690	Kopány 2014. 148–149.
Csesznek	Stone and timber (?)	68	2	1541–1690	Pálffy 1995. 151.
Csesztreg	Timber (?)	75	2		Tolnai 2011. 59 and Benda 1983. 53.
Csobánc	Stone and timber (?)	20	3	1541–1690	Végh 2007. 55–56, Marosi 1991. 22 and Pálffy 1999. passim
Csörötnek	Timber		3	1600–?	Simon 1997. 80.

Name of fortification	Material	Number of garrisons (at the earliest period)	Category	Functioning	Related literature
Csurgó			2	1540–1560	Pálffy 1999. 98, 114, 117, and 131.
Dénesfa	Stone and timber		3	1585–1690	Koppány 2014. 152–153.
Deutsch Schützen (Németlövő/ Lövő)			3	1541–1690	Koppány 2014. 204.
Deutschkreutz (Németkeresz-túr) – Kéthely	Stone and timber	20	3	1555–1590	Pálffy 1995. 149, 171, Jandrics 1999, Vándor 1996. 90, and Koppány 2014. 185.
Devecser	Stone and timber	60	2	1541–1686	Pálffy 1999. 98 and Koppány 2014. 153–155.
Dobri		25	3	1600–1630	Tolnai 2011. 62 and Benda 1983. 53.
Döbrönte	Stone	16	3	1550–1580	Kupovics 2009 and Pálffy 1999. 114–116.
Drassburg (Darufalva)	Stone and timber		3	1551–1690	Koppány 2014.151–152.
Egerszeg (castle)	Stone and timber	309	1	1550–1686	Végh 2010. 30 and Koppány 2014. 158–160.
Egerszeg (town)	Timber		1	1550–1686	Végh 2010. 30 and Koppány 2014. 158–160.
Egervár	Stone and timber	65	2	1540–1686	Marosi 1991. 12 and Koppány 2014. 159–161.
Felsőmarác–Tótfalu	Timber (?)	10	3	1550–1690	Végh 2010. 156, Baráth 2014. 45, and Koppány 2014. 260.
Fityeház	Timber	20	3	1578–1600	Pálffy 1995. 174 and Tolnai 2011. 68.
Fonyód (Fácános Castle)	Tikber (?)	100	2	1544–1575	Magyar–Nováki 2005. 46–47.
Főnyed	Timber		3		Magyar–Nováki 2005. 48.
Gasztony	Timber		3	1600–?	Vadas 2014. 219 and Benczik 1995.
Gattendorf (Lajtakáta)	Stone and timber		3	1629–1690	Koppány 2014. 165–166.
Gétye	Stone and timber		3	1541–1600	Vándor 1996. 91 and Koppány 2014. 166.
Gritsch (Gerese/Grics)	Stone and timber		3	1600–1690	Koppány 2014. 167–168.

Name of fortification	Material	Number of garrisons (at the earliest period)	Category	Functioning	Related literature
Güssing (Németújvár)–Sankóháza	Stone and timber		3	1600–1690	Koppány 2014. 235.
Gyepükaján	Timber (?)		3		Tolnai 2011. 73.
Gyirmót	Timber	4	3		Pálffy 1999. 171.
Gyóró	Timber		3	1600–?	Tolnai 2011. 73.
Győr	Stone and timber (?)	1264	1	1540–1690	Pálffy 1995. 150.
Győr–András-vár	Stone and timber	15	3	1550–1686	Pálffy 1999. 169–172.
Győr–Kis-megyer (Félegyháza)	Timber (?)	25	3	1550–1590 (?)	Pálffy 1999. 171.
Győr–Pataháza	Timber		2	1560–	Villányi 1882. 120 and Tolnai 2011. 158.
Győr–Révfalu	Timber		2	1560–	Villányi 1882. 120 and Tolnai 2011. 159.
Győr–Tarisz-nyavár		25	3		Pálffy 1999. 124, 169 and 171–172.
Győrújbarát	Stone and timber	5	3	1550–1686	Pálffy 1999. 89 and 170–171.
Hahót–Sárkánysziget	Stone and timber		3	1541–1600	Vándor 1996. 91 and Koppány 2014. 235–236.
Hédervár	Stone and timber		3	1543–1686	Marosi 1991. 14 and Kopány 2014. 168–171.
Hídvég	Stone and timber (?)	50	2	1550–1686	Pálffy 1999. 215–219, Simon 1997. 80, and Koppány 2014. 172–173.
Hollós	Timber		3	1590–1686	Pálffy 1999. 215–219.
Hosztót	Timber		3	1560–	Tolnai 2011. 154
Ikervár	Stone and timber		3	1600–?	Tolnai 2011. 154.
Ikrény	Timber	4	3		Pálffy 1999. 170–171.
Ivánc	Timber		3	1570–	Takáts 1915. II. 29, Pálffy 1999. 215, and Tolnai 2011. 77.
Jánosháza	Stone and timber		3	1541–1690	Pálffy 1999. 216 and Koppány 2014. 174–177.
Jánossomorja	Timber		3		Tolnai 2011. 124.
Kacorlak		20	3	1580–1600	Pálffy 1995. 169 and Vándor 1996. 93.

Name of fortification	Material	Number of garrisons (at the earliest period)	Category	Functioning	Related literature
Kanizsa (castle)	Brick and timber	1850 (?)	1	1541–1600	Pálffy 1995. 149.
Kanizsa (watchtower)	Timber (?)	10	3	1541–1600	Pálffy 1995. 169.
Kanizsa–Móricheły	Timber		3	1541–1570	Vándor 1996. 91.
Kanizsa–Palin	Timber	10	3	1570–1600	Kelenik 1995. 167.
Kanizsa–Szentmiklós	Timber (?)	50	2	1580–1600	Pálffy 1995. 168 and Vándor 1996. 93.
Kányavár		15	3	1541–1690	Czigány 2001. 287 and Koppány 2014. 177–178.
Kapornak	Stone and timber	460	1	1541–1686	Pálffy 1995. 150, Simon 1997. 81, and Czigány 2001. 287.
Kaposmérő	Timber		3	1540–1560	Magyar 1991. 102.
Kaposvár–Kaposújvár		50	2	1560–1570	Pálffy 1999. 103, 105 and 118 and Szakály 1975.
Kaposvár–Szentjakab	Stone and timber (?)		3	1540–1560	Tolnai 2011. 123.
Kapuvár	Brick and timber (?)			1560–1686	Marosi 1991. 14.
Kehidakustány–Kehida	Stone and timber		3	1541–1590	Vándor 1996. 94 and Koppány 2014. 178.
Kemendollár–Kemend	Stone	20	3	1570–1686	Pálffy 1995. 170.
Kerecseny			3	1570–1580	Vándor 1996. 90.
Keszthely	Stone and timber	125	2	1560–1686	Végh 2007. 56, Végh 2005, and Pálffy 1999. 79, 134, 145, 160 and 165.
Kilimán	Timber	20	3	1580–1600	Pálffy 1995. 169 and Pálffy 1996.
Klein-mutschen (Pervány)	Stone and timber		3		Koppány 2014. 218.
Körmend	Brick and timber	250	1	1541–1686	Kelenik 1993, Hajmási 1994. 182, Vadas 2013, and Koppány 2014. 194–197.
Kőszeg	Stone (?)			1541–1686	Marosi 1991. 15–16.
Lackenbach (Lakompak)	Stone and timber		3	1540–1690	Koppány 2014. 197–202.
Lakócsa	Timber		3	1680–1690	Tolnai 2011. 88 and Szita 1999. 321.

Name of fortification	Material	Number of garrisons (at the earliest period)	Category	Functioning	Related literature
Lébény	Timber		3	1600–	Tolnai 2011. 88.
Legrad (Légrád)		150	2	1570–1686	Pálffy 1995. 149 and Vándor 1996. 91.
Lendava (Alsólendva)	Stone and timber (?)	100	2	1550–1690	Vándor 1996. 95.
Lenti	Stone and timber	190	2	1540–1686	Marosi 1991. 16 and Végh 2010. 156.
Letenye	Stone and timber (?)		3	1540–1600	Koppány 2014. 203.
Lövő		88	2	1566–1686	Pálffy 1995. 170 and Simon 1997. 81.
Magyarlak	Stone and timber	28	3	1590–1690	Pálffy 1999. 49, 53, 101, 103, 105, 118, and 161–162 and Koppány 2014. 197.
Magyaróvár	Stone and timber			1541–1686	Marosi 1991. 16–17.
Magyarszent-miklós	Timber		3	1650–1690	Tolnai 2011. 126.
Marcali	Stone and timber (?)		3	1540–1570	Tolnai 2011. 91.
Marcaltő	Brick and timber		3	1600–1690	Koppány 2014. 206–207.
Martonfa	Timber		3	1680–1690	Tolnai 2011. 91.
Meggyesko-vácsi	Timber		3	1590–1690	Pálffy 1999. 215 and Koppány 2014. 190.
Ménfőcsanak	Timber	10	3	1610–?	Pálffy 1999. 171.
Ménfőcsanak–Világosvár	Timber	10	3		Pálffy 1999. 170–171.
Mérges	Stone and timber	23	3	1580–?	Pálffy 1999. 170–172 and Tolnai 2011. 92.
Mesteri	Stone and timber		3	1541–1690	Pálffy1999. 216 and Koppány 2014. 207–208.
Mesztegnyő	Stone and timber (?)		3	1540–1560	Tolnai 2011. 92 and Magyar–Nováki 2005. 162–163.
Mihályi	Stone and timber		3	1560–1690	Pálffy 1999. 215 and 219, Koppány 2006. 98–103, and Koppány 2014. 208–211.
Mogyorósd	Timber		3	1550–1650	Vándor 1996. 94 and Koppány 2011. 211–212.
Molnári	Timber		3	1566–1600	Vándor 1996. 91.

Name of fortification	Material	Number of garrisons (at the earliest period)	Category	Functioning	Related literature
Mórichida	Timber (?)		3		Simon 1997. 81.
Murakeresztúr	Timber	60	2	1578–1600	Pálffy 1995. 174 and Vándor 1996. 91.
Muraszemenye	Timber	30	3	1541–1620	Pálffy 1995. 150 and Vándor 1996. 91.
Murska Sobota (Muraszombat)	Stone and timber		3	1541–1690	Koppány 2014. 212–213.
Murska Sobota (Muraszom-bat) – Rakičan (Rakicsány)	Stone and timber		3	1580–1690	Koppány 2014. 223–225.
Nagybajom	Stone and timber		3	1540–1560	Magyar–Nováki 2005. 95 and Tolnai 2011. 95–96.
Nagylózs	Stone (?) and timber		3	1560–1690	Koppány 2014. 203–204.
Nagyrécse–And	Timber (?)		3	1570–1580	Vándor 1996. 91 and 93 and Tolnai 2011. 42.
Neudörfl an der Leitha (Szentmiklós)	Stone and timber		3	1640–1690	Koppány 2014. 249–250.
Nikitsch (Gálosháza)	Stone and timber		3	1569–1690	Koppány 2014. 163–164.
Old	Timber		2	1680–1690	Tolnai 2011. 98 and Szita 1995. 99 and 129.
Old–Lajos-Fortress	Timber (?)		2	1680–1690	Szita 1993. 65 and Tolnai 2011. 88.
Ordacsehi–Orda	Timber		3	1540–1570	Tolnai 2011. 98
Őrimagyarósd		20	3		Simon 1997. 80.
Örményes	Stone and timber (?)		3	1540–1600	Tolnai 2011. 99.
Pacsa–Isabor	Timber	20	3	1570–1590	Pálffy 1995. 170, Vándor 1996. 91, and Tolnai 2011. 77.
Pápa	Stone and timber	750	1	1543–1686	Pálffy 1995. 150 and Marosi 1991. 18.
Pápoc			3		Pálffy 1999. 98 and 123.
Petjanci (Petánc)	Stone and timber		3	1550–1690	Koppány 2014. 219.
Pinkafeld (Pinkafő)	Stone and timber (?)		3	1620–1690	Koppány 2014. 220–221.
Pinkamind-szent	Timber		3	1540–1650	Vándor 1996. 93 and Koppány 2014. 211.

Name of fortification	Material	Number of garrisons (at the earliest period)	Category	Functioning	Related literature
Podturen (Torony)	Stone and timber (?)		3	1541–1690	Koppány 2014. 259–260.
Pornóapáti	Stone and timber (?)		3	1550–1690	Tolnai 2011. 105.
Pölöske	Brick and timber	100	2	1541–1686	Pálffy 1995. 150, Vándor 1996. 90, and Czigány 2001. 287.
Prelog–Oporovec	Timber		3	1540–1550	Tolnai 2011. 42.
Rábapatona	Timber	10	3	1570–1686	Pálffy 1999. 170–172.
Rábapaty (Felsőpaty)	Timber (?)		3		Tolnai 2011. 68.
Rábasebes	Timber		3	1600–	Simon 1997. 75.
Rábaszentandrás	Timber		3	1680–1690	Tolnai 2011. 120.
Rábatöttös (Szent Cecília)	Timber (?)		3		Tolnai 2011. 121.
Rakovac (?)	Timber		3	1600–	Koppány 2014. 226.
Rátót–Tiszamark	Timber		3	1650–1660	Benczik 1995. 45 and Tolnai 2011. 136.
Razkrižje (Ráckanizsa)	Stone and timber (?)		3	1541–1690	Koppány 2014. 233–234.
Rechnitz (Rohonc)	Stone and timber		3	1541–1690	Koppány 2014. 228–233.
Rotenturm (Vörösvár)	Stone and timber (?)		3	1541–1690	Koppány 2014. 267–269.
Rum	Stone and timber (?)		3	1540–1690	Simon 1997. 81, Tolnai 2011. 106, and Koppány 2014. 233.
Sárvár	Stone and brick	800	1	1541–1686	Marosi 1991. 19, Simon 1997. 81, and Pálffy 1999. passim.
Sitke	Stone and timber		3	1541–1690	Pálffy 1999. 114, 216 and 219 and Koppány 2014. 237–239.
Somogysámson–Marót	Timber (?)		3	1550–1560	Tolnai 2011. 91.
Somogyvár	Stone and timber (?)	75	2	1540–1560	Pálffy 1999. 49, 103, and 105.
Somogyvár – Kupa vár	Timber (?)	40	3	1540–1560	Tolnai 2011. 74.
Somogyzsitfa	Timber		3	1590–1600	Magyar–Nováki 2005. 136.

Name of fortification	Material	Number of garrisons (at the earliest period)	Category	Functioning	Related literature
Somogyzsitfa–Szőcsény	Timber		3	1555–1590	Vándor 1996. 90.
Sormás	Timber (?)		3	1570–1580	Vándor 1996. 91.
Sümeg	Stone and timber (?)	32	3	1541–1686	Marosi 1991. 20 and Pálffy 1999. passim.
Szabar	Timber (?)		3	1550–1650	Tolnai 2011. 113.
Szécsisziget (castle)	Stone and timber	150	2	1540–1690	Végh 2010. 156, Simon 1997. 81, and Koppány 245–246.
Szécsisziget (village)	Timber		1	1600–1690	Koppány 2014. 245.
Szecsőd	Timber		3	1600–1686	Vadas 2014. passim
Szentgotthárd	Timber		3	1560–1686	Ilon–Grynaeus–Torma 2001. 38.
Szentgrót		72	2		Pálffy 1995. 170.
Szentgyör-gyvár		46	3	1540–1580	Szatlóczki 2002. 18.
Szentjakab (?)	Timber	2	3	1570–1580	Tolnai 2011. 123.
Szentmárton (Hegyhátszent-márton)	Stone and timber		3	1570–?	Benczik 1992. 47 and Koppány 2014. 249.
Szentmár-tonhely (Pannonhalma)	Stone and timber	167	2		Pálffy 1995. 150 and Marosi 1991. 17.
Szentmi-hály (around Rábapatona [?])	Timber		3	1650–1690	Tolnai 2011. 126.
Szentpéter	Brick and timber	40	3	1550–1686	Marosi 1991. 17 and Simon 1997. 80.
Szenyér	Stone and timber	41	3	1541–1590	Pálffy 1999. 128, Vándor 1996. 90, and Koppány 2014. 250.
Szigliget	Stone (?)	40	3	1540–1686	Végh 2007. 56.
Szombathely–Szőkefölde	Stone and timber (?)		3		Koppány 2014. 58.
Tapolca	Stone and timber	250	1	1541–1686	Pálffy 1999. 77–79 and 83.
Tata	Stone and timber	390	1	1541–1686	Pálffy 1995. 150 and Pálffy 1999. passim.
Tata – Calvary	Timber (?)		3	1540–	Tolnai 2011. 160.

Name of fortification	Material	Number of garrisons (at the earliest period)	Category	Functioning	Related literature
Tihany		76	2	1541–1686	Pálffy 1995. 151.
Türje	Stone and timber (?)	33	3	1550–1690	Tolnai 2011. 138.
Tüskevár (Nagyjenő)	Stone and timber (?)		3	1600–?	Tolnai 2011. 96.
Udvarhely (Somogyud-varhely [?])			3	1560–1580	Tolnai 2011. 138–139.
Ugod		45	3	1550–1686	Pálffy 1999. 50, 114, and 116.
Újudvar	Timber	90	2	1580–1600	Pálffy 1995. 169 and Vándor 1996. 93.
Újudvar (Novidvor)	Brick and timber		3		Koppány 2014. 262.
Újvárfalva–Ko-rotna		250	1	1541–1570	Pálffy 1999. 49, 83, and 90.
Unterloisdorf–Mannersdorf (Répcekét-hely–Tábor)	Stone and timber		3	1540–1690	Koppány 2014. 254–256.
Vág	Brick and timber		3	1560–1640	Koppány 2014. 262–263.
Várgesztes	Stone and timber (?)	32	3		Pálffy 1999. passim
Várkesző	Stone and timber (?)		3	1570–1686	Pálffy 1999. 114–116.
Városlőd	Stone and timber (?)	37	3		Pálffy 1999. 78, 83, 87, 90, 103, and 105.
Várpalota	Stone and timber	400	1		Pálffy 1995. 150.
Vasvár	Stone and timber		3	1541–1690	Pálffy 1999. 216 and 219 and Tolnai 2011.141.
Vasszécsény	Stone and timber		3	1550–1690	Tolnai 2011. 164–165 and Koppány 2014. 243–245.
Vát (Külsővát)	Stone and timber (?)	20	3	1590–1690	Koppány 2014. 265.
Vát (Mersevát)	Stone and timber		3	1650–1690	Koppány 2014. 263–264.
Vázsony	Stone and timber	43	3	1541–1686	Marosi 1991. 17 and Pálffy 1995. 151.
Vértessomló–Vitány	Stone and timber	14	3	1541–1690	Pálffy 1999. 50, 103, 105, and 123.
Veszprém	Stone and timber (?)	600	1	1541–1686	Pálffy 1995. 150.

Name of fortification	Material	Number of garrisons (at the earliest period)	Category	Functioning	Related literature
Vízvár	Timber (?)	34	3	1550–1570	Pálffy 2002. 16.
Zalabér	Stone and timber	10	3	1560–1600	Pálffy 1995. 170, Sorok 1997, Koppány 1999. 49, and Koppány 2014. 129–131.
Zalaistvánd	Timber		3	1600–	Simon 1997. 75.
Zalakomár (Kiskomár)	Stone and timber	400	1	1560–	Pálffy 1995. 149 and Szvitek 2008. 45.
Zalakomár (Nagykomár)	Timber		3	1540–1560	Koppány 2014. 214–215.
Zalaszegvár			3	1600–	Czigány 2004. 145.
Zalaszentba-lázs	Timber (?)		3	1570–1580	Vándor 1996. 91.
Zalatárnok	Timber (?)		1	1560–1630	Benda 1983. 53 and Czigány 2004. 71.
Zalavár	Stone and timber	115	2	1570–1686	Iványi 1960, Pálffy 1995. 150, Vándor 1996. 60, and Czigány 2001. 287.
Zsennye	Stone and timber		3	1550–1690	Koppány 2014. 269–270.
Zsira (Salamonfalva)	Stone and timber (?)		3	1541–1690	Koppány 2014. 234–235.

Appendix 3b. Ottoman fortifications in Transdanubia (c. 1540–1690)

Name of fortification	Material	Number of garrisons (at the earliest period)	Category	Functioning	Related literature
Adony (Dzsánkur-tarán)	Timber (?)		1		Tolnai 2011. 41.
Babócsa	Stone and timber (?)	595	1	1555–1556 and 1566–1686	Hegyi 2007. II. 1312–1319 and III. 1578–1583, and Marosi 1991. 9.
Balatonboglár	Timber		3	1570–?	Tolnai 2011. 54 and Magyar 2004–2005. 44.
Balatonendréd	Timber (?)	73	2	1544–1686	Hegyi 2007. II. 1074–1088.
Balatonföldvár	Timber	38	3	1550–1686	Hegyi 2007. II. 627–639 and Kozák 1970.
Balatonsze-mes–Bolond-vár	Timber (?)	159	2	1570–1686	Hegyi 2007. II. 1239–1243.
Barcs	Timber	200	1	1567–1664	Hegyi 2007. II. 1327–1329 and III. 1590–1594 and Kovács–Sümegi 2011.
Barcs (2)	Timber		3	1680–1690	Tolnai 2011. 49 and Vass 1993. 212.
Bátaszék	Stone and timber (?)	125	2	1550–1686	Hegyi 2007. II. 1140–1152.
Berzence	Stone and timber (?)	396	1	1566–1686	Hegyi 2007. II. 1320–1326 and III. 1583–1589 and Magyar–Nováki 2005. 38–39.
Branjin Vrh (Baranyavár)	Stone and timber	67	2	1543–1686	Hegyi 2007. II. 1180–1186 and Tolnai 2011. 47.
Buda	Stone and timber	2016	1	1541–1686	Hegyi 2007. II. 423–480.
Buda–Gürsz Eljász	Timber (?)	146	2	1593–1686	Hegyi 2007. II. 481–484.
Csókakő	Stone and timber (?)	34	3	1543–1687	Hegyi 2007. II. 1014–1022, Feld et al. 2011. 38–41 and 167, Pálffy 1999. 161 and 222, and Farkas 1989. 39–44.
Darda (Dárda)	Timber (?)	74	2	1600–1686 (?)	Hegyi 2007. II. 1198–1200.

Name of fortification	Material	Number of garrisons (at the earliest period)	Category	Functioning	Related literature
Dombóvár	Stone and timber	150	2	1546–1686	Hegyi 2007. II. 1216–1224, Miklós 2007. 171–180, and Szabó–Csányi 2012.
Döbrököz	Stone (and timber?)	210	1	1545–1686	Hegyi 2007. II. 1111–1120 and Miklós 2007. 190–197.
Dörtesz (?)	Timber (?)		3		Karátson 1908. 38 and Tolnai 2011. 63.
Drávatamási	Timber (?)	44	3	Data only from the 1600s	Hegyi 2007. III. 1602.
Dunaszekcső	Brick and timber	40	3	1541–1686	Hegyi 2007. II. 1172–1179.
Dunaújváros–Pentele	Timber (?)	38	3	1600–?	Hegyi 2007. II. 422.
Ercsi	Timber (?)	200	1	1627–1686	Hegyi 2007. II. 684.
Érd–Hamza bey's seraj	Timber (?)	65	2	1570–1686	Hegyi 2007. II. 666–674 and Marosi 1991. 12.
Esztergom	Stone and timber	2166		1543–1686	Hegyi 2007. II. 686–750.
Esztergom (castle)	Stone and timber (?)	1444	1	1543–1686	Hegyi 2007. II. 686–750.
Esztergom (Tepedelen)	Timber (?)	99	2	1543–1686	Hegyi 2007. II. 686–750.
Esztergom (urban fortification)	Timber (?)	722	1	1543–1686	Hegyi 2007. II. 686–750.
Fehérvár	Stone and timber (?)	2944	1	1543–1688	Hegyi 2007. II. 972–1013.
Fonyód–Bézseny	Timber (?)		3	1570–1580	Pálffy 1999. 140 and 162.
Győr	Stone and timber (?)	2050	1	1594–1598	Hegyi 2007. III. 1495–1498.
Igal	Timber (?)	38	3	17th c.	Hegyi 2007. II. 422.
Kanizsa	Stone and timber (?)	1325	1	1600–1690	Hegyi III. 1535–1551.
Kaposvár	Stone and timber (?)	280	2	1555–1686	Hegyi 2007. II. 1279–1285 and III. 1557–1562.
Karád	Timber (?)	50	2	1555–1686	Hegyi 2007. II. 1225–1229 and Magyar–Nováki 2005. 75.
Korkmaz (?)	Timber	129	2	1546–1686	Hegyi 2007. II. 598–613 and Feld et al. 2011. 21.

Name of fortification	**Material**	**Number of garrisons (at the earliest period)**	**Category**	**Functioning**	**Related literature**
Kőröshegy	Stone and timber		3	1540–	Tolnai 2011. 87.
Madocsa	Stone and timber		3		Tolnai 2011. 89.
Magyareg-regy–Máré	Stone and timber (?)	51	2	Data only from the 1540s	Hegyi 2007. II. 1268 and Marosi 1991. 16.
Mázaszászvár	Stone and timber (?)	159	2	Data only from the 1540s	Hegyi 2007. II. 1266–1267 and Marosi 1991. 16.
Mecseknádasd	Timber (?)	37	3	1570–1686	Hegyi 2007. II. 1288–1289 and III. 1563–1564.
Mohács	Timber (?)	161	2	1560–1686	Hegyi 2007. II. 1187–1196.
Nagyberki (Berkigát)	Timber (?)		3	1650–1660	Magyar–Nováki 2005. 163–164.
Nemeske-Görösgál	Timber (?)	105	2	1544–1559	Hegyi 2007. II. 1269–1272.
Nyergesújfalu	Stone (?)		3		Karátson 1908. 130 and Tolnai 2011. 97.
Osijek (Eszék) (bridge)	Timber (?)	31	3	Data only from the 1610s	Hegyi 2007. III. 1565.
Ozora	Stone and timber	93	2	1545–1686	Hegyi 2007. II. 1089–1098 and Marosi 1991. 17.
Öreglak	Timber (?)	159	2	1566–1686	Hegyi 2007. II. 1230–1238.
Ötvöskónyi	Timber (?)	29	3	Data only from the 1610s	Hegyi 2007. II. 1250.
Ötvösónyi	Stone and timber (?)		3	1556–1650	Magyar–Nováki 2005. 109.
Paks	Timber (?)	69	2	1565–1686	Hegyi 2007. II. 1153–1161 and Miklós 2007. 293–294.
Pápa	Stone and timber (?)	700	1	1594–1597	Hegyi 2007. III. 1499–1500.
Pécs	Stone and timber (?)	870	1	1543–1686	Hegyi 2007. II. 1252–1265 and III. 1551–1556.
Péterhida	Timber (?)		3		Magya–Nováki 2005. 82 and Tolnai 2011. 104.
Pilisvörösvár (Kizilhiszár)	Timber (?)	164	2	1570–1595 and 1630–1686	Hegyi 2007. II. 675–680.

Name of fortification	Material	Number of garrisons (at the earliest period)	Category	Functioning	Related literature
Pincehely	Timber (?)	34	3	1615–1686	Hegyi 2007. II. 1130.
Ságvár	Stone and timber (?)	8	3	1550–1560	Szántó 1980. 34.
Sárbogárd	Timber (?)	44	3	Data from 1608	Hegyi 2007. II. 1128–1129.
Sásd	Timber (?)	55	2	Data only from the 1610s	Hegyi 2007. III. 1566–1567.
Segesd	Stone and timber (?)	400	1	1566–1686	Hegyi 2007. II. 1330–1333 and III. 1590–1595.
Sellye	Timber (?)	30	3	1553–1559	Hegyi 2007. II. 1286–1287.
Siklós	Stone and timber (?)	518	1	1543–1686	Hegyi 2007. II. 1304–1311 and III. 1577.
Simontornya	Stone and timber (?)	156	2	1545–1686	Hegyi 2007. II. 1051–1071.
Sióagárd (Anyavár)	Stone and timber (?)	24	3	1545–1546	Hegyi 2007. II. 1072–1073.
Siófok	Timber (?)	151	2	1600–1686	Hegyi 2007. II. 1244–1249.
Somogycsicsó	Timber (?)		3		Tolnai 2011. 59.
Somogysimo-nyi	Timber (?)		3	1570–1580	Tolnai 2011. 110.
Somogyvá-mos–Reme-teudvar	Timber (?)		3		Sugár 1976. 18. and Tolnai 2011. 106.
Somogyvár–Koppány	Timber (?)	120	2	1550–1686	Hegyi 2007. II. 1201–1215.
Szabadbattyán	Stone and timber	124	2	1567–1687	Hegyi 2007. II. 1030–1039 and Feld et al. 2011. 86–87 and 212.
Szabadhídvég	Timber	50	2	1565–1686	Hegyi 2007. II. 1121–1127 and Feld et al. 2011. 88–90.
Szalánta	Timber (?)		3	?–1660	Magyar–Nováki 2005. 132. and Tolnai 2011. 113.
Szekszárd	Stone and timber	157	2	1541–1686	Hegyi 2007. II. 1131–1139.
Szentlőrinc	Timber (?)	120	2	1541–1559	Hegyi 2007. II. 1273–1275.
Szerdahely (Újpalánk)	Timber (?)	159	2	1600–1686 (?)	Hegyi 2007. II. 1162–1166, Gaál 2003 and Kovács 2015.

Name of fortification	Material	Number of garrisons (at the earliest period)	Category	Functioning	Related literature
Szigetvár (Inner Castle)	Brick and timber	650	1	1566–1689	Hegyi 2007. II. 1292–1303 and III. 1568–1576.
Szigetvár (Outer Castle)	Brick and timber	650	1	1566–1689	Hegyi 2007. II. 1292–1303 and III. 1568–1576.
Szigetvár–Türbék	Stone and timber (?)	60	2	1570–1690	Hegyi 2007. III. 1569–1576 and Tolnai 2011. 128.
Gödreszentmárton (Szentmárton)	Timber (?)	100	2	1552–1599	Hegyi 2007. II. 1276–1278.
Szőcsény	Timber (?)	150	2	1574–1598	Hegyi 2007. II. 1290–1291.
Tamási	Brick and timber	133	2	1545–1686	Hegyi 2007. II. 1099–1110 and 325–334.
Tata	Stone and timber (?)	425	1	1557–1566 and 1594–1598	Hegyi 2007. II. 643–650.
Tihany	Stone and timber (?)	305	1	1594–1598	Hegyi 2007. II. 1048–1049.
Tolna	Stone and timber (?)	127	2	1600–1686 (?)	Hegyi 2007. II. 1167–1171.
Vál	Stone and timber	113	2	1550–1686	Hegyi 2007. II. 614–626 and Feld et al. 2011. 101–102 and 222.
Vârgesztes	Stone and timber	44	3	1557–1598	Hegyi 2007. II. 652–656.
Várpalota	Stone and timber	500	1	1593–1598 and 1605–1687	Hegyi 2007. II. 1040–1047 and Marosi 1991. 24.
Vázsony	Stone and timber (?)	51	2	1594–1598	Hegyi 2007. II. 1050.
Veszprém	Stone and timber (?)	317	1	1552–1566 and 1593–1598	Hegyi 2007. II. 1023–1029.
Visegrád (Lower Castle)	Stone and timber (not one fortification)	101	2	1544–1684	Hegyi 2007. II. 549–569.
Visegrád (Upper Castle)	Stone and timber (not one fortification)	97	2	1544–1684	Hegyi 2007. II. 549–569.

Name of fortification	Material	Number of garrisons (at the earliest period)	Category	Functioning	Related literature
Vitány	Stone and timber (?)	38	3	1557–1566 and 1594–1598	Hegyi 2007. II. 651.
Vörösmart	Timber (?)	80	2	1599–1686 (?)	Hegyi 2007. II. 1197.
Zalakomár	Stone and timber (?)	–		1600–1601	Hegyi 2007. III. 1601.
Zsámbék	Timber (?)	225	1	1546–1686	Hegyi 2007. II. 581–598.

References in Appendix 3

Baráth 2014 Baráth, Zsolt. "A Rába mint védelmi vonal a 17. században. Védelmi munkálatok és létesítmények a folyó Vas megyei szakaszán" [Rába as a defense line – protection measures and constructions in the Vas County-section of the river], *Vasi Honismereti és Helytörténeti Közlemények* no. 1 (2014): 31–53.

Benczik 1995 Benczik, Gyula. "Fentő és góré. A hódoltság kori védelmi munkálatok emlékei a Felső-Rábavidék dűlőneveiben" [Fentő and góré. The memory of the Ottoman-period defensive works in the field names of the Upper Rába region], *Vasi Honismereti és Helytörténeti Közlemények* no. 1 (1995): 42–48.

Benda 1983 Benda, Kálmán. A magyarországi végvári vonal fenntartásának költségei a XVII. század elején [Costs of the upkeep of the frontier defense system in Hungary at the beginning of the 17th century]. In *Magyarországi végvárak a XVI–XVII. században.* (*Tanulmányok.*) [Hungarian frontier fortifications in the 16th–17th centuries (Studies)] (Studia Agriensia, 3), eds. Sándor Bodó and Jolán Szabó, 49–59. Eger: Dobó István Vármúzeum, 1983.

Czigány 2001 Czigány, István. "Hadügyi reformkísérletek a királyi Magyarországon. 1665–1682" [Military reform attempts in royal Hungary, 1665–1682], *Hadtörténelmi Közlemények* 114 (2001): 279–302.

Czigány 2004 — Czigány, István. *Reform vagy kudarc? Kísérletek a magyarországi katonaság beillesztésére a Habsburg Birodalom hadseregébe, 1600–1700* [Reform or defeat? Attempts of the integration of the Hungarian army into the army of the Habsburg Empire, 1600–1700]. Budapest: Balassi, 2004.

Farkas 1989 — Farkas, Gábor. "Csókakővár a törökellenes küzdelmek idején 1543–1687" [Csókakővár in the period of the Ottoman struggles, 1543–1687], *Fejér Megyei Történeti Évkönyv* 19 (1989): 39–44.

Feld et al. 2011 — Feld, István. et al. *Fejér megye várai az őskortól a kuruc korig* [Castles and fortifications in County Fejér, Hungary] (Magyarország várainak topográfiája, 3). Budapest: Castrum Bene Egyesület and Cívertan Bt., 2011.

Gaál 2003 — Gaál, Attila. "Turkish Palisades on the Tolna County Stretch of the Buda to Eszék Road." In *Archaeology of the Ottoman Period in Hungary* (Opuscula Hungarica, 3), eds. Ibolya Gerelyes and Györgyi Kovács, 105–108. Budapest: Hungarian National Museum, 2003.

Hajmási 1994 — Hajmási, Erika. "Körmend a középkorban – néhány szondázó leletmentés tükrében" [Körmend in the Middle Ages in the light of some test excavations]. In *Körmend története* [The history of Körmend], ed. László Szabó, 180–194. Körmend: Körmendi Önkormányzat, 1994.

Hegyi 2007 — Hegyi, Klára. *A török hódoltság várai és várkatonasága*, 3 vols. [Castles and garrisons of Ottoman Hungary] (História Könyvtár. Kronológiák, adattárak, 9). Budapest: História and MTA Történettudományi Intézet, 2007.

Ilon–Grynaeus–Torma 2001 — Ilon, Gábor, András Grynaeus, and Andrea Torma. "A szentgotthárdi török kori palánk kutatásáról" [On the archaeological research of the Ottoman palisade of Szentgotthárd], *Savaria* 31 (2007): 307–328.

Iványi 1960 — Iványi, Béla. "Zalavár és a balatonhídvégi átkelő a török időkben" [Zalavár and the crossing at Balatonhídvég in the Ottoman period]. In *A Göcseji Múzeum jubileumi emlékkönyve 1950–1960* [Jubilee volume of the Göcsej Museum] (A Göcseji Múzeum Közleményei, 8), ed. Imre Szentmihályi, 161–180. Zalaegerszeg: Göcseji Múzeum, 1960.

Ivicsics 1993 Ivicsics Péter. “Kisebb végvárak, véghelyek Vas megyében” [Smaller frontier fortifications and watch towers in Vas County]. In *A végvárak és régiók a XVI–XVII. században. Tudományos tanácskozás előadásai-Noszvaj, 1991. okt. 17–18.* [Borderline castles and regions in the 16th–17th centuries. Proceedings of the conference held at Noszvaj, 17–18 October 1991] (Studia Agriensia, 14), eds. Tivadar Petercsák and Jolán Szabó, 289–310. Eger: Heves Megyei Múzeumok Igazgatósága, 1993.

Jandrics 1999 Jandrics, István. *A kéthelyi Fancsi-vár a környező várak tükrében* [The Fancsy Castle at Kéthely in the light of the neighboring castles]. Kéthely: Kéthelyi Honismereti Egyesület. 1999.

Karátson 1908 Karácson, Imre, ed. *Evlia Cselebi török világutazó magyarországi utazásai, 1664–1666*, vol. 2 [The travels of Evliya Çelebi in Hungary, 1664–1666]. Budapest: Magyar Tudományos Akadémia, 1908.

Kelenik 1993 Kelenik, József. “A nemzetiségi megoszlás, a veszteségek és a fluktuáció mértéke. Tizennégy Kanizsa elleni végvár helyőrségében (1633–1640)” [Ethnic composition, the losses, and the scale of fluctuation. On the example of 14 garrisons against Kanizsa]. In *A végvárak és régiók a XVI–XVII. században. Tudományos tanácskozás előadásai-Noszvaj, 1991. okt. 17–18.* [Borderline castles and regions in the 16th–17th centuries. Proceedings of the conference held at Noszvaj, 17–18 October 1991] (Studia Agriensia, 14), eds. Tivadar Petercsák and Jolán Szabó, 101–121. Eger: Heves Megyei Múzeumok Igazgatósága, 1993.

Koppány 1999 Koppány, Tibor. *A középkori Magyarország kastélyai* [Palaces in medieval Hungary] (Művészettörténeti füzetek, 26). Budapest: Akadémiai, 1999.

Koppány 2006 Koppány, Tibor. *A Castellumtól a kastélyig. Tanulmányok a magyarországi kastélyépítés történetéből* [From castellum to castle. Studies in the history of castle-building in Hungary]. Budapest: Históriaantik Könyvesház Kiadó, 2006.

Koppány 2014 Koppány, Tibor. *Kastélyok a végvárak mögött. Késő reneszánsz és kora barokk kastélyépítészet a 16–17. századi Dunántúlon* [Palaces behind frontier fortifications. Late Renaissance and early Baroque palace architecture in 16th–17th-century Transdanubia]. Budapest: L’Harmattan, 2014.

Kovács 2015 Kovács, Györgyi. “Oszmán erődítmények a Dél-Dunántúlon. Gondolatok Szigetvár–Turbék régészeti kutatása előtt” [Ottoman fortifications at the Southern Transdanubia. Thoughts on the archaeological research of Szigetvár–Turbék], *Mediterrán és Balkán Fórum* no. 2 (2015): 20–33.

Kozák 1970 Kozák, Éva. “Régészeti kutatások a dunaföldvári Öregtoronynál” [Arhcaeological research at Dunaföldvár – Öregtorony], *A Szekszárdi Béri Balogh Ádám Múzeum Évkönyve* 1 (1970): 181–208.

Kupovics 2009 Kupovics, Renáta. “A döbröntei vár kutatása” [Research on the castle of Döbrönte], *Castrum Bene* 9 (2009): 31–68.

Magyar 1991 Magyar, Kálmán. “Somogy várai és véghelyei 1699–1723 között (Különös tekintettel a Kapos-várra)” [The castles and frontier fortifications of Somogy County between 1699 and 1723 (with special regard to Kapos Castle)]. In *A végvárak és végváriak sorsa (1699–1723)* [The fate of frontier fortifications and their garrisons (1699–1723)] (Studia Agriensia, 11), eds. Tivadar Petercsák and Ernő Pető, 99–116. Eger: Dobó István Vármúzeum, 1991.

Magyar 2004–2005 Magyar, Kálmán. “Erődítmények, várak, végvárak Tab környékén: XI–XVIII. századi várkutatás Somogy megye északkeleti vidékén” [Fortifications, castles, frontier forts around Tab: 11th–18th-century castle research in northeast Somogy]. In *Tabi kilátó 2004–2005 Helytörténeti olvasókönyv* [Mirador of Tab, 2004–2005. Local history reader], ed. Béla Bertalan, 8–52. Tab: Tabi Polgármesteri Hivatal, 2004–2005.

Magyar–Nováki 2005 Magyar, Kálmán and Gyula Nováki. *Somogy megye várai a középkortól a kuruc korig* [Castles of Somogy County from the Middle Ages to the *kuruc* period]. Kaposvár: Somogy Megyei Múzeumok Igazgatósága, 2005.

Miklós 2007 Miklós, Zsuzsa. *Tolna megye várai* [Castles of Tolna] (Varia Archaeologia Hungarica, 22). Budapest: Instituti Archaeologici Academiae Scientiarum Hungaricae, 2007.

Pálffy 1995 Pálffy, Géza. “A magyarországi és délvidéki végvárrendszer 1576. és 1582. évi jegyzékei” [The Registers of the Hungarian and Croatian-Slavonian Border Fortresses of 1576 and 1582], *Hadtörténelmi Közlemények* 108, no. 1 (1995): 114–185.

Pálffy 1995b Pálffy, Géza. "A magyarországi török és királyi végvár-rendszer fenntartásának kérdéséhez" [On the question of the management of the Ottoman and Hungarian border defense systems in Hungary], *Keletkutatás* (Spring 1995): 61–86.

Pálffy 1996 Pálffy, Géza. "Egy Zala megyei település nevének keletkezéstörténete és eddig ismeretlen XVI. századi névadója. (Kilimán falu és Andreas Kielman von Kielmansegg)" [The history of the formation of the name of the village in Zala County and its unknown 16th-century eponym (Kilimán village and Andreas Kielman von Kielmansegg)], *Magyar Nyelv* 92 (1996): 163–174.

Pálffy 1999 Pálffy, Géza. *A császárváros védelmében: a győri főkapitányság története, 1526–1598* [In defense of the imperial city: The captainship of Győr, 1526–1598]. Győr: Győr-Moson-Sopron Megye Győri Levéltára, 1999.

Pálffy 2002 Pálffy, Géza. "A Bajcsavárig vezető út. A stájer rendek részvétele a Dél-Dunántúl határvédelmében a 16. században" [The path to Bajcsavár. The participation of the Styrian estates in the frontier protection of Southern Transdanubia]. In *Weitschawar/Bajcsa-Vár. Egy stájer erődítmény Magyarországon a 16. század második felében* [Weitschawar/Bajcsa-Vár. A Styrian fortification in Hungary in the second half of the 16th century], ed. Gyöngyi Kovács, 11–26. Zalaegerszeg: Zala Megyei Múzeumok Igazgatósága, 2002.

Ráth 1865 Ráth, György. "A bodonhelyi erősített kastély Sopron megyében" [The fortified palace of Bodonhely in Sopron County], *Győri Történeti és Régészeti Füzetek* 3 (1865): 276–280.

Sági 1960 Sági, Károly. "A zalacsányi török kori várak" [Ottoman-period castle of Zalacsány]. In *A Göcseji Múzeum jubileumi emlékkönyve 1950–1960* [Jubilee volume of the Göcsej Museum] (A Göcseji Múzeum Közleményei, 8), ed. Imre Szentmihályi, 131–135. Zalaegerszeg: Göcseji Múzeum, 1960.

Simon 1997 Simon, Éva. "Magyar nagybirtokosok tervezetei a Kanizsával szembeni végvidék kiépítéséről" [The plans of the Hungarian aristocracy on the construction of the border defense line against Kanizsa]. In *Zalai történeti tanulmányok* [Studies in the history of Zala County] (Zalai gyűjtemény, 42), ed. Csaba Káli, 61–86. Zalaegerszeg: Zala Megyei Levéltár, 1997.

Sorok 1997 Sorok, János. *Zalabér története* [The history of Zalabér] (Zalai kismonográfiák, 3), ed. Károly Ruzsa. Zalaegerszeg: Zala Megyei Múzeumok Igazgatósága – Zala Megyei Levéltár, 1997.

Sugár 1976 Sugár, István. *Szigetvár és viadala* [Szigetvár and its siege]. Budapest: Zrínyi, 1976.

Szabó–Csányi 2012 Szabó, Géza, and Viktor Csányi. "Werbőczy két Tolna megyei vára: Dombó és Döbrököz az újabb régészeti megfigyelések tükrében" [Two castles of Werbőczy in Tolna County: Dombó and Döbrököz in the light of recent archaeological observations], *A Wosinsky Mór Múzeum Évkönyve* 34 (2012): 179–212.

Szakály 2001 Szakály, Ferenc. "Kaposvár a török időkben" [Kaposvár in the Ottoman period]. In *Kaposvár. Várostörténeti tanulmányok* [Kaposvár. Studies in the history of the town], ed. József Kanyar, 93–138. Kaposvár: Kaposvár Városi Tanács Végrehajtó Bizottság, 1975.

Szántó 1980 Szántó, Imre. *A végvári rendszer kiépítése és fénykora Magyarországon, 1541–1593* [The construction and heyday of the frontier fortification system in Hungary, 1541–1593]. Budapest: Akadémiai, 1980.

Szatlóczki 2002 Szatlóczki, Gábor. "Szentgyörgyvár a török időkben" [Szentgyörgyvár in the Ottoman times]. In *Szentgyörgyvár története* [The history of Szentgyörgyvár] (Zalai Kismonográfiák, 7), ed. Róbert Müller, 33–48. Zalaegerszeg: Zala Megyei Múzeumok Igazgatósága, 2002.

Szita 1993 Szita, László. "Az 1687. évi török elleni hadjárat a haditudósítások tükrében" [The anti-Ottoman military campaign of 1687 in the light of war correspondences]. In *Somogy megye múltjából. Levéltári Évkönyv* [From the history of Somogy County. Archival yearbook], ed. Ferenc Szili, 47–76. Kaposvár: Somogy Megyei Levéltár, 1993.

Szita 1999 Szita, László. "Pécs és régiójának felszabadulása a török megszállás alól" [The liberation of Pécs and its region from the Ottoman occupation]. In *Tanulmányok Pécs történetéből*, vol. 7 [Studies in the history of Pécs, 7], ed. Ferenc Szakály, 305–322. Pécs: Pécs Története Alapítvány, 1999. (2nd ed. Pécs: Kronosz Kiadó – Pécs Története Alapítvány, 2012.)

Szvitek 2008 Szvitek, Róbert József. "Kiskomárom végvár szerepe a dél-dunántúli védelmi rendszerben" [The role of the fortification of Kiskomárom in the defense system of Southern Transdanubia]. PhD diss., Eötvös Loránd Tudományegyetem, 2008.

Takáts 1915 Takáts, Sándor. *Rajzok a török világból*, vol. 2 [Sketches from the Ottoman world, II]. Budapest: Magyar Tudományos Akadémia, 1915.

Toifl 2002 Toifl, Leopold. "Bajcsavár története a stájer levéltári források alapján" [The history of Bajcsavár in the light of Styrian archival data]. In *Weitschawar/Bajcsa-Vár. Egy stájer erődítmény Magyarországon a 16. század második felében* [Weitschawar/Bajcsa-Vár. A Styrian fortification in Hungary in the second half of the 16th century], ed. Gyöngyi Kovács, 27–40. Zalaegerszeg: Zala Megyei Múzeumok Igazgatósága, 2002.

Tolnai 2011 Tolnai, Gergely. *Palánkvárak Magyarországon* [Earth and wood fortifications in Hungary]. Budapest: Martin Opitz, 2011.

Vadas 2013 Vadas, András. *Körmend és a vizek. Egy település és környezete a kora újkorban* [Körmend and the waters. A settlement and its environment in the early modern period] (ELTE BTK Történelemtudományok Doktori Iskola. Tanulmányok – konferenciák, 5). Budapest: Történelemtudományok Doktori Iskola, 2013.

Vadas 2014 Vadas, András. "Vízgazdálkodás és háborús védekezés. Csákány és a Vas megyei Rába-mente a kora újkorban (1600–1658)" [Water management and border protection at Csákány along the Rába is Vas County between 1600 and 1658]. In *Víz és társadalom Magyarországon a középkortól a XX. század végéig* [Water and society in Hungary from the Middle Ages to the end of the 20th century], ed. Gergely Krisztián Horváth, 207–245. Budapest: Balassi, 2014.

Vándor 1996 Vándor, László. "A zalai végvárrendszer a 16–17. században" [The frontier fortification system of Zala County in the 16th–17th centuries]. In *Zala megye ezer éve. Tanulmánykötet a magyar államalapítás milleniumának tiszteletére* [Millennium of Zala County. Studies in honor of the millennium of the Hungarian state foundations], ed. idem, 89–96. Zalaegerszeg: Zala Megyei Múzeumok Igazgatósága, 1996.

Vass 1993 Vass, Előd. "Szigetvár város és a szigetvári szandzsák jelentősége az Oszmán-Török Birodalomban 1566–1689" [The role of the town of Szigetvár and the sanjak of Szigetvár in the Ottoman Empire, 1566–1689]. In *Tanulmányok a török hódoltság és a felszabadító háborúk történetéből. A szigetvári történész konferencia előadásai a város és vár felszabadításának 300. évfordulóján, 1989* [Studies in the history of Ottoman Hungary and the liberation wars. Proceedings of the historical conference held at Szigetvár on the 300th anniversary of the liberation of the town, 1989], ed. László Szita, 193–218. Pécs: Baranya Megyei Levéltár and Magyar Történelmi Társulat Dél-dunántúli Csoportja, 1993.

Végh 2005 Végh, Ferenc. "Keszthely praesidiális mezőváros térszerkezete és látképe a XVII. század második felében. A történeti városkép-rekonstrukció alternatív módszerei" [The spatial organization and image of the fortified market town of Keszthely in the second half of the 17th century], *Castrum. A Castrum Bene Egyesület Hírlevele* 1, no. 1 (2005): 31–46.

Végh 2007 Végh, Ferenc. *Birodalmak határán – Balaton partján: Keszthely végvárváros a XVI–XVII. században* [On the border of Empires – by the Lake Balaton: Keszthely the fortified frontier town in the 16th–17th centuries] (Disszertációk a társadalomtudományok köréből, 1). Budapest: Históriaantik Könyvesház, 2007.

Végh 2010 Végh, Ferenc. *Egerszeg végvár és város a 17. században* [The borderline fortification of Egerszeg and the town in the 17th century] (Zalaegerszegi füzetek, 10). Zalaegerszeg: Millecentenáriumi Közalapítány: 2010.

Bibliography

Archival Sources

Batthyanyisches Herrschaftsarchiv Güssing [BHG] (Austria)

Hadtörténeti Intézet és Múzeum [HIM] [Military History Institute and Museum] (Budapest, Hungary)
B IX a – Osztrák Birodalom, Osztrák-Magyar Monarchia, Magyarország általános politikai, topográfiai térképei, valamint Magyarország megye térképei [General political, topographic maps of Hungary and the Austro-Hungarian Monarchy, and county maps of Hungary]

Landesarchiv Baden-Württemberg: Generallandesarchiv Karlsruhe [LBW GK] (Germany)
HFK – Hausfideikommission

Magyar Nemzeti Levéltár Országos Levéltára [MNL OL] [State Archive of the Hungarian National Archives] (Budapest, Hungary)
Magyar Kancelláriai Levéltár [Archives of the Hungarian Chancery]
A 14 – Insinuata Consilii Bellici
Magyar Kamara Archívuma [Archives of the Hungarian Chamber]
E 15 – Expeditiones camerales
E 21 – Benignae resolutiones
E 41 – Litterae ad cameram exaratae
E 142 – Acta Publica
E 156 – Urbaria et Conscriptiones
E 185 – Archivum Familiae Nádasdy
E 211 – Lymbus

Családi Levéltárak [Family archives]
P 107 – Erdődy család vöröskői levéltára [Archive of the Erdődy family at Červený Kameň]
P 632 – Szentiványi család levéltára. A család által rendezett iratok [Archive of the Szentiványi Family. Documents arranged by the family]
P 1313 – A herceg Batthyány család levéltára, a Batthyány család törzslevéltára [Archive of the Batthyány family. Central archive of the family]

P 1314 – A herceg Batthyány család levéltára, Missiles [Archive of the Batthyány family. Letters]
P 1322 – A herceg Batthyány család levéltára, a Batthyány család körmendi központi igazgatósága [Archive of the Batthyány family. Central administration of Körmend of the family]
Collectio Antemohácsiana
DL – Diplomatikai Levéltár [Collection of diplomatic documents]
Collection of diplomatic photographs
DF – Diplomatikai Fényképgyűjtemény [Collection of diplomatic photographs]
Map collection
S 12 – Helytartótanácsi térképek [Maps of the Consilium Regium Locumtenentiale Hungaricum]
S 78 – Kataszteri térképek [Cadastral maps]
Collection of microfiches
X 767 – Canonica visitationes
X 5136 – Insinuata Consilii Bellici

Magyar Nemzeti Levéltár Vas Megyei Levéltára [MNL VML] [Hungarian National Archive. Vas County Archive] (Szombathely, Hungary)
XII.1. a. – Vasvár–Szombathelyi Székeskáptalan Hiteleshelyi Levéltára. Jegyzőkönyvek [Archive of the Vasvár-Szombathely cathedral chapter's place of authenticity. Minutes]

Magyar Nemzeti Levéltár Szabolcs-Szatmár-Bereg Megyei Levéltára [MNL SzSzBML] [Hungarian National Archive. Szabolcs-Szatmár-Bereg County Archive] (Nyíregyháza, Hungary)
IV. – Megyei törvényhatóságok [County authorities]

Országos Széchényi Könyvtár [Széchényi National Library] [OSZK] (Budapest, Hungary)
Kézirattár [Collection of Manuscripts] [K]
App. M. – Apponyi Metszet [Etchings of the Apponyi Collection]
Fol. Lat. – Folia Latina
Térképtár [Map collection] [T]

Österreichisches Staatsarchiv [ÖSta] (Vienna, Austria)
Finanz- und Hofkammerarchiv [FHKA]
Sammlungen und Selekte Patente

Haus-, Hof- und Staatsarchiv [HHSta]
Csáky – Familienarchiv Csáky at Szepesmindszent
Erdődy – Familienarchiv Erdődy
Kriegsarchiv [KA]
AFA – Alte Feldakten
HKR – Akten des Wiener Hofkriegsrates

Rába Helytörténeti Múzeum Adattára [RHMA] [Data collection of the Rába Local History Museum] (Körmend, Hungary)

Slovenský Národný Archív [SNA] [National Archive of Slovakia] (Bratislava, Slovakia)
Esterházy – Archív česneckej línie rodu Esterházi [Archive of the Csesznek branch of the Esterházy family]

Szombathelyi Egyházmegyei Könyvtár [SZEK] [Szombathely Diocesan Library] (Hungary)
Can. vis. – Canonica visitationes

Printed Primary Sources

Anon., "Regéczi uradalmi erdőrendtartás" [Forest-use instruction from the domain of Regéc], *Magyar Gazdaságtörténelmi Szemle* 5 (1899): 462–464.

Bártfai Szabó, László. *Pest megye történetének okleveles emlékei, 1002–1599* [The history of Pest County in documentary sources, 1002–1599]. Budapest: [author's edition], 1938.

Bel, Matthias. *Notitia Hungariae novae historico geographica*, 6 vols., eds. Tóth Gregorius, Ladislaus Glück, Bernadett Benei, Zoltanus Gőzsy, and Rudolphus Jarmalov. Budapest: Magyar Országos Levéltár and Magyar Tudományos Akadémia Történettudományi Intézete, 2011–2020.

Bél, Mátyás. "A kunok és jászok avagy filiszteusok kerületei" [District of the Cumans, and Jazyians or Phylistines], eds. Bálint Illyés and Rudolf Szőts. In *Bács-Kiskun megye múltjából*, vol. 1 [From the past of Bács-Kiskun County, vol. 1], ed. Tibor Iványosi-Szabó, 7–52. Kecskemét: Bács-Kiskun Megyei Levéltár, 1975.

Fejér, Georgius, ed. *Codex diplomaticus Hungariae ecclesiasticus ac civilis*, 11 vols. in 44 parts. Buda: Typis Typogr. Regia Universitatis Ungaricae, 1829–1844.

Fekete, Lajos. *A hódoltság török levéltári forrásai nyomában* [In pursuit of the archival sources of Ottoman Hungary], ed. Géza Dávid (Budapest Oriental

Reprints, Ser., A 6). Budapest: Körösi Csorna Társaság and Magyar Tudományos Akadémia Könyvtára, 1993.

Fraknói, Vilmos, ed. *Magyar országgyűlési emlékek történeti bevezetésekkel*, III. (1546–1556.) [Hungarian dietary records with introductions] (Magyar történelmi emlékek 3. osztály. Országgyűlési emlékek, 3). Budapest: Ráth and Magyar Tudományos Akadémia, 1876.

Házi, Jenő. *Sopron szabad királyi város története. II. rész, 2. kötet, Végrendeletek, közgyűlési jegyzőkönyvek, polgárkönyvi feljegyzések és különféle számadások 1400-tól 1541-ig* [History of the free royal town of Sopron. II/2. Last wills, town council minutes, town book entries and different account books]. Sopron: Székely és Társa Könyvnyomdája, 1931.

Heimler, Károly. *Payr György és Payr Mihály krónikája 1584–1700* [The chronicle of György Payr and Mihály Payr]. Sopron: Soproni Városszépítő Egyesület, 1942.

Hornyik, János. *Kecskemét város története, oklevéltárral*, vol. 2 [History of the town of Kecskemét, with a cartulary]. Kecskemét: [author's edition], 1864.

Ila, Bálint. "Körmend város 1649. évi összeírása" [Conscription of the town of Körmend from 1649], *Ethnographia* 78 (1967): 556–568.

Istvánffy, Miklós. *Magyarok dolgairól írt históriája Tállyai Pál XVII. századi fordításában*, vol. I/2 [The history of Miklós Istvánffy in the 17th-century translation of Pál Tállyai], ed. Péter Benits. Budapest: Balassi, 2009.

Iványosi-Szabó, Tibor, ed. *Kecskeméti szabályrendeletek 1659–1849* [Kecskemét town statutes]. Kecskemét: Bács-Kiskun Megyei Levéltár, 1991.

Káldy-Nagy, Gyula. *A csanádi szandzsák 1567. és 1579. évi összeírása* [Conscription of the Sanjak of Cenad, 1567, and 1579] (Dél-aföldi évszázadok, 15). Szeged: Csongrád Megyei Levéltár, 2000.

———. *A szegedi szandzsák települései, lakosai és török birtokosai 1570-ben* [The settlements, population, and Ottoman owners of the Sanjak of Szeged in 1570] (Dél-alföldi évszázadok, 24). Szeged: Csongrád Megyei Levéltár, 2008.

Karácson, Imre, ed. *Evlia Cselebi török világutazó magyarországi utazásai, 1660–1664* [The travels of Evliya Çelebi in Hungary, 1660–1664]. Budapest: Magyar Tudományos Akadémia, 1904.

Kenyeres, István and Péter Kis, eds. *XVI. századi uradalmi utasítások. Utasítások a kamarai uradalmak prefektusai, udvarbírái és ellenőrei részére*, 2 vols. [16th-century estate instructions. Instructions to the administrators of the estates of the Hungarian Chamber] (Fons Könyvek, 2). Budapest: Szentpétery Imre Történettudományi Alapítvány, 2002.

Kisari Balla, György. *Karlsruhei térképek a török háborúk korából / Kriegskarten und Pläne aus der Türkkenzeit in Karlsruher Sammlungen*. Budapest: Kisari Balla György, 2000.

———. *Marsigli tábornok térképei / Le mappe del Generale Marsigli*. Budapest: [n. p.], 2005.

Kondicsné Kovács, Éva, ed. *Szemelvénygyűjtemény Körmend történetének tanulmányozásához*, 2 vols. [Sources to the history of Körmend]. Körmend: Jubileumi Bizottság, 1993–2002.

Mályusz, Elemér, Iván Borsa, Norbert C. Tóth, Tibor Neumann, Bálint Lakatos, and Gábor Mikó, eds. *Zsigmondkori oklevéltár*, 14 vols. [1387–1427] [Sigismundian cartulary] [A Magyar Országos Levéltár kiadványai II. Forráskiadványok 1, 3–4, 22, 25, 27, 32, 37, 39, 41, 43, 49, 52, 55, 59], Budapest: Akadémiai and Magyar Országos Levéltár, 1951–2020.

Mályusz, Elemér. "Az Országos Levéltár Nádasdy-levéltárának magyar levelei: IV. közlemény" [Hungarian-language letters of the Nádasdy archive at the Hungarian National Archive], *Levéltári Közlemények* 3, no. 1 (1925): 68–85.

Márkus, Dezső, ed. *Magyar törvénytár. 1608–1657 évi törvényczikkek* [Corpus Juris Hungarici. Laws of Hungary]. Budapest: Franklin, 1900.

Mátyás-Rausch, Petra. "Az ismeretlen földjén – Thorda Zsigmond jelentése a liptói bányászatról (1560)" [On the land of the unknown – The report of Zsigmond Thorda on mining in Liptó County], *Lymbus – Magyarságtudományi Forrásközlemények* 17 (2019): 29–84.

Merényi, Lajos. "Kismartoni uradalmi faárszabás 1632-ből" [Price limitations of timber and wood from the domain of Eisenstadt from 1632], *Magyar Gazdaságtörténelmi Szemle* 12 (1905): 280.

Nagy, Imre. *Sopron vármegye története. Oklevéltár.* I. kötet: 1156–1411 [History of Sopron County. Cartulary]. Sopron: Litfass Károly Könyvnyomdája, 1889.

Nagy, Imre, Iván Nagy, and Dezső Véghely, eds. *A zichy és vásonkeői gróf Zichy-család idősb ágának okmánytára*, 12 vols. [Cartulary of the ancient branch of the Zichy family of Zich and Vásonkeő]. Pest: Magyar Történelmi Társulat, 1871–1931.

Olahus, Nicolaus. *Hungaria – Athila* (Bibliotheca Scriptorum Medii Recentisque Aevorum. Saeculum XVI), eds. Colomannus Eperjessy and Ladislaus Juhász. Budapest: K.M. Egyetemi Nyomda, 1938.

Őze, Sándor, ed. *500 magyar levél a XVI. századból*, 2 vols. [500 letters from the 16th century]. Budapest: MNM, 1996.

Schefer, Charles, ed. *Le voyage d'outremer de Bertrandon de la Broquiere, premier écuyer tranchant et conseiller de Philippe le Bon, duc de Bourgogne* (Recueil de voyages et de documents pour servir à l'histoire de la géographie depuis le XIIIe jusqu'à la fin du XVIe siècle, 12). Paris: Ernest Leroux, 1892.

Szalay, Ágoston, ed. *Négyszáz magyar levél a XVI. századból, 1504–1560* [Four-hundred letters from the 16th century]. Pest: [Landerer és Heckenast], 1861.

Tagányi, Károly, ed. *Magyar erdészeti oklevéltár*, 3 vols. [Cartulary of Hungarian forestry]. Budapest: Pátria, 1896.

Thallóczy, Lajos and Antal Hodinka, eds. *Magyarország melléktartományainak oklevéltára = Codex diplomaticus partium Regno Hungariae adnexarum*

(Monumenta Hungariae historica. Első osztály: okmánytárak, 31). Budapest: Magyar Tudományos Akadémia, 1903.

Thaly, Kálmán. "Az Rába rectificatiójáról való determinatió, 1699" [Decision on the regulation of the Rába, 1699], *Történelmi Tár* 8 (1885): 154–164.

Tóth, Péter, ed. *Vas vármegye közgyűlési jegyzőkönyveinek regesztái*, vol. II. 1601–1620., 1631–1641 [Regestas of the noble gathering of Vas County] (Vas megyei levéltár füzetek, 5). Szombathely: Vas Megyei Levéltár, 1992.

———, ed. *Sopron vármegye közgyűlési jegyzőkönyveinek regesztái*, vol. I. 1579–1589 [Regestas of the noble gathering of Sopron County]. Sopron: Sopron Város Levéltára, 1994.

Turbuly, Éva, ed. *Sopron vármegye közgyűlési jegyzőkönyveinek regesztái*, vol. II. 1595–1608 [Regestas of the noble gathering of Sopron County]. Sopron: Győr-Moson-Sopron Megyei Levéltár Soproni Levéltára, 2002.

Von Kunits, Michael. *Topographische Beschreibungen des Königreiches Ungarn und seiner einverleibten Provinzen*. Pesth: Ludwig Landerer Edlen v. Fűskút, 1824.

Wagner, Hans and Irmtraut Lindeck-Pozza, eds. *Urkundenbuch des Burgenlandes und der angrenzenden Gebiete der Komitate Wieselburg, Ödenburg und Eisenburg*, 3 vols. (Publikationen des Instituts für Österreichische Geschichtsforschung, 7). Graz and Cologne: Böhlau, 1955–1979.

Wenzel, Gusztáv, ed. *Árpádkori új okmánytár / Codex Diplomaticus Arpadianus continuatus*, 12 vols. Pest: Eggenberger, 1860–1874.

Zivuska, Jenő. *A besztercebányai m. kir. erdőigazgatóság okiratainak tartalomjegyzéke* [Summary of the documents of the Hungarian Royal Forestry of Banská Bystrica]. Banská Bystrica: Magyar Királyi Erdőigazgatóság 1906.

Secondary Literature

Abulafia, David and Nora Berend, eds. *Medieval Frontiers: Concepts and Practices*. Burlington, VT: Routledge, 2002.

Ágoston, Gábor. *A hódolt Magyarország* [Ottoman-period Hungary]. [n.p.]: Adams Kiadó, 1992.

———. "Ottoman Conquest and the Ottoman Military Frontier in Hungary." In *A Millennium of Hungarian Military History* (Atlantic Studies on Society in Change, 114 = East European Monographs, 621 = War and Society in East Central Europe, 37), eds. Béla Király and László Veszprémy, 85–110. Boulder, CO: Atlantic Research and Publications, 2002.

———. *Guns for the Sultan: Technology, Industry, and Military Power in the Ottoman Empire*. Cambridge: Cambridge University Press, 2004.

———. "Birodalom és információ: Konstantinápoly, mint a koraújkori Európa információs központja" [Empire and information: Constantinople as the information center of early modern Europe]. In *Az értelem bátorsága. Tanulmányok Perjés Géza emlékére* [The courage of intellect. Essays offered in memory of Géza Perjés], ed. Gábor Hausner, 31–60. Budapest: Argumentum, 2005.

———. "Mohács." In *The Seventy Great Battles of All Time*, ed. Jeremy Black, 100–112. London: Thames & Hudson, 2005.

———. "Information, Ideology, and Limits of Imperial Policy: Ottoman Grand Strategy in the Context of Ottoman-Habsburg Rivalry." In *The Early Modern Ottomans: Remapping the Empire*, eds. Virginia H. Aksan and Daniel Goffman, 75–103. Cambridge: Cambridge University Press, 2007.

———. "Where Environmental and Frontier Studies Meet: Rivers, Forests, Marshes and Forts along the Ottoman–Hapsburg Frontier in Hungary." In *The Frontiers of the Ottoman World* (Proceedings of the British Academy, 156), ed. Andrew C.S. Peacock, 57–79. Oxford: Oxford University Press, 2009.

———. "Defending and Administering the Frontier: The Case of Ottoman Hungary." In *The Ottoman World*, ed. Christine Woodhead, 220–236. Milton Park, Abingdon, and Oxon: Routledge, 2012.

———. "The Ottoman Empire and Europe." In *The Oxford Handbook of Early Modern European History: 1350–1750. Volume II. Cultures and Power*, ed. Scott M. Hamish, 612–637. Oxford: Oxford University Press, 2015.

Ágoston, Gábor and Teréz Oborni. *A tizenhetedik század története* [History of the 17th century] (Magyar századok). Budapest: Pannonica, 2000.

Albion, Robert G. *Forests and Sea Power: The Timber Problem of the Royal Navy, 1652–1862*. Cambridge, MA: Harvard University Press, 1936.

Almási, Gábor. *The Uses of Humanism: Johannes Sambucus (1531–1584), Andreas Dudith (1533–1589), and the Republic of Letters in East Central Europe* (Brill's Studies in Intellectual History, 185). Leiden and Boston: Brill, 2010.

Ambrózy, Pál. "A Felső-Rába vízgyűjtőjének éghajlati jellegzetességei" [Climatic specificities of the catchment area of the Upper Rába], *Vízügyi Közlemények* 79 (1997): 498–517.

Andrásfalvy, Bertalan. *A Duna mente népének ártéri gazdálkodása Tolna és Baranya megyében az ármentesítés befejezéséig* [The flood plain economy of the peoples of the Interl valley in Tolna and Baranya counties before the completion of the regulation works] (Tanulmányok Tolna megye történetéből, 7). Szekszárd: Tolna Megyei Levéltár, 1975.

———. *A Duna mente népének ártéri gazdálkodása: ártéri gazdálkodás Tolna és Baranya megyében az ármentesítési munkák befejezése előtt* [The flood plain economy of the peoples of the Danube valley in Tolna and Baranya Counties before the completion of the regulation works]. [Budakeszi]: Ekvilibrium, 2007.

Anon. "A haltermelés és -fogyasztás alakulása a világon és Magyarországon" [The fish production and consumption in the World and Hungary], *Statisztikai Tükör*

6, no. 84 (2013): 1–4. Online document: https://www.ksh.hu/docs/hun/xftp/stattukor/halaszat.pdf (last accessed: 3 April 2020)

Appuhn, Karl. "Inventing Nature: Forests, Forestry, and State Power in Renaissance Venice," *The Journal of Modern History* 72 (2000): 851–889.

———. *A Forest on the Sea: Environmental Expertise in Renaissance Venice.* Baltimore, MD: Johns Hopkins University Press, 2009.

Arroyo, Felix Labrador and Koldo Trápaga Monchet. "Forestry, Territorial Organization, and Military Struggle in the Early Modern Spanish Monarchy," *Environmental History* 23 (2018): 318–341.

Bagi, Zoltán. "Egy kudarc okai: Kanizsa 1601. évi ostroma" [Reasons of a failure: The siege of Kanizsa in 1601], *Aetas* 28, no. 1 (2013): 5–30.

Bagi, Zoltán Péter. "A folyóvíz, a csapadék és az áradások mint a hadakozást befolyásoló tényezők a tizenöt éves háború időszakában" [Rivers, precipitation, and floods as agents of military campaigns in the period of the Fifteen Years' War]. In *Víz és társadalom Magyarországon a középkortól a XX. század végéig* [Water and society in Hungary from the Middle Ages to the end of the 20th century], ed. Gergely Krisztián Horváth, 189–206. Budapest: Balassi, 2014.

———. "The Life of Soldiers during the Long Turkish War (1593–1606)," *The Hungarian Historical Review* 4 (2015): 384–417.

Bajdó, Bettina. "Előkerült a régi malom." [The old mill has been discovered] [interview with the archaeologist, Gyula Siklósi]. Online document: http://www.vasnepe.hu/ fokuszban/20070615_elokerult_a_regi_malom/print (last accessed: 1 May 2020)

Bak, János M. "Servitude in the Medieval Kingdom of Hungary." In *Forms of Servitude in Northern and Central Europe Decline, Resistance, and Expansion* (Medieval Texts and Cultures of Northern Europe, 9), eds. Paul Freedman and Monique Bourin, 387–400. Turnhout: Brepols, 2005.

Bálint, Marianna. "Az Árpád-kori településhálózat rekonstrukciója a Dorozsma-Majsai Homokhát területén" [Reconstruction of the Árpádian-Age Settlement Network in the Dorozsma-Majsa Sandy Ridge]. PhD diss., Eötvös Loránd Tudományegyetem, 2006.

Balogh, Lajos and Ferenc Ördög, eds. *Győr-Moson-Sopron megye földrajzi nevei 1. A kapuvári járás* [Place names in Győr-Moson-Sopron County, I. The surroundings of Kapuvár]. Győr: Apáczai Csere János Tanítóképző Főiskola, 1998.

Bamford, Paul. *Forests and French Sea Power, 1660–1789*. Toronto: Toronto University Press, 1956.

Bándi, Zsuzsanna. *Körmend a középkorban* [Körmend in the Middle Ages]. Körmend: Körmend Város Tanácsa Végrehajtó Bizottsága, 1987.

Bankoff, Greg. "Wood for War: The Legacy of Human Conflict on the Forests of the Philippines, 1600–1946." In *War and the Environment: Military Destruction in the Modern Age*, ed. Charles Closmann, 32–48. College Station, TX: Texas A&M University Press, 2009.

Baráth, Zsolt. "A Rába mint védelmi vonal a 17. században. Védelmi munkálatok és létesítmények a folyó Vas megyei szakaszán" [Rába as a defense line – protection measures and constructions in the Vas County-section of the river], *Vasi Honismereti és Helytörténeti Közlemények* no. 1 (2014): 31–53.

———. "A Rába védelmi vonal és Szentgotthárd térsége a 17. század derekán" [The Rába defense line and the surroundings of Szentgotthárd in the mid-17th century]. In *1664 – A szentgotthárdi csata* [1664 – The battle of Szentgotthárd], 257–295. Szombathely: Szülőföld Kiadó, 2015.

———. "Oszmán terjeszkedés a Dunántúlon és az 1594. évi Rába felmérés" [Ottoman expansion at the Transdanubia and the 1594 Rába River survey], *Győri Tanulmányok* 39 (2018): 53–76.

Barca, Stefania. *Enclosing Water: Nature and Political Economy in a Mediterranean Valley, 1796–1916*. Cambridge: White Horse Press, 2010.

Bartha, Dénes. "Történeti erdőhasználatok Magyarországon" [Historical forest exploitation in Hungary], *Magyar Tudomány* 48 (2003): 1566–1577.

Bartha, Dénes and Sándor Oroszi. "Magyar erdők" [Hungarian forests]. In *Pannon enciklopédia. Magyarország növényvilága* [Pannonian encyclopedia. Flora of Hungary], ed. Magda Járainé Komlódi, 221–231. Budapest: Dunakanyar 2000 Könyvkiadó, 2000.

Bartholy, Judit, Rita Pongrácz, and Zsófia Molnár. "Classification and Analysis of Past Climate Information based on Historical Documentary Sources for the Carpathian Basin," *International Journal of Climatology* 24 (2004): 1759–1776.

———. "Extremes and Millennial Trends in the Carpathian Basin Using the Rethly Documentary Collection," *Bulletin of the American Meteorological Society* 12 (2004): 3791–3802.

Bartosiewicz, László. *Animals in the Urban Landscape in the Wake of the Middle Ages: A Case Study from Vác* (BAR International Series, 609). Oxford: Tempus Reparatum, 1995.

Bartosiewicz, László and Erika Gál. "Ottoman Period Animal Exploitation in Hungary." In *Archeology of the Ottoman Period in Hungary* (Opuscula Hungarica, 3), eds. Ibolya Gerelyes and Gyöngyi Kovács, 365–376. Budapest: Magyar Nemzeti Múzeum, 2004.

Batizi, Zoltán. "Mining in Medieval Hungary." In *The Economy of Medieval Hungary* (East Central and Eastern Europe in the Middle Ages, 450–1450, 49), eds. József Laszlovszky et al., 166–181. Leiden and Boston: Brill, 2018.

Batta, István. "Középkori bányászatunk és kohászatunk a Metercián" [Mining and metallurgy in medieval Hungary as represented on the *Metercia*], *Bányászati és Kohászati Lapok. Kohászat* 121 (1988): 277–285.

Bechmann, Roland. *Trees and Man: The Forest in the Middle Ages*. Translated by Katharyn Duncan. New York: Random House, 1990.

Belitzky, János. *Sopron vármegye története*, vol. 1 [The history of Sopron County, 1]. Budapest: Stephaneum, 1938.

Bello, David A. *Across Forest, Steppe, and Mountain: Environment, Identity, and Empire in Qing China's Borderlands*. Cambridge: Cambridge University Press, 2016.

Bellon, Tibor. *A Tisza néprajza. Ártéri gazdálkodás a tiszai Alföldön* [The ethnography of Tisza. Flood plain economy at the Tisza in the Great Hungarian Plain]. Budapest: Timp, 2003.

Benczédi, László. "A Habsburg-abszolutizmus indítékai és megvalósulása az 1670-es évek Magyarországán" [The motivations and execution of Habsburg absolutism in Hungary in the 1670s], *Történelmi Szemle* 21 (1978): 535–556.

Benczik, Gyula. "Fentő és góré. A hódoltság kori védelmi munkálatok emlékei a Felső-Rábavidék dűlőneveiben" [*Fentő* and *góré*. The memory of the Ottoman-period defensive works in the field names of the Upper Rába Region], *Vasi Honismereti és Helytörténeti Közlemények* no. 1 (1995): 42–48.

Benda, Borbála. *Étkezesi szokások a magyar főúri udvarokban a kora újkorban* [Dietary practices in Hungarian aristocratic courts in the early modern period] (Archivum Comitatus Castriferrei, 6). Szombathely: Magyar Nemzeti Levéltár Vas Megyei Levéltára, 2014.

Bende, Lajos. "A törökkori magyar végvárakról" [On Ottoman-period Hungarian borderline fortifications], *Hadtörténelmi Közlemények* 18, no. 3 (1971): 501–527.

Benkő, Elek. *A középkori Székelyföld*, 2 vols. [Székely Lands in the Middle Ages]. Budapest: MTA Bölcsészettudományi Kutatóközpont Régészeti Intézet, 2012.

Benkő, Elek and Teréz Oborni, eds. *Székelyföld története*, 3 vols. [History of the Székely Lands]. Székelyudvarhely: MTA Bölcsészettudományi Kutatóintézet, Erdélyi Múzeum-Egyesület, and Haáz Rezső Múzeum, 2016.

Berend, Nora. "Preface." In *Medieval Frontiers: Concepts and Practices*, eds. David Abulafia and Nora Berend, x–xv. Burlington, VT: Routledge, 2002.

Bergmann, Heinz et al. *Hydrologische Monographie des Einzugsgebietes der Oberen Raab – A Felső-Rába vízgyűjtőjének hidrológiai monográfiája* (Schriftenreihe zur Wasserwirtschaft, 23). Graz and Budapest: Technisches Universität Graz, 1996.

Biró, Marianna. "A történeti térképekre alapuló vegetációrekonstrukció és alkalmazásai a Duna-Tisza közén" [The possibilities and applications of using historical maps in vegetation reconstructions in the Danube-Tisza Interfluve area]. PhD diss., Pécsi Tudományegyetem, 2006.

Bíró, Mariann and Zsolt Molnár. "Az Alföld erdei a folyószabályozások és az alföldfásítás előtti évszázadban." [The forests of the Great Hungarian Plain in the century before the river regulations and the reforestation] In *Környezettörténet.*

Az utóbbi 500 év környezeti eseményei történeti és természettudományi források tükrében [Environmental history. The environmental events of the last 500 years in the light of historical and scientific data] (Környezettörténet, 1), ed. Miklós Kázmér, 169–206. Budapest: Hantken, 2009.

Blanchard, Ian. "The Continental European Cattle Trades, 1400–1600," *The Economic History Review* [New Series] 39 (1986): 427–460.

Born, Robert. "The Ottoman Expansion and the Development of Cartography in East-Central Europe (15th–18th Centuries)," *Revue des Études Sud-Est Européennes* 54 (2017): 121–152.

Bostan, Ídris. "A szultáni ágyúöntő műhelyben (Tophâne-i Âmire) folyó tevékenység a 16. század elején" [Cannon founding at the Sultans Foundry (Tophâne-i Âmire) at the beginning of the 16th century], *Aetas* 18, no. 2 (2003): 5–20.

Bothe, Jan Philipp. "How to 'Ravage' a Country. Destruction, Conservation and Assessment of Natural Environments in Early Modern Military Thought," *Hungarian Historical Review* 7 (2018): 510–540.

Brady, Lisa M. "Life in the DMZ: Turning a Diplomatic Failure into an Environmental Success," *Diplomatic History* 32 (2008): 585–611.

———. *War upon the Land: Military Strategy and the Transformation of Southern Landscapes during the American Civil War.* Foreword by Paul S. Sutter. Athens, GA: University of Georgia Press. 2012.

———. "War from the Ground Up: Integrating Military and Environmental Histories." In *A Field on Fire: The Future of Environmental History*, eds. Mark D. Hersey and Ted Steinberg, 250–262. Tuscaloosa, AL: The University of Alabama Press, 2019.

Brázdil, Rudolf et al. "Flood Events of Selected European Rivers in the Sixteenth Century," *Climatic Change* 43 (1999): 239–285.

———. "Documentary Data and the Study of Past Droughts: A Global State of the Art," *Climate of the Past* 14 (2018): 1915–1960.

Brázdil, Rudolf, Hubert Valášek, and Oldřich Kotyza. "Meteorological Records of Michel Stueler of Krupka and their Contribution to the Knowledge of the Climate of the Czech Lands in 1629–1649." In *Czech Geography at the Dawn of the Millenium*, ed. Dušan Drbohlav, 95–112. Olomouc: Palacký University, 2004.

Brown, Kate. *Manual for Survival: A Chernobyl Guide to the Future.* New York: W.W. Norton, 2020.

Brummett, Palmira. "The River Crossing: Breaking Points (Metaphorical and 'Real') in Ottoman Mutiny." In *Rebellion, Repression, Reinvention: Mutiny in Comparative Perspective*, ed. Jane Hathaway, 215–231. Westport, CT and London: Praeger, 2001.

———. *Mapping the Ottomans. Sovereignty, Territory, and Identity in the Early Modern Mediterranean.* Cambridge: Cambridge University Press, 2015.

Büntgen, Ulf et al. "Combined Dendro-Documentary Evidence of Central European Hydroclimatic Springtime Extremes over the Last Millennium," *Quaternary Science Reviews* 30 (2011): 3947–3959.

Büntgen, Ulf, Rudolf Brázdil, Petr Dobrovolný, Mirek Trnka, and Tomáš Kyncl. "Five Centuries of Southern Moravian Drought Variations Revealed from Living and Historic Tree Rings," *Theoretical and Applied Climatology* 105 (2011): 167–180.

Camenisch, Chantal et al., "The 1430s: A Cold Period of Extraordinary Internal Climate Variability during the Early Spörer Minimum with Social and Economic Impacts in North-Western and Central Europe," *Climate of the Past* 12 (2016): 2107–2126.

Campbell, Bruce M.S., James A. Galloway, Derek Keene, and Margaret Murphy. *A Medieval Capital and Its Grain Supply: Agrarian Production and Distribution in the London Region, c. 1300* (Historical Geography Research Series, 40). [n. p.]: Historical Geography Research Group, Institute of British Geographers, 1993.

Casale, Giancarlo, *The Ottoman Age of Exploration*. New York: Oxford University Press, 2010.

Castro, Fernando Domínguez, Juan I. Santisteban, Mariano Barriendos, and Rosa Mediavilla. "Reconstruction of Drought Episodes for Central Spain from Rogation Ceremonies Recorded at the Toledo Cathedral from 1506 to 1900: A Methodological Approach," *Global and Planetary Change* 63 (2008): 230–242.

Crosby, Alfred W. *Children of the Sun. A History of Humanity's Unappeasable Appetite for Energy*. New York: W.W. Norton, 2006.

Csapody, István. "Sopron város (és volt úrbéres községei) erdeinek története (XII–XX. század)" [History of the forests of the town of Sopron (and its belonging villages), 12th to 20th centuries], *Erdészettörténeti Közlemények* no. 3–4 (1968): 3–16.

Cserményi, Vajk and Endre Tóth. "Der Abschnitt der Bernsteinstrasse in Ungarn," *Savaria* 16 (1982): 283–290.

Csiffáry, Gergely. "Végvárak és hadiipari létesítmények" [Borderline fortifications and military industrial complexes]. In *Végvár és ellátás* [Borderline fortifications and supply] (Studia Agriensia, 22), eds. Tivadar Petercsák and Mátyás Berecz, 107–129. Eger: Heves Megyei Múzeumi Szervezet, Dobó István Vármúzeum, 2001.

Curta, Florin. "Introduction." In *Borders, Barriers, and Ethnogenesis: Frontiers in Late Antiquity and the Middle Ages* (Studies in the Early Middle Ages, 12), ed. idem, 1–9. Turnhout: Brepols, 2010.

———. "Linear Frontiers in the 9th century: Bulgaria and Wessex," *Quaestiones Medii Aevi Novae* 16 (2011): 15–32.

Curtis, Daniel. *Coping with Crisis. The Resilience and Vulnerability of Pre-Industrial Settlements*. Burlington, VT: Ashgate, 2014.

Dávid, Géza. *A simontornyai szandzsák a 16. században* [The sanjak of Simontornya in the 16th century]. Budapest: Akadémiai, 1982.

———. "Magyarország népessége a XVI–XVII. században" [The population of Hungary in the 16th–17th centuries]. In *Magyarország történeti demográfiája (896–1996)* [Historical demography of Hungary], ed. József Kovacsics, 141–171. Budapest: KSH, 1997.

Dávid, Géza and Pál Fodor, eds. *Ransom Slavery along the Ottoman Borders (Early Fifteenth–Early Eighteenth Centuries)* (The Ottoman Empire and its Heritage: Politics, Society and Economy, 37). Leiden and Boston: Brill, 2007.

Degroot, Dagomar. "The Frigid Golden Age: Experiencing Climate Change in the Dutch Republic, 1560–1720." PhD diss., York University, 2014.

———. *The Frigid Golden Age: Climate Change, the Little Ice Age, and the Dutch Republic, 1560–1720*. Cambridge: Cambridge University Press, 2018.

Dénes, József. "Körmend a középkorban" [Körmend in the Middle Ages]. In *Körmend története* [The history of Körmend], ed. László Szabó, 32–98. Körmend: Önkormányzat, [1994].

Deryabina, T.G. et al. "Long-Term Census Data Reveal Abundant Wildlife Populations at Chernobyl," *Current Biology* 25, no. 19 (2015): R824–R826.

Détshy, Mihály. "A sárospataki ágyúöntőház története" [History of the cannon foundry of Sárospatak], *Technikatörténeti Szemle* 5 (1968–1970): 69–115.

Dobrovolný, Petr et al. "Monthly and Seasonal Temperature Reconstructions for Central Europe Derived from Documentary Evidence and Instrumental Records since AD 1500," *Climatic Change* 101 (2010): 69–107.

Dóka, Klára. "A Rába-szabályozás kérdése 1786-ban" [The question of the Rába-regulation in 1786], *Soproni Szemle* 30 (1976): 55–60.

———. "A Rába szabályozása, 1762–1895" [The regulation of the River Rába], *Technikatörténeti Szemle* 11 (1979): 85–101.

Dominkovits, Péter. *„Egy nemzetek lévén…" A Nyugat-Dunántúl Bocskai István 1605. évi hadjárata idején* ["Being one nation…" The Western Transdanubia in the period of István Bocskai's military campaign]. Budapest: Martin Opitz, 2006.

———. "Folyóvizek és a XVII. századi vármegyei közigazgatás, bíráskodás" [Rivers and 17th-century county administration and jurisdiction]. In *Víz és társadalom magyarországon a középkortól a XX. század végéig* [Water and society in Hungary from the Middle Ages to the end of the 20th century], ed. Gergely Krisztián Horváth, 155–188. Budapest: Balassi, 2014.

Domokos, György. "A kassai királyi hadszertár fegyverzete és felszerelése a XVI–XVII. századi inventáriumok tükrében" [Weapons and munitions of the royal armory of Košice based on inventories in the 16th and 17th centuries], *Hadtörténelmi Közlemények* 110, no. 4. (1997): 667–747.

———. "Egy itáliai várfundáló mester Magyarországon a XVI. század második felében. Ottavio Baldigara élete és tevékenysége" [An Italian castle-builder in

Hungary in the second half of the 16th century. The life and the activity of Ottavio Baldigara], *Hadtörténelmi Közlemények* 111, no. 4 (1998): 767–856.

———. *Ottavio Baldigara: egy itáliai várfundáló mester Magyarországon* [An Italian castle-builder in Hungary, Ottavio Baldigara]. Budapest: Balassi, 2000.

Dowling, Abigail P. and Richard Keyser, eds. *Conservation's Roots: Managing for Sustainability in Preindustrial Europe, 1100–1800.* (Environment in History: International Perspectives, 19). New York: Berghahn Books, 2020.

Drake, Brian Allen, ed. *The Blue, the Gray, and the Green. Toward an Environmental History of the Civil War.* Athens, GA: University of Georgia Press, 2015.

Emiralioğlu, Pinar. *Geographical Knowledge and Imperial Culture in the Early Modern Ottoman Empire.* Burlington, VT: Ashgate, 2014.

Endrei, Walter. "A magyar malomipar, 1550–1800" [Hungarian milling industry]. In *Műszaki innovációk sorsa Magyarországon: malomipar, vaskohászat, textilipar* [Fate of technological innovations in Hungary: Milling industry, iron smelting and textile industry], ed. idem, 48–65. Budapest: Akadémiai, 1995.

Engel, Pál. "Magyarország és a török veszély Zsigmond korában, 1387–1437" [Hungary and the Ottoman threat in the age of King Sigismund, 1387–1437], *Századok* 128 (1994): 273–287.

———. "A török dúlások hatása a népességre: Valkó megye példája" [The impact of the Ottoman plundering on the population of Valkó County], *Századok* 134 (2000): 267–321.

———. *The Realm of St. Stephen. A History of Medieval Hungary, 895–1526.* Translated by Tamás Pálosfalvi. London and New York: I.B. Tauris, 2001.

Esser, Raingard and Steven G. Ellis. "Introduction." In *Frontier and Border Regions in Early Modern Europe* (The Formation of Europe, 7), eds. Raingard Esser and Steven G. Ellis, 7–16. Hannover: Wehrhahn Verlag, 2013.

———, eds. *Frontier and Border Regions in Early Modern Europe* (The Formation of Europe, 7). Hannover: Wehrhahn Verlag, 2013.

Evans, Julian. "Coppice Forestry – An Overview," In *Ecology and Management of Coppice Woodlands*, ed. G.P. Buckley, 18–27. New York: Chapman and Hall, 1992.

Evenden, Matthew. *Fish versus Power: An Environmental History of the Fraser River.* Cambridge: Cambridge University Press, 2004.

Falkowski, Mateusz. "Fear and Abundance: Reshaping of Royal Forests in Sixteenth-Century Poland and Lithuania," *Environmental History* 22 (2017): 618–642.

F. Romhányi, Beatrix. "Szempontok a Kárpát-medence térszervezésének változásaihoz (5–14. század)" [Changes in the early medieval settlement structure of the Carpathian Basin (5th–14th centuries)]. In *Hatalmi központok az Avar Kaganátusban – Power Centres of the Avar Khaganate*, eds. Csilla Balogh, József Szentpéteri, and Erika Wicker, 399–420. Kecskemét: Katona József Múzeum and MTA BTK MŐT, 2019.

F. Romhányi, Beatrix, Zsolt Pinke, and József Laszlovszky. "Environmental Impacts of Medieval Uses of Natural Resources in the Carpathian Basin," *Hungarian Historical Review* 9 (2020): 241–283.

Fara, Andrea. "An Outline of Livestock Production and Cattle Trade from Hungary to Western Europe in Late Middle Ages and Early Modern Period (XIVth–XVIth centuries)," *Crisia* 45 (2015): 87–95.

———. "Il commercio di bestiame ungherese verso la Penisola italiana tra tardo Medioevo e prima Età moderna (XIV–XVI secolo)," *Mélanges de l'École française de Rome – Moyen Âge*", 47, no. 2 (2015). doi: 10.4000/mefrm.2709.

Faragó, Tamás and Péter Őri. *Az 1784–1787 évi népszámlálás II. Az Alföld, a Délvidék és a Dunántúl népességi adatai* [The census of 1784–1787, II. The population data of the Great Hungarian Plain, the Southern Frontier and Transdanubia]. Budapest: [n. p.], 2008. Available online: http://real.mtak.hu/2144/1/46348_ZJ1.pdf (last accessed: 7 July 2021).

Febvre, Lucien. "Frontière," *Bulletin du Centre international de synthèse. Section de synthèse historique* no. 5 (1928): 31–44.

———. *A Geographical Introduction to History*. London and New York: Kegan Paul, Trench, Trubner and Co., and Alfred A. Knopf, 1932.

Fekete, Lajos. "A hódoltság-kori törökség Magyarországra vonatkozó földrajzi ismeretei" [Geographical knowledge of Ottomans on Hungary], *Hadtörténelmi Közlemények* 31, nos. 1, and 2 (1930): 1–17, and 134–154.

Ferenczi, László. "Water Management in the Medieval Hungary," In *The Economy of Medieval Hungary* (East Central and Eastern Europe in the Middle Ages, 49), eds. József Laszlovszky et al., 238–252. Leiden and Boston: Brill, 2018.

Fiege, Mark. "Gettysburg and the Organic Nature of the American Civil War." In *Natural Enemy, Natural Ally: Toward an Environmental History of War*, eds. Richard P. Tucker and Edmund P. Russell, 93–109. Corvallis, OR: Oregon State University Press, 2004.

Filipovic, Emir O. "The Key to the Gate of Christendom? The Strategic Importance of Bosnia in the Struggle against the Ottomans." In *The Crusade in the Fifteenth Century: Converging and Competing Cultures* (Crusades – Subsidia, 8), ed. Norman Housley, 151–168. London and New York: Routledge, 2016.

Finkel, Caroline. *The Administration of Warfare: The Ottoman Military Campaigns in Hungary, 1593–1606*. Vienna: VWGÖ, 1988.

Fleet, Kate. *The Ottomans and the Sea* (Oriento Moderno, XX/1). Rome: Skilliter Centre for Ottoman Studies, 2001.

Fodor, Pál. "Török várerődítési munkák Magyarországon a XVI–XVII. században" [Ottoman fortification works in 16th–17th-century Hungary]. *Hadtörténelmi Közlemények* 26, no. 3 (1979): 375–398.

———. "Bauarbeiten der Türken an den Burgen in Ungarn im 16–17. Jahrhundert," *Acta Orientalia Academiae Scientiarum Hungaricae* 35 (1981): 55–88.

———. *The Unbearable Weight of Empire. The Ottomans in Central Europe – A Failed Attempt at Universal Monarchy (1390–1566)*. Budapest: MTA BTK Történettudományi Intézet, 2015.

Frank, Norbert. "A kocsányos tölgy (Quercus robur L.) erdőművelési tulajdonságai" [The forestry characteristics of pedunculate oak], *Erdészeti Lapok* 150 (2015): 314–315

Fügedi, Erik. "Das mittelalterliche Königreich Ungarn als Gastland." In *Die deutsche Ostsiedelung des Mittelalters als Problem der europäischen Geschichte* (Vorträge und Forschungen, 18), ed. Walter Schlesinger, 471–507. Sigmaringen: Thorbecke, 1974.

———. *Castle and Society in Medieval Hungary*. Translated by János M. Bak (Studia historica Academiae Scientiarum Hungaricae, 187). Budapest: Akadémiai, 1986.

Gaál, Attila. "Turkish Palisades on the Tolna-County Stretch of the Buda-to-Eszék Road." In *Archeology of the Ottoman Period in Hungary* (Opuscula Hungarica, 3), eds. Ibolya Gerelyes and Gyöngyi Kovács, 104–108. Budapest: Magyar Nemzeti Múzeum, 2004.

Galloway, James, Derek Keene, and Margaret Murphy. "Fuelling the City: Production and Distribution of Firewood and Fuel in London's Region, 1290–1400," *Economic History Review* 49 (1996): 455–458.

Gecsényi, Lajos. "Elképzelések a Rábaköz önvédelmének megszervezésére a XVII. század közepén" [Ideas about the defense of the Rábaköz region in the mid-17th century], *Soproni Szemle* 45 (1991): 343–347.

———. "Győr erődváros kiépítése a 16. század második felében" [The formation of Győr as a garrison town in the second half of the 16th century]. In idem, *Gazdaság, társadalom, igazgatás. Tanulmányok a kora újkor történetéből* [Economy, society, administration. Studies in the history of the early modern period], 445–452. Győr: Győr-Moson-Sopron Megye Győri Levéltára and Győr Megyei Jogú Város Levéltára, 2008.

Gellén, Zsolt. "A Rába védelmében" [In defense of the Rába]. In *Táj és történelem. Tanulmányok a történeti ökológia világából* [Landscape and history. Studies in historical ecology], ed. Ágnes R. Várkonyi, 232–254. Budapest: Osiris, 2000.

Géra, Eleonóra. *"Kőhalomból (fő)város." Buda város hétköznapjai a 18. század elején* ['(Capital) city from a pile of stone.' The everyday life of Buda at the beginning of the 18th century]. Budapest: L'Harmattan, 2013.

Gerő, László. "A török elleni harcokban átépített, vagy újonnan épített bástyás várak kialakulása" [The new fortifications with bastions that were built or rebuilt during the period of the Ottoman wars]. In *Várépítészetünk* [Castle building in Hungary], ed. idem, 325–344. Budapest: Műszaki, 1976.

Glaser, Lajos. "A Dunántúl középkori úthálózata, I–II" [Medieval road network in the Transdanubia], *Századok* 63 (1929): 138–167, and 257–285.

Glaser, Rüdiger, *Klimageschichte Mitteleuropas – 1200 Jahre Wetter, Klima, Katastrophen*. 3rd ed. Darmstadt: Primus, 2013.

Glaser, Rüdiger et al. "The Variability of European Floods since AD 1500," *Climatic Change* 101 (2010): 235–256.

Gleitsmann, Rolf-Jürgen. "Rohstoffmangel und Lösungsstrategien: Das Problem der vorindustriellen Holzknappheit," *Technologie und Politik* 16 (1980): 104–154.

Göckenjan, Hansgerd. *Hilfsvölker und Grenzwächter im mittelalterlichen Ungarn* (Quellen und Studien zur Geschichte des östlichen Europa). Wiesbaden: Franz Steiner, 1972.

Göcsei, Imre. *Kapuvári-Rábaköz földrajza* [Geography of the Rábaköz area around Kapuvár] (Értekezések a Magyar Királyi Horthy Miklós Tudományegyetem Földrajzi Intézetéből, 5 [35]). Szeged: [n. p.], 1943.

Goda, László and Vilmos Vasvári. "A Felső-Rába vízjárásának statisztikai jellemzése" [Statistical analysis of the water-regime of the Upper Rába], *Vízügyi Közlemények* 79 (1997): 518–538.

Győrffy, István. *A feketekőrös-völgyi magyarság települése: az erdélyi magyarság eredete* [The Hungarians of the valley of the Fekete-Körös: the origin of the Hungarians in Transylvania]. Budapest: Fritz Ny., 1914.

H. Németh, István. "Háború és népesség a kora újkori Magyarországon (16–17. század)" [War and population in early modern Hungary, 16th–17th centuries]. In *Történeti demográfiai évkönyv 2001* [Historical demography yearbook, 2001], eds. Tamás Faragó and Péter Őri, 129–141. Budapest: KSH Népességtudományi Kutatóintézet, 2001.

Hausner, Gábor and András Németh, eds. *Zrínyi-Újvár. A Seventeenth-Century Frontier Defensive System on the Edge of the Ottoman Empire*. Budapest: Dialóg Campus, 2020.

Heckenast, Gusztáv. *A magyarországi vaskohászat története a feudalizmus korában* [History of iron-smelting in Hungary in the age of feudalism]. Budapest: Akadémiai, 1991.

Hegyi, Klára. *A török hódoltság várai és várkatonasága*, 3 vols. [Castles and garrisons of Ottoman Hungary] (História Könyvtár. Kronológiák, adattárak, 9). Budapest: História and MTA Történettudományi Intézet, 2007.

Henshaw, Robert E. and Frances F. Dunwell. *Environmental History of the Hudson River: Human Uses that Changed the Ecology, Ecology that Changed Human Uses*. Albany, NY: State University of New York Press, 2011.

Herényi, István. *A nyugati gyepü erődítményei: erődítmények, várak Nyugat-Magyarországon* [Constructions of the Western *gyepű*: fortifications and castles in Western-Hungary]. Budapest: Heraldika, 2007.

Hoffmann, Richard C. "Footprint Metaphor and Metabolic Realities Environmental Impacts of Medieval European Cities." In *Natures Past. The Environment and Human History*, ed. Paolo Squatriti, 288–325. Ann Arbor, MI: University of Michigan Press, 2007.

———. *An Environmental History of Medieval Europe*. Cambridge: Cambridge University Press, 2014.

Holub, József. "Ete város története. (Adalékok a Tolna megyei Sárköz település- és gazdaságtörténetéhez)" [The history of the town of Ete (Data on the settlement and economic history of the Sárköz in Tolna County)], *Történeti Statisztikai Közlemények* 2, no. 3–4 (1958): 28–46.

Hóman Bálint and Gyula Szekfű, *Magyar történet*, vol. 5 [Hungarian history]. Budapest: Királyi Magyar Egyetemi Nyomda, 1928.

Horváth, Gergely Krisztián, ed. *Víz és társadalom Magyarországon a középkortól a XX. század végéig* [Water and society in Hungary from the Middle Ages to the end of the 20th century]. Budapest: Balassi, 2014.

Hupy, Joseph P. "The Environmental Footprint of War," *Environment and History* 14 (2008): 405–421.

Husain, Faisal H. "Changes in the Euphrates River: Ecology and Politics in a Rural Ottoman Periphery, 1687–1702," *Journal of Interdisciplinary History* 47 (2016): 1–25.

———. *Rivers of the Sultan: The Tigris and Euphrates in the Ottoman Empire*. New York: Oxford University Press, 2021.

Ihrig, Dénes, ed. *A magyar vízszabályozás története* [History of water regulation in Hungary]. Budapest: Országos Vízügyi Hivatal, 1973.

Illik, Péter, "Török kártételek a nyugat-dunántúli hódoltsági peremvidéken a 17. század első felében" [Ottoman raids in Transdanubia. Plundering on the fringes of the Ottoman Empire's Transdanubian parts in the first half of the 17th century]. PhD diss., Pázmány Péter Katolikus Egyetem, 2009.

Illyés, Eszter and János Bölöni, eds. *Lejtősztyepek, löszgyepek és erdőssztyeprétek Magyarországon* [Slope steppes, loess steppes and forest steppe meadows of Hungary]. Budapest: [MTA Ökológiai és Botanikai Kutatóintézet], 2007.

Ilon, Gábor, András Grynaeus, and Andrea Torma. "A szentgotthárdi török kori palánk kutatásáról" [On the archaeological research of the Ottoman palisade of Szentgotthárd], *Savaria* 31 (2007): 307–328.

Imber, Colin. "The Navy of Süleyman the Magnificent," *Archivum Ottomanicum* 6 (1980): 211–282.

Ivanics, Mária. *A Krími Kánság a tizenöt éves háborúban* [The Crimean Tatar Khanate in the Fifteen Years' War] (Kőrösi Csoma kiskönyvtár, 22). Budapest: Akadémiai, 1994.

Iványi, Béla. "A tüzérség története Magyarországon kezdettől 1711-ig, III" [History of artillery in Hungary from the beginnings to 1711], *Hadtörténelmi Közlemények* 27, no. 3 (1926): 259–289.

———. *A körmendi levéltár memorabiliái 1352–1698. / Acta Memorabilia in tabulario gentis principum de Batthyány reperibilia* (Körmendi füzetek, 2). Körmend: Rábavidék, 1942.

———. "Gróf Batthyány Ádám, a levéltárrendező" [Ádám Batthyány, the archivist], *Levéltári Közlemények* 20–23 (1942–1945): 290–309.

———. *Képek Körmend multjából / Ex praeteritis oppidi Körmend* (Körmendi füzetek, 4). Körmend: "Rábavidék" nyomda és lapkiadóvállalat, 1943.

———. *Részletek a magyarországi fertőző betegségek történetéből. Adatok a körmendi levéltárból, a pestis XVI–XVII. századi történetéhez /1510–1692/* [To the history of contagious diseases in Hungary. Data to the 16th–17th-century history of pestilence in the archive of Körmend (1510–1692)] (Communicationes ex Bibliotheca Historiae Medicae Hungarica. Supplementum, 3). Budapest: Országos Orvostörténeti Könyvtár, 1965.

———. *Körmend története a római kortól a mohácsi vészig* [History of Körmend from the Roman period to the battle of Mohács]. Körmend: Csaba József Honismereti Egyesület, 1999.

Janecek, Andrzej. "Frontiers and Borderlands in Medieval Europe. Introductory Remarks," *Quaestiones Medii Aevi Novae* 16 (2011): 5–14.

Jankovich B., Dénes. "Adatok a Körösvidék középkori vízrajzához és a vizek hasznosításához" [Data to the medieval hydrography and water management in the Körösvidék], *A Békés Megyei Múzeumok Közleményei* 16 (1996): 305–349.

Juhász, Lajos. "A Vas megyei Farkaserdő a XVII–XVIII. században" [Farkaserdő in Vas County in the 17th–18th century], *Századok* 71 [Supplementum] (1937): 553–575 [37–69].

K. Németh, András. "Vizek és vízgazdálkodás a középkori Tolna megyében I. Vízimalmok" [Waters and water management in medieval Tolna County, I. Water mills], *A Wosinsky Mór Múzeum Évkönyve* 35 (2013): 121–151.

———. "Fák és erdők a középkori Tolna megyében" [Trees and woodlands in medieval Tolna County], *A Wosinsky Mór Múzeum Évkönyve* 38 (2016): 45–70.

Kaán, Károly. *Alföldi kérdések. Erdők és vizek az Alföld kérdéseiben* [Question of the Great Hungarian Plain. Forests and waters in the questions of the Great Hungarian Plain]. Budapest: Stádium, 1939.

Kalász, Elek. *A szentgotthárdi apátság birtokviszonyai és a ciszterci gazdálkodás a középkorban* [Estate structure of the Szentgotthárd abbey and Cistercian economy in the Middle Ages] (Tanulmányok a magyar mezőgazdaság történetéhez, 5). Budapest: Sárkány Nyomda, 1932.

Káldy-Nagy, Gyula. *A budai szandzsák 1546–1590. évi összeírásai. Demográfiai és gazdaságtörténeti adatok* [The conscriptions of the sanjak of Buda, 1546–1590. Demographical and economic historical data] (Pest megye múltjából, 6). Budapest: Pest Megyei Levéltár, 1985.

Károlyi, Zoltán and Sándor Somogyi. "Felszíni vízfolyások" [Above ground waters]. In *A Kisalföld és a nyugat-magyarországi peremvidék* [Little Hungarian Plain and the Western Hungarian Highlands] (Magyarország tájföldrajza, 3), 98–115. Budapest: Akadémiai, 1975.

Károlyi, Zsigmond. *A vízhasznosítás, vízépítés és vízgazdálkodás története Magyarországon.* (Vázlat) [History of water use, water construction, and water management in Hungary (A sketch)]. Budapest: Tankönyvkiadó, 1960.

Karp, Hans-Jürgen. *Grenzen in Ostmitteleuropa während des Mittelalters. Ein Beitrag zur Entstehungsgeschichte der Grenzlinie aus dem Grenzsaum* (Forschungen und Quellen zur Kirchen- und Kulturgeschichte Ostdeutschlands, 9). Cologne and Vienna: Böhlau, 1972.

Kelenik, József. "Körmend a hadtörténelemben 1526–1711" [Körmend in military history 1526–1711]. In *Körmend a hadtörténelemben* [Körmend in military history], eds. László Veszprémy, József Kelenik, Róbert Hermann, and László Bencze, 51–139. Körmend: Önkormányzat, 1992.

———. "A nemzetiségi megoszlás, a veszteségek és a fluktuáció mértéke. Tizennégy Kanizsa elleni végvár helyőrségében (1633–1640)" [Ethnic composition, the losses, and the scale of fluctuation. On the example of 14 garrisons against Kanizsa]. In *A végvárak és régiók a XVI–XVII. században. Tudományos tanácskozás előadásai-Noszvaj, 1991. okt. 17–18* [Borderline castles and regions in the 16th–17th centuries. Proceedings of the conference held at Noszvaj, 17–18 October 1991] (Studia Agriensia, 14), eds. Tivadar Petercsák and Jolán Szabó, 101–121. Eger: Heves Megyei Múzeumok Igazgatósága, 1993.

———. "A kanizsai övezet és természetföldrajzi adottságai a XVI. század 70-es éveinek végén" [The Kanizsa region and its geographical conditions in the 1570s]. In *Végvár és környezet. Tudományos tanácskozás előadásai – Noszvaj, 1993. okt. 14–15.* [Borderline castles and the environment. Proceedings of the conference, Noszvaj, 14–15 October 1993] (Studia Agriensia, 15), eds. Tivadar Petercsák and Ernő Pető, 163–174. Eger: Heves Megyei Múzeumi Szervezet, 1995.

———. "Tata helye és szerepe a végvári rendszerben a 16. század utolsó harmadában" [The place of Tata in the borderline defense in the last third of the 16th century]. In *Tata a tizenötéves háborúban. Tatán 1997. május 23-án megtartott tudományos ülésszakon elhangzott előadások anyaga* [Tata in the Fifteen Years' War. Papers read at the conference held at Tata, 23 May 1997] (Annales Tataienses, 1), eds. János Fatuska, Éva Mária Fülöp, and László ifj. Gyuszi, 45–58. Tata: Mecénás Közalapítvány, 1998.

Kenyeres, István. "Kamarai uradalmak igazgatása a XVI. században" [Administration of the domains of the Hungarian Chamber in the 16th century]. In *XVI. századi uradalmi utasítások. Utasítások a kamarai uradalmak prefektusai, udvarbírái és ellenőrei részére*, 2 vols. [16th-century instructions. Instructions to the prefects, provisors and checkers of the Chamber domains] (Fons Könyvek, 2), eds. István Kenyeres and Péter Kis, vol. I, 15–90. Budapest: Szentpétery Imre Történettudományi Alapítvány, 2002.

———. "A királyi Magyarország bevételei és kiadásai a 16. században" [The incomes and expenses of Hungary in the 16th century], *Levéltári Közlemények* 74, no. 1–2 (2003): 59–103.

———. "A török elleni küzdelem finanszírozása Buda elestétől a drinápolyi békéig" [The financing of the anti-Ottoman struggles from the fall of Buda to the treaty of Adrianople]. In *Mozgó frontvonalak. Háború és diplomácia a várháborúk időszakában, 1552–1568* [Moving front lines. War and diplomacy in the period of the fortress wars, 1552–1568] (Studia Agriensia, 35), eds. Györgyi Bujdosné Pap, Ingrid Fejér, and Ágota H. Szilárd, 19–40. Eger: Dobó István Vármúzeum, 2017.

———. "The Economy of Castle Estates in the Late Medieval Kingdom of Hungary." In *The Economy of Medieval Hungary* (East Central and Eastern Europe in the Middle Ages, 49), eds. József Laszlovszky et al., 394–416. Leiden and Boston: Brill, 2018.

Keresztesi, Béla. "Az akác erdőművelési tulajdonságai és erdőgazdasági jelentősége a Magyar Alföldön" [The forestry characteristics of acacia and its role in forest management in the Great Hungarian Plain], *Az Erdő* 3 (1954): 181–189.

Kertész, Róbert, András Morgós, Dénes Nagy, and Zsuzsanna Szántó. "Tisza-hidak a török hódoltság korából radiokarbon és dendrokronológiai vizsgálatok tükrében" [Radiocarbon and dendrochronological investigations of Tisza bridges built during the Ottoman period]. In *Az erdő és a fa régészete és néprajza (kézművesipar-történeti megközelítésben) / Archaeology and Ethnography of Forest and Wood (in Approximation of Handicraft History)* (Az anyagi kultúra a Kárpát-medencében, 2), ed. János Gömöri, 145–178. Sopron: MTA VEAB Iparrégészeti, Archaeometriai Munkabizottság és Kézművesipar-történeti Munkabizottság, 2007.

Király, János. *A pozsonyi nagy-dunai vám- és révjog története* [The history of the customs and toll at the Danube by Bratislava]. Bratislava: Drodtleff, 1890.

Kiss, Andrea. "Historical Climatology in Hungary: Role of Documentary Evidence in the Study of Past Climates and Hydrometeorological Extremes," *Időjárás* 113 (2009): 315–339.

———. "Floods and Long-Term Water-Level Changes in Medieval Hungary." PhD diss., Central European University, 2011.

———. *Floods and Long-Term Water-Level Changes in Medieval Hungary* (Springer Water). Cham: Springer International Publishing, 2019.

———. "The Great 1506–1507 Drought and its Consequences in Hungary in a (Central) European Context," *Regional Environmental Change* 20 (2020): no. 50. doi: 10.1007/s10113-020-01634-5.

Kiss, Gábor and Balázs Zágorhidi Czigány. "Víznevek névalkotó szerepéről – Vas megyei példákon" [To the role of hydronyms in forming place names – on examples from Vas County], *Helynévtörténeti Tanulmányok* 3 (2008): 111–120.

———. "A Lapincs–Rába-vonaltól délre eső terület Árpád-kori történeti földrajzához: a megyetörténet műhelyéből. I. rész" [To the Árpádian-period historical geography of the area south of the Lapincs–Rába-line – from the workshop of county history], *Vasi Szemle* 64 (2010): 711–721.

Kiss, Gábor and Endre Tóth. "A vasvári 'Római sánc' és a 'Katonák útja' időrendje és értelmezése. Adatok a korai magyar gyepűrendszer topográfiájához. I. rész" [The chronology and interpretation of the 'Roman rampart" and the 'Soldiers' Road' at Vasvár. Data on the topography of the *gyepű* system], *Communicationes Archaeologicae Hungariae 1987*, 101–137.

Kołodziejczyk, Dariusz. "Between Universalistic Claims and Reality: Ottoman Frontiers in the Early Modern Period." In *The Ottoman World*, ed. Christine Woodhead, 471–529. London and New York: Routledge, 2012.

Koltai, András. *Batthyány Ádám. Egy magyar főúr és udvara a XVII. század közepén* [Ádám Batthyány. A Hungarian aristocrat and his court in the mid-17th century] (A Győri Egyházmegyei Levéltár Kiadványai. Források – Feldolgozások, 14). Győr: Győri Egyházmegyei Levéltár, 2012.

Komáromy, András. "Sopron vármegye végzései a malmokról és a Rábaköz védelmezéséről" [The decrees of Sopron County on the mills and the defense of the Rábaköz], *Történelmi Tár* [New Series] 9 (1908): 61–71.

Koppány, Tibor. *Körmend városának építéstörténete* [Construction history of the town of Körmend]. Körmend: Körmend város tanácsa és végrehajtó bizottsága, 1986.

Koroknay, Gyula. "Kálló véghely két ostroma (1670, 1672)" [The two sieges of the borderline fortification of Kálló], *Hadtörténelmi Közlemények* 103, no. 3 (1990): 72–84.

Kovács, Gyöngyi and Csilla Zatykó, eds. *"per sylvam et per lacus nimios." The Medieval and Ottoman Period in Southern Transdanubia, Southwest Hungary: The Contribution of the Natural Sciences*. Budapest: Institute of Archaeology, Research Centre for the Humanities, Hungarian Academy of Sciences, 2016.

Kovács, Gyöngyi and Pál Sümegi. "Palánkvárak, fák, erdők. Régészeti és környezettörténeti adatok a török kori palánkvárak faanyag-felhasználásához" [Earth and wood fortifications, trees and forests. Archaeological and environmental history data to the timber consumption of Ottoman-period fortifications]. In *Várak nyomában. Tanulmányok a 60 éves Feld István tiszteletére* [On the trail of castles. Studies in honor of István Feld on his 60th birthday], eds. György Terei et al., 113–120. Budapest: Castrum Bene Egyesület, 2011.

Kraker, Adriaan M.J. "Flood Events in the Southwestern Netherlands and Coastal Belgium, 1400–1953," *Hydrological Sciences-Journal-des Sciences Hydrologiques* 51 (2006): 913–929.

———. "War, Climatic Stress and Environmental Degradation during the 15th and 16th Centuries. The Case of the North Flemish Coastal Landscape in the Estuary of the Western Scheldt." In *The Dance of Death in Late Medieval and Renaissance Europe: Environmental Stress, Mortality and Social Response*, eds. Andrea Kiss and Kathleen Pribyl, 66–85. New York: Routledge, 2019.

Kubinyi, András. "A magyar királyság népessége a XV. század végén" [The population of the Kingdom of Hungary at the end of the 15th century]. In *Magyarország történeti demográfiája (896–1996)* [Historical demography of Hungary], ed. József Kovacsics, 93–110. Budapest: KSH, 1997.

———. *Városfejlődés és vásárhálózat a középkori Alföldön és az Alföld szélén* [Urbanization and the Network of Markets in the Great Hungarian Plain in the Middle Ages] (Dél-alföldi évszázadok, 14). Szeged: Csongrád Megyei Levéltár, 2000.

Laakkonen, Simo et al., eds. *The Resilient City in World War II Urban Environmental Histories*. Cham: Springer International Publishing and Palgrave MacMillan, 2019.

Lacko, Miroslav. "Frühneuzeitlicher Bergbau und Umwelt in Mitteleuropa. Probleme und Perspektiven der Forschung." In *Bergbau und Umwelt. 15. Internationaler Montanhistorischer Kongress: Sterzing/Hall in Tirol/Schwaz*, eds. Wolfgang Ingenhaeff and Johann Bair, 191–211. Wattens: Berenkamp Buch- und Kunstverlag, 2017.

Laczlavik, György. "Várday Pál helytartói működése, 1542–1549" [Pál Várday as procurator, 1542–1549], *Levéltári Közlemények* 83, no. 1–2 (2012): 3–67.

Ladányi-Benedikt, Ildikó. "Mindennapi élet a körmendi várban – Francsics Gáspár körmendi kapitány levelei 1648–1650-ben" [Everyday life in the castle of Körmend: the letter of Gáspár Francsics, captain of Körmend, 1648–1650]. In *A Batthyányak évszázadai. Tudományos konferencia Körmenden 2005. október 27–29.* [Centuries of the Batthyány family. Conference held at Körmend, 27–29 October 2005], ed. Zoltán Nagy, 205–216. Körmend and Szombathely: Körmend Város Önkormányzata, 2006.

Langdon, John. *Horses, Oxen and Technological Innovation. The Use of Draught Animals in English Farming from 1066–1500*. Cambridge: Cambridge University Press, 1986.

———. "The Use of Animal Power from 1200 to 1800." In *Economia e energia, secc. XIII–XVIII*, ed. Simonetta Cavaciocchi, 213–221. Florence: Le Monnier, 2003.

Lászlóffy, Woldemár, ed. *A Kárpát-medence vízborította és vízjárta területei az ármentesítő és lecsapoló munkálatok megkezdése előtt.* [Map] [Water covered areas and wetlands in the Carpathian Basin before the beginning of flood protection and drainage works]. Budapest: Vízrajzi Intézet, 1938.

Laszlovszky, József et al. "Contextualizing the Mongol Invasion of Hungary in 1241–42: Short and Long-Term Perspectives," *Hungarian Historical Review* 7 (2018): 419–450.

Lavaud, Sandrine and Patrick Fournier, eds. *Eaux et conflicts dans l'Europe médiévale et moderne: actes des XXXII*[es] *Journées Internationales d'Histoire de l'Abbaye de Flaran, 8 et 9 octobre 2010*. [Toulouse]: Presses Universitaires du Mirail, 2012.

Le Maresquier-Kesteloot, Yvonne-Hélène. "L'approvisionnement de Paris en bois (XIVe–XVe siècles)," *Franco-British Studies* 20 (1995): 69–83.

Lee, John S. "Postwar Pines: The Military and the Expansion of State Forests in Post-Imjin Korea, 1598–1684," *The Journal of Asian Studies* 77 (2018): 319–332.

Lenke, Walter. *Klimadaten von 1621–1650 nach Beobachtungen des Landgrafen Hermann IV. von Hessen* (*Uranophilus Cyriandrus*). Offenbach am Main: Dt. Wetterdienst, 1960.

Lewis, A.R. "The Closing of the Medieval Frontier 1250–1350," *Speculum* 33 (1958): 475–485.

Magyar, Eszter. *A feudalizmus kori erdőgazdálkodás az alsó-magyarországi bányavárosokban, 1255–1747* [Feudal-age forest management in the mining towns of Lower Hungary, 1255–1747] (Értekezések a Történeti Tudományok Köréből [New Series], 101). Budapest: Akadémiai, 1983.

Magyar, Kálmán and Gyula Nováki. *Somogy megye várai a középkortól a kuruc korig* [The castles of Somogy County from the Middle Ages until the *kuruc* period]. Kaposvár: Somogy Megyei Múzeumok Igazgatósága, 2005.

Makkai, László. "Östliches Erbe und westliche Leihe in der ungarischen Landwirtschaft der frühfeudalen Zeit," *Agrártörténeti Szemle* 16 [Supplementum] (1974): 1–53.

Maksay, Ferenc. "Urbáriumok" [Terriers]. In *A történeti statisztika forrásai* [Sources of historical statistics], ed. József Kovacsics, 119–144. Budapest: Közgazdasági és Jogi könyvkiadó, 1957.

Malanima, Paolo. "The Energy Basis for Early Modern Growth, 1650–1820." In *Early Modern Capitalism Economic and Social change in Europe, 1400–1800*, ed. Maarten Prak, 51–68. London and New York: Routledge, 2001.

Maliniak, Pavol. "Mlynárstvo na strednom Pohroní v stredoveku a na začiatku novoveku" [Milling in the Central Hron Regon in the Middle Ages and at the beginning of the modern age]. In *Vydavateľstvo: Z histórie technicko-hospodárskeho využitia vodných tokov na strednom Pohroní* [From the history of technological and economic use of water courses in Central Hron Region], eds. J. Žilák and P. Hronček, 30–45. Banská Bystrica: Centrum vedy a výskumu FHV UMB, 2011.

Manners, Ian. *European Cartographers and the Ottoman World* (Oriental Institute Museum Publications, 27). Chicago: Oriental Inst. Museum of the University of Chicago, 2007.

Marosi, Endre. *XVI. századi váraink* [16th-century castles in Hungary]. Budapest and Miskolc: Hungária-Európa Alapítvány and Borsod-Abaúj-Zemplén Megyei Levéltár, 1991.

Marton, Szabolcs. "Gondolatok középkori katonaságunk szeszesital-ellátásával kapcsolatban" [Notes on the alcohol consumption of the garrisons in medieval Hungary], *Hadtörténelmi Közlemények* 120, no. 2. (2007): 577–589.

Máté Ágnes and Teréz Oborni, eds. *Isabella Jagiellon, Queen of Hungary (1539–1559). Studies*. Budapest: Bölcsészettudományi Kutatóközpont, 2020.

Máté, Gábor. "Landscape Reconstruction of the Southern Transdanubian Puszta (1683–1735) Based on 18th-Century Border Litigations," *Acta Ethnographica Hungarica* 62 (2017): 105–134.

Mátyás-Rausch, Petra. "A liptói nemesércbányászat a 16. század második felében" [Precious metal mining in Liptó County in the second half of the 16th century], *Urbs. Magyar Várostörténeti Évkönyv* 12 (2018): 107–123.

Mauch, Christof and Christian Pfister, eds. *Natural Disasters, Cultural Responses. Case Studies toward a Global Environmental History*. Lanham, MD: Lexington Books, 2009.

Mauch, Christof and Thomas Zeller, eds. *Rivers in History: Perspectives on Waterways in Europe and North America*. Pittsburgh: University of Pittsburgh Press, 2008.

Mayer, Hannes. *Wälder des Ostalpenraumes; Standort, Aufbau und waldbauliche Bedeutung der wichtigsten Waldgesellschaften in den Ostalpen samt Vorland*. Stuttgart: Gustav Fischer, 1974.

McNeill, John. "Woods and Warfare in World History," *Environmental History* 9 (2004): 388–410.

———. *Mosquito Empires: Ecology and War in the Greater Caribbean, 1620–1914*. New York: Cambridge University Press, 2012.

Mihalik, Béla Vilmos, Dániel Kálmán, and Áron Zarnóczki. *A veszprémi káptalan 1727. és 1755. évi urbáriumai* [Terriers of the chapter of Veszprém from 1727 and 1755] (A Veszprémi Egyházmegye Múltjából, 24). Veszprém: Veszprémi Érseki és Főkáptalani Levéltár and Veszprémi Érseki Könyvtár, 2012.

Mikhail, Alan, *Nature and Empire in Ottoman Egypt: An Environmental History*. Cambridge: Cambridge University Press, 2011.

———. "Anatolian Timber and Egyptian Grain: Things that Made the Ottoman Empire." In *Early Modern Things: Objects and Their Histories, 1500–1800*, ed. Paula Findlen, 274–294. New York: Routledge, 2013.

———. "Oriental Democracy," *Global Environment* 7 (2014): 381–404.

———. *Under Osman's Tree: The Ottoman Empire, Egypt and Environmental History.* Chicago: University of Chicago Press, 2017.

Miklós, Zsuzsa. "Beiträge zur Siedlungsgeschichte des mittelalterlichen Marktfleckens," *Acta Archaeologica Academiae Scientiarum Hungaricae* 53 (2002): 195–254.

———. *Tolna megye várai* [Castles of Tolna County]. Budapest: Históriaantik Könyvesház Kiadó, 2007.

Mócsy, András and Mária Szilágyi. "Úthálózat" [Road network]. In *Pannónia régészeti kézikönyve* [Archaeology handbook of Pannonia], 118–124. Budapest: Akadémiai, 1990.

Mollay, Károly. "Pájer Imre: Rábaköz népének védekezése az áradások ellen 1870–1889. Csorna, 1990" [Imre Pájer: The fight of the inhabitants of the Rábaköz against the floods, 1870–1889] [Book review]. *Soproni Szemle* 46 (1992): 383.

Molnár, József. "Török emlékek: Eszék–Dárdai híd a XVII. században" [Ottoman monuments: Osijek – Bridge of Dárda in the 17th century], *Művészettörténeti Értesítő* 7 (1958): 259–261.

Molnár, Zsolt. "A Duna-Tisza köze és a Tiszántúl növényzete a 18–19. század fordulóján I. Módszertan, erdők, árterek és lápok" [Vegetation of the Danube–Tisza Interfluve and Transtisza Regions at the turn of the 18th and 19th centuries, I: methods, woodlands, floodplains, and fens], *Botanikai Közlemények* 95 (2008): 11–38.

Molnár, Zsolt and András Kun, eds. *Alföldi erdőssztyepp-maradványok Magyarországon* [Forested steppe remains in Hungary] (WWF Füzetek, 15). Budapest: WWF Magyarország, 2000.

Molnár, Zsolt, Gábor Fekete, Marianna Biró, and András Kun. "A Duna-Tisza közi homoki sztyepprétek történeti tájökológiai jellemzése" [Historical ecological analysis of steppe meadows in the Danube-Tisza Interfluve]. In *Talaj-vegetáció-klíma kölcsönhatások. Köszöntjük a 70 éves Láng Editet* [Soil – vegetation – climate. Honorary volume on the 70th birthday of Edit Láng], eds. György Kröel-Dulay, Tibor Kalapos, and Andrea Mojzes, 39–56. Vácrátót: MTA Ökológiai és Biológiai Kutatóintézete, 2008.

Mordovin, Maxim. "Szécsény városának kora újkori palánkerődítése (A szécsényi Pintér-háznál feltárt maradványok alapján)" [The early modern plank fortification of Szécsény (in the light of the excavations at the Pintér house)]. In *Várak nyomában. Tanulmányok a 60 éves Feld István tiszteletére* [On the trail of castles. Studies in honor of István Feld on his 60th birthday], eds. György Terei et al., 149–159. Budapest: Castrum Bene Egyesület, 2011.

———. "The Post-Medieval Fortifications of Earth and Timber in Hungary." In *Études de castellologie médiévale. Château et frontière* (Château Gaillard, 26), eds. Peter Ettel, Anne-Marie Flambard Héricher, and Kieran O'Conor, 273–282. Caen: Presses Universitaires de Caen, 2014.

Moreno, Eduardo Manzano. "The Creation of a Medieval Frontier: Islam and Christianity in the Iberian Peninsula, Eighth to Eleventh Centuries." In *Frontiers in Question: Eurasian Borderlands, 700–1700*, eds. Daniel Power and Naomi Standen, 32–54. Basingstoke: Macmillan, 1999.

Nagy, Balázs, ed. *Tatárjárás* [Mongol invasion] (Nemzet és emlékezet). Budapest: Osiris, 2003.

Niederkorn, Jan Paul. *Die europäischen Mächte und der 'Lange Türkenkrieg' Kaiser Rudolfs II. 1593–1606* (Archiv für Österreichische Geschichte, 135). Vienna: Verlag der Österreichischen Akademie der Wissenschaften, 1993.

Niklas, Thomas. *Um Macht und Einheit des Reiches: Konzeption und Wirklichkeit der Politik bei Lazarus von Schwendi, 1522–1583*. Husum: Matthiesen, 1995.

Nógrády, Árpád. "A középkor végi Magyarország mindennapi kenyere (Beregszász lakosságának gabonavásárlásai 1530-ban és a Jagelló-kori malomvámok)" [The everyday bread of late medieval Hungary. The grain provisioning of the population of Berehove in the 1530s and the mill customs in the Jagiellonian period], *A Jósa András Múzeum Évkönyve* 42 (2000): 155–169.

Nünning, Ansgar. "Krise als Erzählung und Metapher: Literaturwissenschaftliche Bausteine für eine Metaphorologie und Narratologie von Krise." In *Krisengeschichte(n): 'Krise' als Leitbegriff und Erzählmuster in kulturwissenschaftlicher Perspektive* (Beihefte der Vierteljahrschrift für Sozial- und Wirtschaftsgeschichte, 210), eds. Carla Meyer, Katja Patzel-Mattern, and Gerrit Jasper Schenk, 117–144. Stuttgart: Steiner, 2013.

Nyári, Diána and Tímea Kiss. "Blown Sand Movement at Kiskunhalas on the Danube-Tisza Interfluve, Hungary," *Journal of Environmental Geography* 2, no. 3–4 (2009): 31–36.

Nyári, Diána, Tímea Kiss, and György Sipos. "Investigation of Holocene Blown-Sand Movement Based on Archaeological Findings and OSL Dating, Danube-Tisza Interfluve, Hungary," *Journal of Maps. Student Edition 2007*, 46–57.

O'Reilly, William. "Border, Buffer and Bulwark. The Historiography of the Military Frontier, 1521–1881." In *Frontiers and the Writing of History, 1500–1850*, eds. Steven G. Ellis and Raingard Esser, 229–244. Hanover: Wehrhahn, 2006.

Oborni, Teréz. "From Province to Principality: Continuity and Change in Transylvania in the First Half of the Sixteenth Century." In *Fight Against the Turk in Central-Europe in the First Half of the 16th Century*, ed. István Zombori, 165–180. Budapest: METEM, 2004.

———. "Between Vienna and Constantinople: Notes on the Legal Status of the Principality of Transylvania." In *The European Tributary States of the Ottoman Empire in the Sixteenth and Seventeenth Centuries* (The Ottoman Empire and Its Heritage: Politics, Society and Economy, 53), eds. Gábor Kármán and Lovro Kunčević, 67–89. Leiden and Boston: Brill, 2013.

Őriné Bilkei, Irén. "A zalavári és a kapornaki konventek hiteleshelyi tevékenysége és ügyfelei, a megyei nemesség a Mohács utáni évtizedekben" [The activity of the chapters of Zalavár and Kapornak as places of authenticity, and its clients, the nobility of the county in the decades after the battle of Mohács]. PhD diss., Eötvös Loránd Tudományegyetem, 2007.

Oross, András. *A Magyar Királyság törökellenes határvédelmi rendszerének felszámolása és átszervezése* [The dissolution and reorganization of the anti-Ottoman border defense system of the Kingdom of Hungary] (Fons Könyvek, 4). Budapest: Szentpétery Imre Történettudományi Alapítvány, 2013.

Ortvay, Tivadar. *Magyarország régi vízrajza a 13. század végéig*, 2 vols. [Ancient hydrography of Hungary until the 13th century]. Budapest: Magyar Tudományos Akadémia, 1882.

———. *Pozsony város története*, vol. II/2 [The history of the town of Bratislava]. Bratislava: Stampfel Károly, 1898.

Pálffy, Géza. "A főkapitányi hadiipari műhely kiépülése Kassán és nyersanyagellátó forrásai" [The formation and the raw material provisioning of the chief captainship at Košice]. In *Végvár és környezet. Tudományos tanácskozás előadásai – Noszvaj, 1993. okt. 14–15.* [Borderline castles and the environment. Proceedings of the conference, Noszvaj, 14–15 October 1993] (Studia Agriensia, 15), eds. Tivadar Petercsák and Ernő Pető, 183–221. Eger: Heves Megyei Múzeumi Szervezet, 1995.

———. "A magyarországi és délvidéki végvárrendszer 1576. és 1582. évi jegyzékei" [The Registers of the Hungarian and Croatian–Slavonian Border Fortresses of 1576 and 1582], *Hadtörténelmi Közlemények* 108, no. 1 (1995): 114–185.

———. "A magyarországi török és királyi végvárrendszer fenntartásának kérdéséhez" [On the question of the management of the Ottoman and Hungarian border defense systems in Hungary], *Keletkutatás* (Spring 1995): 61–86.

———. "A török elleni védelmi rendszer szervezetének története a kezdetektől a 18. század elejéig. (Vázlat egy készülő nagyobb összefoglaláshoz)" [The history of the anti-Ottoman defense from the beginnings to the early 18th century. (Sketch of an overview in progress)], *Történelmi Szemle* 38 (1996): 163–217.

———. "Pápa szerepe a XVI. századi végvárrendszerben" [Role of Pápa in the 16th-century defense system]. In *Tanulmányok Pápa város történetéből*, vol. 2 [Studies in the history of the town of Pápa], ed. István Hermann, 81–98. Pápa: Pápa Város Önkormányzata, 1996.

———. "A rabkereskedelem és rabtartás gyakorlata és szokásai a XVI–XVII. századi török–magyar határ mentén. (Az oszmán–magyar végvári szokásjog történetéhez)" [Practices and customs of captive-trade and captive keeping along the 16th–17th-century Ottoman-Hungarian frontier (On the customary law of the Ottoman-Hungarian frontier region)], *Fons. Forráskutatás és Történeti Segédtudományok* 4 (1997): 5–78.

———. *Pápa a hosszú török háborúban. A végvár története az 1594–1597. esztendőkben, különös tekintettel a töröktől való visszafoglalására* [Pápa in the Long Turkish War. The history of the border fortress from 1594 to 1597]. Pápa: Jókai Mór Városi Könyvtár, 1997.

———. *A császárváros védelmében: a győri főkapitányság története, 1526–1598* [In defense of the imperial city: The captainship of Győr, 1526–1598]. Győr: Győr-Moson-Sopron Megye Győri Levéltára, 1999.

———. *Európa védelmében: haditérképészet a Habsburg Birodalom magyarországi határvidékén a 16–17. században* [In defense of Europe. Military cartography at the Hungarian borders of the Habsburg Empire in the 16th–17th centuries]. 2nd ed. Pápa: Jókai Városi Könyvtár, 2000.

———. "The Origins and Development of the Border Defence System Against the Ottoman Empire in Hungary. (Up to the Early Eighteenth Century)." In *Ottomans, Hungarians, and Habsburgs in Central Europe: The Military Confines in the Era of the Ottoman Conquest* (The Ottoman Empire and its Heritage, Politics, Society and Economy, 20), eds. Géza Dávid and Pál Fodor, 3–69. Leiden, Boston, and Cologne: Brill, 2000.

———. "Egy szlavóniai köznemesi família két ország szolgálatában: a budróci Budor család a XV–XVIII. században" [A Slavonian lesser noble family in the service of two countries: The Budor family of Budróc in the 15th–18th centuries], *Hadtörténelmi Közlemények* 115, no. 4 (2002): 923–1007.

———. "Elképzelések a török hódoltság elpusztításáról a XVI–XVII. században. (A Habsburg Birodalom magyarországi hadszínterének néhány főbb sajátosságáról)" [Ideas on the destruction of the Ottoman Empire in Hungary in the 16th–17th centuries (Some specificities of the Hungarian military scene of the Habsburg Empire)]. In *"Quasi liber et pictura" Tanulmányok Kubinyi András hetvenedik születésnapjára / Studies in Honour of András Kubinyi on his Seventieth Birthday*, ed. Gyöngyi Kovács, 387–403. Budapest: ELTE Régészettudományi Intézet 2004.

———. "Scorched-Earth Tactics in Ottoman Hungary: On a Controversy in Military Theory and Practice on the Habsburg-Ottoman Frontier," *Acta Orientalia Academiae Scientiarum Hungaricae* 61 (2008): 181–200.

———. *The Kingdom of Hungary and the Habsburg Monarchy in the Sixteenth Century* (East European Monographs, 735 = Center for Hungarian Studies and Publications Series, 18). Translated by J. Thomas and Helen D. DeKornfeld. Boulder, CO and Wayne, NJ: Social Science Monographs, Center for Hungarian Studies and Publications, Inc., 2009.

———. *Die Anfänge der Militärkartographie in der Habsburgermonarchie. Die regelmäßige kartographische Tätigkeit der Burgbaumeisterfamilie Angielini an den kroatisch-slawonischen und den ungarischen Grenzen in den Jahren 1560–1570.* Budapest: Ungarisches Nationalarchiv, 2011.

———. "Egy rendkívüli forrás a magyar politikai elit 16. századi földrajzi ismereteiről. Az 1526 és 1556 között török kézbe került magyarországi városok, várak és kastélyok összeírása a Német-római Birodalom rendjei számára" [A unique document on the geographical knowledge of the political elite of the Kingdom of Hungary in the middle of the 16th century: The register of the Hungarian and Slavonian towns, fortresses and castles occupied by the Ottomans from 1526 until 1556 for the estates of the Holy Roman Empire]. In *Várak nyomában. Tanulmányok a 60 éves Feld István tiszteletére* [On the trail of castles. Studies in honor of István Feld on his 60th birthday], eds. György Terei et al., 177–194. Budapest: Castrum Bene Egyesület, 2011.

———. "Un penseur militaire alsacien dans la Hongrie au XVIe siècle: Lazare baron von Schwendi (1522–1583)." In *La pensée militaire hongroise à travers les siècles*, eds. Hervé Coutau-Bégarie and Ferenc Tóth, 41–59. Paris: Economica, 2011.

———. "The Habsburg Defense System in Hungary Against the Ottomans in Sixteenth Century: A Catalyst of Military Development in Central Europe." In *Warfare in Eastern Europe, 1500–1800* (History of Warfare, 72), ed. Brian J. Davies, 35–61. Leiden and Boston: Brill, 2012.

———. "A Magyar Királyság új fővárosa: Pozsony a XVI. században" [The new capital of the Kingdom of Hungary: Bratislava in the 16th century], *Fons. Forráskutatás és Történeti Segédtudományok* 20 (2013): 3–76.

———. "Mohács radikálisabb változásokat hozott, mint Trianon" [Mohács brought more radical changed than did Trianon]. Online document: Transindex, 2015. http://vilag.transindex.ro/?cikk=25515 (last accessed: 29 June 2020)

Palkó, István. "Körmend névtudományi vizsgálata" [Onomastic study of Körmend]. In *Körmend története* [The history of Körmend], ed. László Szabó, 11–17. Körmend: Önkormányzat, [1994].

Pálosfalvi, Tamás. *From Nicopolis to Mohács: A History of Ottoman-Hungarian Warfare, 1389–1526* (The Ottoman Empire and Its Heritage, 63). Leiden and Boston: Brill, 2018.

Pánya, István. "Az eszéki Dráva-híd" [The bridge over the Dráva at Osijek], *Várak, kastélyok, templomok. Évkönyv, 2018*, 166–170.

Parker, Geoffrey. *Global Crisis: War, Climate Change and Catastrophe in the Seventeenth Century*. New Haven, CT: Yale University Press, 2013.

Pataki, Vidor. "A XVI. századi várépítés Magyarországon" [Castle-building in 16th-century Hungary], *A Bécsi Magyar Történeti Intézet Évkönyve* 1 (1931): 98–133.

Paulinyi, Oszkár. "A Garam-vidéki bányavárosok lakosságának lélekszáma a XVI. század derekán" [Population of the mining towns along the River Hron in the mid-16th century], *Történelmi Szemle* 1 (1958): 351–378.

Peacock, A.C.S., ed. *The Frontiers of the Ottoman World* (Proceedings of the British Academy, 156). Oxford: Oxford University Press, 2009.

Pearson, Chris, Peter A. Coates, and Tim Cole, eds. *Militarized Landscapes: From Gettysburg to Salisbury Plain*. London: Continuum, 2010.

Perjés, Géza. *The Fall of the Medieval Kingdom of Hungary: Mohács 1526 – Buda 1541* (War and Society in East Central Europe, 26 = Atlantic Studies on Society in Change, 56 = East European Monographs, 255). Boulder, CO: East European Monographs, 1989.

Petercsák, Tivadar and Ernő Pető, eds. *Végvár és környezet. Tudományos tanácskozás előadásai – Noszvaj, 1993. okt. 14–15.* [Borderline castles and the environment. Proceedings of the conference, Noszvaj, 14–15 October 1993] (Studia Agriensia, 15). Eger: Heves Megyei Múzeumi Szervezet, 1995.

Petercsák, Tivadar and Jolán Szabó, eds. *A végvárak és régiók a XVI–XVII. században. Tudományos tanácskozás előadásai-Noszvaj, 1991. okt. 17–18.* [Borderline castles and regions in the 16th–17th centuries. Proceedings of the conference held at Noszvaj, 17–18 October 1991] (Studia Agriensia, 14). Eger: Heves Megyei Múzeumok Igazgatósága, 1993.

Péterfi, Bence. "Debates Concerning the Regulation of Border Rivers in the Late Middle Ages: The Case of the Mura River," *Hungarian Historical Review* 8 (2019): 313–335.

Pinke, Zsolt. "Alkalmazkodás és felemelkedés – modernizáció és leszakadás: Kis jégkorszaki kihívások és társadalmi válaszok a Tiszántúlon" [Adaptation and Rise – Modernization and Decline: Little Ice Age Challenges and Social Responses on the Trans-Tisza Region (Hungary)]. PhD diss., Pécsi Tudományegyetem, 2014.

———. "Modernization and Decline: An Eco-Historical Perspective on Regulation of the Tisza Valley, Hungary," *Journal of Historical Geography* 45 (2014): 92–105.

Pinke, Zsolt and Beatrix Szabó. "Analysis of the Map of the Ministry of Agriculture: Water Covered Areas and Wetlands in the Carpathian Basin Before the Commencement of Flood Protection and Draining." In *A 2. Nemzetközi és 8. Országos Interdiszciplináris Grastyán konferencia előadásai* [Papers presented at the 2nd International and 8th National Grastyán Conference], eds. Virág Rab and Melinda Szappanyos, 207–217. Pécs: PTE Grastyán Endre Szakkollégium, 2010.

Pinke, Zsolt, László Ferenczi, Gyula Gábris, and Balázs Nagy. "Settlement Patterns as Indicators of Water Level Rising? Case Study on the Wetlands of the Great Hungarian Plain," *Quaternary International* 415 (2016): 204–215.

Póka, Ágnes. "A Batthyány-birtokkomplexum igazgatása Batthyány Ádám alatt (1632–1659)" [The administration of the Batthyány estate complex in the period of Ádám Batthyány]. In *Tanulmányok Badacsonyból. A Fiatal Levéltárosok Egyesületének konferenciája, Badacsony, 2010. július 9–10.* [Studies from Badacsony. Conference of the Association of Young Archivists, 9–10 June 2010] (Fiatal Levéltárosok Egyesületének kiadványai, 1), eds. Béla Vilmos Mihalik and Áron Zarnóczki, 44–59. Budapest: ELTE, 2011.

Power, Daniel. "Introduction." In *Frontiers in Question: Eurasian Borderlands, 700–1700*, eds. idem and Naomi Standen, 1–12. Basingstoke: Macmillan, 1999.

Purcell, Nicholas and Peregrine Horden. *The Corrupting Sea: A Study of Mediterranean History*. Oxford: Blackwell, 2000.

Rackham, Oliver. *Trees and Woodland in the British Landscape*. London: J.M. Dent, 1976.

Rácz, Lajos. *Climate History of Hungary since 16th Century: Past Present and Future* (Discussion Papers, 28). Pécs: Centre for Regional Studies of the Hungarian Academy of Sciences, 1999.

———. *Magyarország éghajlattörténete az újkor idején* [Climate history of Hungary in the Modern times]. Szeged: Juhász Gyula Felsőoktatási Kiadó, 2001.

———. "Környezeti változások a kora újkori Magyarországon – környezettörténeti vázlat" [Environmental changes in early modern Hungary – environmental history sketch]. In *Környezettörténet. Az utóbbi 500 év környezeti eseményei történeti és természettudományi források tükrében* [Environmental history. The environmental events of the last 500 years in the light of historical and scientific data] (Környezettörténet, 1), ed. Miklós Kázmér, 157–165. Budapest: Hantken, 2009.

———. "The Price of Survival. Transformations in Environmental Conditions and Subsistence Systems in Hungary in the Age of Ottoman Occupation," *Hungarian Studies* 24, no. 1 (2010): 21–39.

———. *The Steppe to Europe. An Environmental History of Hungary in the Traditional Age*. Cambridge: White Horse Press, 2013.

Radkau, Joachim. "Holzverknappung und Krisenbewußtsein im 18. Jahrhundert," *Geschichte und Gesellschaft* 9 (1983): 513–543.

———. "Zur angeblichen Energiekrise des 18. Jahrhunderts. Revisionistische Betrachtungen zur 'Holznot'," *Vierteljahrschrift für Sozial- und Wirtschaftsgeschichte* 73 (1986): 1–37.

———. *Wood: A History*. Translated by Patrick Camiller. Cambridge: Polity, 2012.

Radó, Dezső. "Bel- és külterületi fasorok EU-módszer szerinti értékelése" [Assessment of esplanades in settlements and in outskirts based on an EU scheme], *Lélegzet* no. 7–8 [Supplement] (1999): 1–12.

Ráth, Károly. "A Győr vármegyei hódoltságról" [On the Ottoman-period in Győr], *Magyar Történelmi Tár* 5, no. 1 (1860): 3 –91.

Rauscher, Peter. "Kaiser und Reich. Die Reichstürkenhilfen von Ferdinand I. bis zum Beginn des Langen Türkenkriegs (1548–1593)." In *Finanzen und Herrschaft. Materielle Grundlagen fürstlicher Politik in den habsburgischen Ländern und im Heiligen Römischen Reich im 16. Jh.* (Veröffentlichungen des Instituts für Österreichische Geschichtsforschung, 38), eds. Friedrich Edelmayer, Maximilian Lanzinner, and Peter Rauscher, 45–83. Vienna: Oldenbourg, 2003.

Rázsó, Gyula. "Military Reforms in the Fifteenth Century." In *A Millennium of Hungarian Military History* (Atlantic Studies on Society in Change, 114 = East European Monographs, 621 = War and Society in East Central Europe, 37), eds. Béla Király and László Veszprémy, 54–82. Boulder, CO: Atlantic Research and Publications, 2002.

Restás, Attila. "A kora újkori Magyarország-toposzok történetéhez Paczoth Ferenc (1617) és Johann Fechner (1650) beszédei" [On the history of tropes in early modern Hungary: The speeches of Ferenc Paczoth (1617) and Johann Fechner (1650)], *Irodalomtörténeti Közlemények* 122, no. 4 (2018): 490–507.

Réthly, Antal. *Időjárási események és elemi csapások Magyarországon 1700-ig* [Weather events and natural disasters in Hungary until 1700]. Budapest: Akadémiai, 1962.

Richards, John F. *The Unending Frontier: An Environmental History of the Early Modern World*. Berkeley, CA: University of California Press, 2005.

Rieber, Alfred J. *The Struggle for the Eurasian Borderlands: From the Rise of Early Modern Empires to the End of the First World*. New York: Cambridge University Press, 2014.

Ringer, István, Péter Barkóczy, and Árpád Kovács. "A sárospataki ágyúöntő műhely régészeti kutatása és a régészeti leletanyag metallurgiai vizsgálata" [Archaeological research of the Sárospatak cannon foundry and the metallurgical study of the finds], *Archeometriai Műhely* no. 4 (2011): 351–370.

Rodrigo, F.S., M.J. Esteban-Parra, and Y. Castro-Diez. "On the Use of the Jesuit Order Private Correspondence Records in Climate Reconstructions: A Case Study from Castille (Spain) for 1634–1648 A.D.," *Climatic Change* 40 (1998): 625–645.

Rodríguez, Francisco Zamora. "'Quando el Agua llegare aquí Sevilla…' La avenida del río Guadalquivir en 1626 según un documento de la Biblioteca da Ajuda (Portugal)," *Historia, instituciones, documentos* 41 (2013): 407–431.

Rohr, Christian. "Measuring the Frequency and Intensity of Floods of the Traun River (Upper Austria), 1441–1574," *Hydrological Sciences Journal* 51 (2006): 834–847.

Roll, Christine. "Grenzen und Grenzüberschreitungen in der Frühen Neuzeit – eine Einführung in die Forschung." In *Grenzen und Grenzüberschreitungen. Bilanz und Perspektiven der Frühneuzeitforschung* (Frühneuzeit-Impulse, 1), eds. Christine Roll, Frank Pohle, and Matthias Myrczek, 13–22. Vienna, Cologne, and Weimar: Böhlau, 2010.

Roll, Christine, Frank Pohle, and Matthias Myrczek, eds. *Grenzen und Grenzüberschreitungen. Bilanz und Perspektiven der Frühneuzeitforschung* (Frühneuzeit-Impulse, 1). Vienna, Cologne, and Weimar: Böhlau, 2010.

Sabján, Tibor. "Ship Mills in Historical Hungary." In *Ruralia 5: Water Management in Medieval Rural Economy. Les usages de l'eau en milieu rural au Moyen Âge*, ed.

Jan Klápště, 242–250. Prague: Institute of Archaeology, Academy of Sciences of the Czech Republic, 2005.

Sági, Károly. "A zalacsányi török kori várak" [Ottoman-period castle of Zalacsány]. In *A Göcseji Múzeum jubileumi emlékkönyve 1950–1960* [Jubilee volume of the Göcsej Museum] (A Göcseji Múzeum Közleményei, 8), ed. Imre Szentmihályi, 131–135. Zalaegerszeg: Göcseji Múzeum, 1960.

Sahin-Tóth, Péter. "La France et les français face à la 'longue guerre' de Hongrie (1591–1606)." 2 vols. PhD diss., Université François-Rabelais, 1997.

Sandberg, Brian. "Going Off to the War in Hungary: French Nobles and Crusading Culture in the Sixteenth Century," *Hungarian Historical Review* 4 (2015): 346–383.

Sárosi, Edit. *Deserting Villages – Emerging Market Towns. Settlement Dynamics and Land Management in the Great Hungarian Plain 1300–1700* (Series Minor, 39). Budapest: Archaeolingua, 2016.

Sarusi Kiss, Béla. "Vasgyártás és vasgazdálkodás Murányban a XVI. században" [Iron production and iron economy at Muránsky Hrad in the 16th century], *Fons. Forráskutatás és Történeti Segédtudományok* 4, no. 1 (1997): 79–98.

Schmale, Wolfgang and Reinhard Stauber, eds. *Menschen und Grenzen in der Frühen Neuzeit.* (Innovationen. Bibliothek zur Neueren und Neuesten Geschichte, 2). Berlin: Verlag Spitz, 2000.

Schmid, Martin. "The Environmental History of Rivers in the Early Modern Period." In *An Environmental History of the Early Modern Period: Experiments and Perspectives*, eds. Martin Knoll and Reinhold Reith, 19–26. Berlin: LIT Verlag, 2014.

Schönach, Paula. "River Histories: A Thematic Review," *Water History* 9 (2017): 233–257.

Şen, Gül. *Jordan as an Ottoman Frontier Zone in the Sixteenth-Eighteenth Centuries* (Ulrich Haarmann Memorial Lecture, 15). Berlin: EB-Verlag Dr. Brandt, 2018.

Siklósi, Gyula. *A középkori Körmend védelmi rendszere* [The defense system of medieval Körmend] (Testis Temporis, 15). Körmend: Körmend Város Önkormányzata and Siklósi Gyula, 2006.

———. "Körmend, Mónus Illés u. 15" [Körmend, 15 Mónus Illés Street]. In *Régészeti kutatások Magyarországon 2007* [Archaeological excavations in Hungary, 2007], ed. Judit Kisfaludi, 239–240. Budapest: Kulturális Örökségvédelmi Hivatal, 2008.

Simon, Éva. "Magyar nagybirtokosok tervezetei a Kanizsával szembeni végvidék kiépítéséről" [The plans of the Hungarian aristocracy on the construction of the border defense line against Kanizsa]. In *Zalai történeti tanulmányok* [Studies in the history of Zala County] (Zalai gyűjtemény, 42), ed. Csaba Káli, 61–86. Zalaegerszeg: Zala Megyei Levéltár, 1997.

Simonkay, Márton. "Csorbuló önrendelkezés vagy állami siker? Érdekek és vélemények a Rába-szabályozás kapcsán (1886–1893)" [Confined autonomy or state success? Interests and opinions regarding the regulation of the River Rába]. In

Víz és társadalom Magyarországon a középkortól a XX. század végéig [Water and society in Hungary from the Middle Ages to the end of the 20th century], ed. Gergely Krisztián Horváth, 343–376. Budapest: Balassi, 2014.

Skelton, Leona. *Tyne after Tyne: An Environmental History of a River's Battle for Survival, 1530–2015*. Cambridge: White Horse Press, 2017.

Skorka, Renáta. "Pozsony a bécsi közvetítőkereskedelem árnyékában" [Bratislava in the shadow of intermediary trade]. In *Tiszteletkör. Történeti tanulmányok Draskóczy István egyetemi tanár 60. születésnapjára* [Lap of honor. Studies in honor of the 60th birthday of István Draskóczy], eds. Gábor Mikó, Bence Péterfi, and András Vadas, 301–311. Budapest: ELTE Eötvös Kiadó, 2012.

———. "On Two Sides of the Border. The Hungarian-Austrian Border Treaty of 1372," *Hungarian Historical Review* 8 (2019): 290–312.

Slavin, Philip. *Bread and Ale for the Brethren: The Provisioning of Norwich Cathedral Priory, 1260–1536* (Studies in Regional and Local History, 11). Hatfield: University of Hertfordshire Press, 2012.

———. "Warfare and Ecological Destruction in Early Fourteenth-Century British Isles," *Environmental History* 19 (2014): 528–550.

———. "Ecology, Warfare and Famine in Early Fourteenth-Century British Isles: A Small Prolegomenon to a Big Topic." In *Guerra y carestía en la Edad Media*, ed. Pere Benito i Monclús, 85–99. Lleida: Milenio, 2015.

———. *Experiencing Famine: A Fourteenth-Century Environmental Shock in the British Isles* (Environmental Histories of the North Atlantic World [EHNAW], 4). Turnhout: Brepols, 2019.

Solymosi, László. "Az Ernuszt-féle számadáskönyv és a középkor végi népességszám" [The account book of Ernuszt and the late medieval population], *Történelmi Szemle* 28 (1985): 414–436.

Sonnlechner, Christoph, Severin Hohensinner, and Gertrud Haidvogl. "Floods, Fights and a Fluid River: The Viennese Danube in the Sixteenth Century," *Water History* 5 (2013): 173–194.

Soós, Elemér. "A tokaji vár története" [The history of the castle of Tokaj], *Hadtörténelmi Közlemények* 14, no. 1 and 2 (1913): 65–88, and 167–194.

Sperl, Karin, Martin Scheutz, and Arno Strohmeyer, eds. *Die Schlacht von Mogersdorf/St. Gotthard und der Friede von Eisenburg/Vasvár*. Eisenstadt: Amt der Burgenländischen Landesregierung, 2016.

Srodecki, Paul. *Antemurale Christianitatis: Zur Genese der Bollwerksrhetorik im östlichen Mitteleuropa an der Schwelle vom Mittelalter zur Frühen Neuzeit*. Husum: Matthiesen Verlag, 2015.

Standen, Naomi. "Introduction. Nine Case Studies of Pre-Modern Frontiers." In *Frontiers in Question: Eurasian Borderlands, 700–1700*, eds. Daniel Power and Naomi Standen, 13–31. Basingstoke: Macmillan, 1999.

Starlander, Jakob. "Conflict and Negotiation: Management of Forest Commons in Seventeenth-Century Northern Finland," *Scandinavian Economic History Review* 69 (2021): 177–194.

Stefánik, Martin. "Die Anfänge der slowakischen Bergstädte. Das Beispiel Neusohl." In *Stadt und Bergbau* (Städteforschung A/64), eds. Karl Heinrich Kaufhold and Wilfried Reininghaus, 295–312. Cologne: Böhlau, 2004.

Stegena, Lajos, ed. *Lazarus Secretarius: The First Hungarian Mapmaker and His Work*. Budapest: Akadémiai, 1982.

Stein, Mark L. *Guarding the Frontier: Ottoman Border Forts and Garrisons in Europe*. London and New York: I.B. Tauris, 2007.

Strumia, Giorgio. "Tree-Ring Based Reconstruction of Precipitation in Eastern Austria." PhD diss., BOKU, 1999.

Sturm, Katrin et al. "Hochwasser in Mitteleuropa seit 1500 und ihre Beziehung zur atmosphärischen Zirkulation," *Petermanns Geographische Mitteilungen* 145, no. 6 (2001): 14–23.

Sugár, István. "Az egri vár építőanyagainak beszerzési helyei 1548–1564" [Origin of the building materials of the castle of Eger from 1548 to 1564]. In *Végvár és környezet. Tudományos tanácskozás előadásai – Noszvaj, 1993. okt. 14–15* [Borderline castles and the environment. Proceedings of the conference, Noszvaj, 14–15 October 1993] (Studia Agriensia, 15), eds. Tivadar Petercsák and Ernő Pető, 175–182. Eger: Heves Megyei Múzeumi Szervezet, 1995.

Sümegi, Pál et al. "The Environmental History of Southern Transdanubia during the Medieval and the Ottoman Period in the Light of Palaeoecological and Geoarchaeological Research." In *"per sylvam et per lacus nimios." The Medieval and Ottoman Period in Southern Transdanubia, Southwest Hungary: The Contribution of the Natural Sciences*, eds. Gyöngyi Kovács and Csilla Zatykó, 15–74. Budapest: Institute of Archaeology, Research Centre for the Humanities, Hungarian Academy of Sciences, 2016.

Sümegi, Pál, Gusztáv Jakab, Péter Majkut, Tünde Törocsik, and Csilla Zatykó. "Middle Age Paleoecological and Paleoclimatological Reconstruction in the Carpathian Basin," *Időjárás* 113 (2009): 265–298.

Szabó, Géza and Viktor Csányi. "Werbőczy két Tolna megyei vára: Dombó és Döbrököz az újabb régészeti megfigyelések tükrében" [Two castles of Werbőczy in Tolna County. Dombó and Döbrököz in the light of archaeological observations], *A Wosinsky Mór Múzeum Évkönyve* 34 (2012): 179–212.

Szabó, János, B. and Ferenc Tóth. *Mohács 1526. Soliman le Magnifique prend pied en Europe central*. Paris: Économica, 2009.

Szabó, Péter. "'There is hope for a tree': Pollarding in Hungary," *Medium Aevum Quotidianum* 44 (2001): 41–60.

———. "Sources for the Historian of Medieval Woodland." In *People and Nature in Historical Perspective* (CEU Medievalia, 5), eds. József Laszlovszky and Péter Szabó, 265–288. Budapest: CEU Press and Archaeolingua, 2003.

———. *Woodland and Forests in Medieval Hungary* (Archaeolingua – Central European Series, 2 = BAR International Series, 1348). Oxford: Archaeopress, 2005.

———. "Changes in Woodland Cover in the Carpathian Basin." In *Human Nature: Studies in Historical Ecology and Environmental History*, eds. idem and Radim Hédl, 106–115. Brno: Institute of Botany of the ASC, 2008.

———. "Erdők a kora újkorban: történelem, régészet, ökológia" [Forests in the early modern period: History, archaeology, ecology]. In *Környezettörténet. Az utóbbi 500 év környezeti eseményei történeti és természettudományi források tükrében* [Environmental history. The environmental events of the last 500 years in the light of historical and scientific data] (Környezettörténet, 1), ed. Miklós Kázmér, 137–156. Budapest: Hantken, 2009.

Szabó, Péter, Jana Müllerová, Silvie Suchánková, and Martin Kotacka. "Intensive Woodland Management in the Middle Ages: Spatial Modelling Based on Archival Data," *Journal of Historical Geography* 48 (2015): 1–10.

Szakály, Ferenc. "Tolna megye negyven esztendeje a mohácsi csata után, 1526–1566" [Forty years of Tolna County after the battle of Mohács], *Tanulmányok Tolna Megye Történetéből* 2 (1969): 5–85.

———. "Phases of Turco-Hungarian Warfare before the Battle of Mohács (1365–1526)," *Acta Orientalia Academiae Scientiarum Hungaricae* 33 (1979): 65–111.

———. *Magyar adóztatás a török hódoltságban* [Hungarian taxation in Ottoman Hungary]. Budapest: Akadémiai, 1981.

———. "A török–magyar küzdelem szakaszai a mohácsi csata előtt, 1365–1526" [Stages of the Ottoman–Hungarian struggles before the battle of Mohács]. In *Mohács. Tanulmányok a mohácsi csata 450. évfordulója alkalmából* [Mohács. Studies in the memory of the 450th anniversary of the battle of Mohács], eds. Lajos Ruzsás and Ferenc Szakály, 11–57. Budapest: Akadémiai, 1986.

Szalay, Gergely and Endre Szilágyi, *Magyarország vizeinek műszaki-hidrológiai jellemzése: Mosoni-Dunaág, Rába* [The technological – hydrological analysis of the waters of Hungary. Mosoni Danube, Rába]. Budapest: Vízgazdálkodási Intézet, 1989.

Szamota, István. *Magyar oklevél-szótár. Régi oklevelekben és egyéb iratokban előforduló magyar szók gyűjteménye* [Hungarian charter dictionary. Collection of Hungarian language words in old charters and other documents]. Budapest: Hornyánszky Viktor Könyvkereskedése, 1902–1906.

Szántó, Imre. "A végvári rendszer kiépítése Magyarországon" [Formation of the borderline defense in Hungary], *Acta Universitatis Szegediensis: Acta Historica* 38 (1971): 3–44.

Szaszkóné Sin, Aranka, ed. *Magyarország történeti helységnévtára 1773–1808 – Pest-Pilis-Solt megye és a Kiskunság* [Historical settlement names of Hungary, 1773–1808 – Pest-Pilis-Solt County and the Kiskunság]. Budapest: Központi Statisztikai Hivatal Könyvtár és Dokumentációs Szolgálat, 1988.

Szatlóczki, Gábor. "Szentgyörgyvár a török időkben" [Szentgyörgyvár in the Ottoman period]. In *Szentgyörgyvár története* [History of Szentgyörgyvár] (Zalai Kismonográfiák, 7), ed. Róbert Müller, 33–48. Zalaegerszeg: [n. p.], 2002.

Szende, Katalin. "Towns Along the Way. Changing Patterns of Long-Distance Trade and the Urban Network of Medieval Hungary." In *Towns and Communication. 2. Communication between Towns. Proceedings of the Meeting of the International Commission for the History of Towns (ICHT) London 2007 – Leeds 2008*, eds. Hubert Houben and Kristjan Toomaspoeg, 161–226. Galatina: Mario Congedo Editore, 2011.

———. "Iure Theutonico?' German Settlers and Legal Frameworks for Immigration to Hungary in an East-Central European Perspective," *Journal of Medieval History* 45 (2019): 360–379.

———. "Mills and Towns: Textual Evidence and Cartographic Conjectures from Hungarian Towns in the Preindustrial Period." In *Extra muros Vorstädtische Räume in Spätmittelalter und Früher Neuzeit. Espaces suburbains au bas Moyen Âge et à l'époque moderne* (Städteforschung, 91), eds. Guy Thewes and Martin Uhrmacher, 485–516. Cologne: Böhlau, 2019.

Szende, Katalin and Magdolna Szilágyi. "Town Typology in the Context of Historic Towns Atlases: A Target or a Tool?" In *Political Functions of Urban Spaces and Town Types through the Ages. Making Use of the Historic Towns Atlases in Europe*, eds. Roman Czaja et al., 267–302. Cracow, Toruń, and Vienna: Böhlau, 2019.

Szilágyi, Emőke Rita. "Zur Überlieferungsgeschichte von Nicolaus Olahus' Hungaria." In *Wiener Archivforschungen: Festschrift für den ungarischen Archivdelegierten in Wien, István Fazekas* (Publikationen der Ungarischen Geschichtsforschung in Wien, 10), eds. Zsuzsanna Cziráki et al., 69–75. Vienna: Institut für Ungarische Geschichtsforschung in Wien and Ungarische Archivdelegation beim Haus-, Hof- und Staatsarchiv, Wien, 2014.

———. "Az önkéntes száműzetés alakzatai és trópusai Oláh Miklós Brüsszelben" [Figures and tropes of voluntary exile. Miklós Oláh at Brussels]. In *Börtön, exilium és szenvedés: Bethlen Miklós élettörténetének kora újkori kontextusai kontextusai* [Prison, exile, suffer. The early modern contexts of the life of Miklós Bethlen], eds. Anna Fajt, Emőke Rita Szilágyi, and Zsombor Tóth, 51–60. Budapest: Reciti, 2017.

Szilágyi, Magdolna. "Árpád Period Communication Networks: Road Systems in Western Transdanubia." PhD diss., Central European University, 2012.

———. "Városok, utak, kereskedelem. Az úthálózat szerepe Vas megye városi fejlődésében a 13–14. században" [Towns, roads, trade. The role of the road network in the urban development of Vas County in the 13th–14th centuries], *Savaria* 36 (2013): 223–241.

———. "Római utak a középkori Dunántúlon. Az utak nevei és szerepük a középkori térszervezésben" [Roman roads in medieval Transdanubia. The name of the roads and their role in medieval spatial organization], *Történelmi Szemle* 56 (2014): 1–25.

Szilágyi, Miklós. *Halászóvizek – halásztársadalom – halászati technika. A tiszai halászat történeti-néprajzi elemzése* [Fishing waters – fishing society – fishing techniques. Historical-ethnological analysis of fishing at Tisza]. Debrecen: KLTE Néprajzi Tanszék, 1992.

Szvitek, Róbert József. "Kiskomárom végvár építéstörténete" [The building history of the borderline fortification of Kiskomárom], *Castrum Bene* 14 (2011): 41–55.

Tagányi, Károly. "Gyepü és gyepüelve, I–IV" [*Gyepű* and *gyepűelve*], *Magyar Nyelv* 9 (1913): 97–104, 145–152, 201–206, and 254–266.

Takács, Károly. "Néhány észrevétel Györffy György Árpád-kori történeti földrajzának legújabb kötetéhez" [Some notes to the new volume of György Györffy's Árpádian-Period Historical Geography], *Aetas* 14, no. 3 (1999): 101–107.

———. "Medieval Hydraulic Systems in Hungary: Written Sources, Archaeology and Interpretation." In *People and Nature in Historical Perspective* (CEU Medievalia, 5), eds. József Laszlovszky and Péter Szabó, 289–312. Budapest: CEU Press and Archaeolingua, 2003.

Takács, Katalin and Zoltán Kern. "Multidecadal Changes in the River Ice Regime of the Lower Course of the River Drava since AD 1875," *Journal of Hydrology* 529 (2015): 1890–1900.

Takács, Katalin, Zoltán Kern, and Balázs Nagy. "Impacts of Anthropogenic Effects on River Ice Regime: Examples from Eastern Central Europe," *Quaternary International* 293 (2013): 275–282.

Takács, Lajos. *Egy irtásfalu földművelése* [The agriculture of a clearance village]. Budapest: Akadémiai, 1987.

———. *Határjelek, határjárás a feudális kor végén Magyarországon* [Boundary marks and perambulation at the end of the feudal age in Hungary]. Budapest: Akadémiai, 1987.

Takáts, Sándor. "A komáromi vizahalászat a XVI. században, I–II" [Sturgeon fishing at Komárom in the 16th century], *Magyar Gazdaságtörténelmi Szemle* 4 (1897): 425–445, and 485–509.

———. "Dunai hajózás a XVI. és XVII. században. III. közlemény" [Navigation on the Danube in the 16th–17th centuries, III], *Magyar Gazdaságtörténelmi Szemle* 7 (1900): 193–222.

———. "A magyar vár" [Hungarian castles], *Századok* 41 (1907): 726–741 and 815–837.

———. "A magyar faragómolnárok I–II" [Hungarian building millwrights]. In Sándor Takáts, *Rajzok a török világból*, 4 vols. [Sketches from the Ottoman world], II, 422–463. Budapest: Magyar Tudományos Akadémia, 1915–1917.

———. *Emlékezzünk eleinkről*, 2 vols. [Let's remember the ancestors of the Hungarians]. [Budapest]: Genius, [1929].

———. *Művelődéstörténeti tanulmányok a XVI–XVII. századból* [Studies in 16th–17th-century intellectual history]. Budapest: Gondolat, 1961.

Tarkó, Ilona. "Rabkereskedelem és anyagi kultúra a XVI–XVII. században a Batthyány család levéltára alapján" [The Ottoman Slave Trade and Material Culture in the 16th–17th century according to the Family Archive of the Batthyánys]. PhD diss., Pázmány Péter Katolikus Egyetem, 2012.

Terei, György, Gyula Nováki, Zsolt Mráv, István Feld, and Sebestyén Sárközy, *Fejér megye várai az őskortól a kuruc korig* [The castles of Fejér County from Prehistory to the *kuruc* period] (Magyarország várainak topográfiája, 3). Budapest: Castrum Bene Egyesület and Cívertan Bt., 2011.

Thompson, James Westfall. "Profitable Fields of Investigation in Medieval History," *American Historical Review* 18 (1913): 490–504.

Tilcsik, György. "Eine unbekannte topographische Beschreibung der Fürst Philipp Batthyányschen Herrschaft in Körmend aus der ersten Hälfte der 1820er Jahre," *Zeitschrift des Historischen Vereins für Steiermark* 101 (2010): 179–194.

Tillman, David A. *Wood as an Energy Resource*. New York: Academic Press, 1978.

Tóber, Márta. "Mennyiben tükrözi Bertrandon de la Brocquière útleírása a középkori Homokhátság természeti viszonyait?" [How much does the travel account of Bertrandon de la Brocquière reflect the natural conditions of the medieval Homokhátság?]. In *A táj változásai a Kárpát-medencében: történelmi emlékek a tájban* [Landscape changes in the Carpathian Basin: historical monuments in the landscape], ed. György Füleky, 309–314. Gödöllő: Környezetkímélő Agrokémiáért Alapítvány, 2013.

Tóber, Márta and Andrea Kiss. "Landscape History of the Medieval Sand Ridge Area in Central Hungary: Examples of Sand and Arboreal Vegetation in Medieval Documentation Compared to the Results of Natural Scientific and Archaeological Investigations," *Siedlungsforschung. Archäologie – Geschichte – Geographie* 31 (2014): 247–269.

Toda, Oana. "Economic and Material Aspects of the Late Medieval Bridges from Transylvania: The Written Sources," *Banatica* 27 (2017): 361–397.

Toifl, Leopold. "Bajcsavár története a stájer levéltári források alapján" [The history of Bajcsavár in the light of archival data from Styria]. In *Weitschawar/Bajcsa-Vár. Egy stájer erődítmény Magyarországon a 16. század második felében* [Weitschawar/Bajcsa-Vár. A Styrian fortification in Hungary in the second half of the 16th century], ed. Gyöngyi Kovács, 27–40. Zalaegerszeg: Zala Megyei Múzeumok Igazgatósága, 2002.

Tolnai, Gergely. *Palánkvárak Magyarországon* [Earth and wood fortifications in Hungary]. Budapest: Martin Opitz, 2011.

Török, Zsolt. "Renaissance Cartography in East-Central Europe c. 1450–1650," In *Cartography in the European Renaissance: The History of Cartography*, vol. 3, part 2, ed. David Woodward, 1806–1851. Chicago: University of Chicago Press, 2007.

Tóth, Albert, ed. *„Áldás és átok a víz" – Tudományos emlékülés a Mirhó-gát megépítésének 200. évfordulóján* ['Blessing or curse is the water' – Scientific conference on the 200th anniversary of the construction of the Mirhó Dam]. Kisújszállás: Városi Tanács, 1987.

Tóth, Endre. "Körmend vidéke római kori történetéhez" [On the history of the Roman period of Körmend and its neighborhood], *Vasi Szemle* 33 (1979): 342–346.

———. *Itineraria Pannonica. Római utak a Dunántúlon* [Roman roads in the Transdanubia]. Budapest: MNM, 2006.

Tóth, Gergely. "Bél Mátyás 'Notitia Hungariae novae…' című művének keletkezéstörténete és kéziratainak ismertetése." 2 vols. [The History of the Formation of the 'Notitia Hungariae novae…' of Matthias Bel and its manuscripts]. PhD diss., Eötvös Loránd Tudományegyetem, 2007.

Tóth, István György. "Körmend a kora újkorban (1526–1809)" [Körmend in the early modern period (1526–1809)]. In *Körmend története* [The history of Körmend], ed. László Szabó, 98–167. Körmend: Önkormányzat, [1994].

Tóth, István György. *Jobbágyok, hajdúk, deákok: a körmendi uradalom társadalma a 17. században* [Tenant peasants, heyduk, literates: The society of Körmend in the 17th century] (Értekezések a Történeti Tudományok Köréből [New Series], 115). Budapest: Akadémiai, 1992.

Totman, Conrad D. *The Green Archipelago: Forestry in Pre-Industrial Japan*. Berkeley, CA: University of California Press, 1989.

Tringli, István. "A magyar szokásjog a malomépítésről" [Hungarian customary law of mill construction] In *Tanulmányok a középkorról* [Studies on medieval history] (Analecta Medievalia, 1), ed. Tibor Neumann, 251–267. Budapest and Piliscsaba: PPKE, 2001.

Tucker, Richard P. "The Impact of Warfare on the Natural World: A Historical Survey." In *Natural Enemy, Natural Ally: Toward an Environmental History of War*, eds. Richard P. Tucker and Edmund P. Russell, 15–42. Corvallis, OR: Oregon State University Press, 2004.

———. "War and the Environment." In *A Companion to Global Environmental History*, eds. John R. McNeill and Erin Stewart Mauldin, 319–339. Chichester: Wiley Blackwell, 2012.

Tucker, Richard, Tait Keller, J.R. McNeill, and Martin Schmid, eds. *Environmental Histories of the First World War*. Cambridge: Cambridge University Press, 2018.

Turner, Frederick Jackson. "The Significance of the Frontier in American History." *Proceedings of the State Historical Society of Wisconsin at its Forty-First Annual Meeting, Held December 14, 1893*, 79–112. Madison, WI: Democrat Print. Co., 1894.

———. *The Frontier in American History.* New York: Henry Holt and Company, 1920.

Unger, Richard W. *Beer in the Middle Ages and the Renaissance.* Philadelphia: University of Pennsylvania Press, 2004.

———. "Thresholds for Market Integration in the Low Countries and England in the Fifteenth Century." In *Money, Markets and Trade in Late Medieval Europe Essays in Honour of John H.A. Munro*, eds. Lawrin Armstrong, Ivana Elbl, and Martin M. Elbl, 349–380. Leiden and Boston: Brill, 2007.

Vadas, András. "Documentary Evidence on the Weather Conditions and a Possible Crisis in 1315–1317: Case Study from the Carpathian Basin," *Journal of Environmental Geography* 2, no. 3–4 (2009): 23–29.

———. "Late Medieval Environmental Changes of the Southern Great Hungarian Plain – A Case Study," *Annual of the Medieval Studies at CEU* 17 (2011): 41–60.

———. *Körmend és a vizek. Egy település és környezete a kora újkorban* [Körmend and the waters. A settlement and its environment in the early modern period] (ELTE BTK Történelemtudományok Doktori Iskola. Tanulmányok – konferenciák, 5). Budapest: Történelemtudományok Doktori Iskola, 2013.

———. "'A Dunára én bizon nem megyek, mert még nem akarok meghalnom' – a Duna jégjelenségei a kora újkorban (1530–1650)" ["I am not going on the Danube's ice since I don't want to die" The Ice-Regime of Danube in the early modern period (1530–1650)]. In *Micae mediaevales III. Fiatal történészek dolgozatai a középkori Magyarországról és Európáról* [Studies of young medievalists on medieval Hungary and Europe] (ELTE Történelemtudományok Doktori Iskola. Tanulmányok – Konferenciák, 6), eds. Judit Gál et al., 219–235. Budapest: ELTE BTK Történettudományok Doktori Iskola, 2013.

———. "The 'waters leave their beds frequently' – A Western-Hungarian Town and the Flooding of the Rába/Raab River in the Seventeenth Century (1600–1659)," *Water History* 5 (2013): 267–286.

———. "Vízgazdálkodás és háborús védekezés. Csákány és a Vas megyei Rába-mente a kora újkorban (1600–1658)" [Water management and border protection at Csákány and long the Rába is Vas County between 1600 and 1658]. In *Víz és társadalom Magyarországon a középkortól a XX. század végéig* [Water and society in Hungary from the Middle Ages to the end of the 20th century], ed. Gergely Krisztián Horváth, 207–245. Budapest: Balassi, 2014.

———. "A Rába-mente környezeti viszonyai a 16. század közepén egy 1543–44-es folyófelmérés tükrében" [The environment of the Rába-valley in the mid-16th century as reflected in the river survey of 1543–44], *Soproni Szemle* 69 (2015): 16–40.

———. "Városárkok és vízgazdálkodás a késő-középkori Közép-Európa városaiban" [Urban moats and water management in the late medieval towns of Central Europe], *Urbs. Magyar Várostörténeti Évkönyv* 10–11 (2015–2016): 323–353.
———. "Some Remarks on the Legal Regulations and Practice of Mill Construction in Medieval Hungary." In *Wasser in der mittelalterlichen Kultur / Water in Medieval Culture Gebrauch – Wahrnehmung – Symbolik / Uses, Perceptions, and Symbolism* (Das Mittelalter. Perspektiven mediävistischer Forschung, 4), eds. Gerlinde Huber-Rebenich, Christian Rohr, and Michael Stolz, 291–304. Berlin: de Gruyter Verlag, 2017.
———. "Vízépítés és munkaszervezési formák a késő középkori és kora újkori Magyar Királyságban" [Water construction and labor organization in late medieval and early modern Hungary]. In *Techné: a mesterségbeli tudás átadásának lehetséges színterei az ó- és középkorban* [Techné. The scenes of knowledge transfer in the Antiquity and the Middle Ages], eds. Márta Munding, Kornél Szovák, and László Takács 477–510. Piliscsaba: Avicenna Avicenna Közel-Kelet Kutatások Intézete, 2017.
———. "Border by the River – But Where is the River? Hydrological Changes and Borders in Medieval Hungary," *Hungarian Historical Review* 8 (2019): 336–360.
———. "Technologies on the Road between West and East: The Spread of Water Mills and the Christianization of East Central Europe." In *The Medieval Networks in East Central Europe: Commerce, Contacts, Communication*, eds. Balázs Nagy, Felicitas Schmieder, and András Vadas, 123–138. New York: Routledge, 2019.
———. "Who Stole the Water? The Control and Appropriation of Water Resources in Medieval Hungary." PhD diss., Central European University, 2020.
———. *Egy határfolyó környezettörténete. Háború és vízgazdálkodás a kora újkori Rába-völgyben* [The environmental history of a frontier river. War and water management in the early modern Rába River valley]. Budapest: Bölcsészettudományi Kutatóközpont Történettudományi Intézet, 2021.
———. "A Batthyány uradalomrendszer és a 17. század első felének pestisjárványai" [The Batthyány Estate Complex and the Plagues in the First Half of the 17th Century], *Századok* 156 (2022): 25–46.
———. "For the Benefit of Generations to Come or for the Sake of Survival? Measures for Protecting Forests in Early Modern Hungary," *Historical Studies on Central Europe* 2, no. 1 (2022): 4–26.
Vadas, András and Lajos Rácz. "Climatic Changes in the Carpathian Basin during the Middle Ages. The State of Research," *Global Environment* 12 (2013): 198–227.
Vági, István. "Van-e hazánkban ezeréves puszta, vagy azt a török hódoltság okozta. Megváltozott-e a Nagy-Alföld éghajlata a török hódoltság miatt aszályosabb irányban, továbbá a talajok is alig javíthatóan megromlottak-e a valóságban" [Was there a millennial puszta in Hungary or was it caused by the Ottomans. Did the climate of the Great Hungarian Plain became drier in consequence of

the Ottoman Empire and were soils indeed devastated to a barely amendable state], *Erdészeti Lapok* 73 (1934): 670–682.

Vándor, László. "Kanizsa története a honfoglalástól a város török alóli jelszabadulásáig" [The history of Kanizsa from the Hungarian conquest to its liberation from the Ottoman occupation]. In *Nagykanizsa. Városi monográfia,* vol. 1. [Nagykanizsa. Town monograph], ed. József Béli, 217–424. Nagykanizsa: Nagykanizsa Megyei Jogú Város Önkormányzata, 1994.

———. "A zalai végvárrendszer a 16–17. században" [The frontier fortification system of Zala County in the 16th–17th centuries]. In *Zala megye ezer éve. Tanulmánykötet a magyar államalapítás millenniumának tiszteletére* [Millennium of Zala County. Studies in honor of the millennium of the Hungarian state foundations], ed. idem, 89–96. Zalaegerszeg: Zala Megyei Múzeumok Igazgatósága, 1996.

———, ed. *Zala megye ezer éve. Tanulmánykötet a magyar államalapítás millenniumának tiszteletére* [Millennium of Zala County. Studies in honor of the millennium of the Hungarian state foundations]. Zalaegerszeg: Zala Megyei Múzeumok Igazgatósága, 1996.

Varga J., János. *Jobbágyrendszer a magyarországi feudalizmus kései századaiban, 1556–1767* [The tenant peasant-system in Hungary in the late feudal period]. Budapest: Akadémiai, 1969.

Várkonyi, Ágnes R. *Pelikán a fiaival* [Pelican with his sons]. Budapest: Liget Műhely Alapítvány, 1992.

———. "Környezet és végvár. Végvárrendszer és a történeti ökológia kérdései a 16–17. századi Magyarországon" [Environment and borderline castles. Border defense system and questions of historical ecology in 16th–17th-century Hungary]. In *A végvárak és régiók a XVI–XVII. században. Tudományos tanácskozás előadásai-Noszvaj, 1991. okt. 17–18.* [Borderline castles and regions in the 16th–17th centuries. Proceedings of the conference held at Noszvaj, 17–18 October 1991] (Studia Agriensia, 14), eds. Tivadar Petercsák and Jolán Szabó, 7–27. Eger: Heves Megyei Múzeumok Igazgatósága, 1993.

———. "A párbeszéd esélyei, (A végvárrendszer-kutatások humánökológiai megközelítéséről)" [The chances of dialogue (on the human ecological approach to the borderline fortification research]. In *Végvár és környezet. Tudományos tanácskozás előadásai – Noszvaj, 1993. okt. 14–15.* [Borderline castles and the environment. Proceedings of the conference, Noszvaj, 14–15 October 1993] (Studia Agriensia, 15), eds. Tivadar Petercsák and Ernő Pető, 7–32. Eger: Heves Megyei Múzeumi Szervezet, 1995.

———. "Történeti ökológia" [Historial ecology]. In *A történelem segédtudományai* [Auxiliary sciences of history] (A történettudomány kézikönyve, 1), ed. Iván Bertényi, 44–65. Budapest: Osiris Kiadó, 2001.

———. "Természet és társadalom. A történeti ökológia regionális lehetőségei" [Nature and society. The regional perspectives of environmental history], *A Nógrád Megyei Múzeumok Évkönyve* 26 (2002): 345–373.

———. "'A természet majd az értelemmel...' Történeti ökológia és a XVIII. századi Magyarország környezeti válsága" ['Nature with conscience...' Historical ecology and the environmental crisis of Hungary in the 18th century]. In *Környezettörténet. Az utóbbi 500 év környezeti eseményei történeti és természettudományi források tükrében* [Environmental history. The environmental events of the last 500 years in the light of historical and scientific data] (Környezettörténet, 1), ed. Miklós Kázmér, 21–54. Budapest: Hantken, 2009.

———. "Az öltözködés filozófiájáról" [On the philosophy of clothing], *Történelmi Szemle* 53 (2011): 503–536.

Végh, Ferenc. *Egerszeg végvár és város a 17. században* [The borderline fortification of Egerszeg and the town in the 17th century] (Zalaegerszegi füzetek, 10). Zalaegerszeg: Millecentenáriumi Közalapítány: 2010.

———. "A balatoni 'hadiflotta' a török korban" [The 'fleet' on Lake Balaton in the Ottoman Period], *Hadtörténelmi Közlemények* 129, no. 1 (2016): 27–56.

Warde, Paul. *Ecology, Economy, and State Formation in Early Modern Germany.* Cambridge: Cambridge University Press, 2006.

———. *Energy Consumption in England & Wales 1560–2000* (Energy Consumption, 2). [n. p.]: Consiglio Nazionale delle Ricerche Istituto di Studi sulle Società del Mediterraneo, 2007.

———. *The Invention of Sustainability: Nature and Destiny, c. 1500–1870.* Cambridge: Cambridge University Press, 2018.

Webb, Robert H. *Requiem for the Santa Cruz: An Environmental History of an Arizona River.* Tucson: University of Arizona Press, 2014.

Weidlein, János. "A dűlőnévkutatás történeti vonatkozásai" [Historical applications of research into field names], *Századok* 69 (1935): 665–692.

Weiss, Gerhard. "Mountain Forest Policy in Austria: A Historical Policy Analysis on Regulating a Natural Resource," *Environment and History* 7 (2001): 335–355.

Weisz, Boglárka. "Az alsó-magyarországi bányavárosok kiváltságai a Zsigmond-korban" [The privileges of the Lower Hungarian mining towns in the age of Sigismund], *Urbs. Magyar Várostörténeti Közlemények* 12 (2018): 21–48.

Wellmann, Imre. "Népesség és mezőgazdaság a XVII. és a XVIII. század fordulóján" [Population and agriculture at the turning of the 17th century], *Történelmi Szemle* 18, no. 4 (1978): 701–730.

Wenzel, Gusztáv. *Magyarország mezőgazdaságának története* [History of Hungarian agriculture]. Budapest: Magyar Tudományos Akadémia, 1887.

Werther, Lukas et al. "On the Way to the Fluvial Anthroposphere—Current Limitations and Perspectives of Multidisciplinary Research," *Water* 13, no. 2188 (2021). doi: 10.3390/w13162188.

Westing, Arthur H. "The Environmental Aftermath of Warfare in Viet Nam," *Natural Resources Journal* 23 (1983): 365–389.

Wetter, Oliver et al. "The Largest Floods in the High Rhine Basin since 1268 Assessed from Documentary and Instrumental Evidence," *Hydrological Sciences Journal* 56 (2011): 733–758.

White, Sam. *The Climate of Rebellion in the Early Modern Ottoman Empire*. Cambridge: Cambridge University Press, 2011.

Wilson, Robert J.S., Brian H. Luckman, and Jan Esper. "A 500 Year Dendroclimatic Reconstruction of Spring–Summer Precipitation from the Lower Bavarian Forest Region, Germany," *International Journal of Climatology* 25 (2005): 611–630.

Wing, John T. *Roots of Empire. Forests and State Power in Early Modern Spain, c. 1500–1750* (Brill's Series in the History of the Environment, 4). Leiden and Boston: Brill, 2015.

Winiwarter, Verena, Martin Schmid, and Gert Dressel. "Looking at Half a Millennium of Co-Existence: The Danube in Vienna as a Socio-Natural Site," *Water History* (2013): 101–119.

Zombori, István, ed. *Fight Against the Turk in Central-Europe in the First Half of the 16th Century*. Budapest: METEM, 2004.

Zsoldos, Attila. "A vasi várispánság felbomlása" [Dissolution of the royal county district of Vas], *Vasi Szemle* 54, no. 1 (2000): 27–46.

Websites, databases

Az Első Katonai Felmérés 1763–1785 [First Military survey, 1763–1785]. Available online: http://mapire.eu (last accessed: 28 April 2022).

Climate and Environmental History Collaborative Research Environment Database. http://tambora.org (last accessed: 2 May 2021).

Environment and War Bibliography, ed. Richard Tucker. http://environmentandwar.com/bibliographies/ (last accessed: 17 April 2021).

Index of Geographic Names[1]

1 Hungary, Ottoman Empire, Körmend, and Rába were omitted since these appear in too many instances.

Index of Personal Names[1]

1 For the dignitaries of the Batthyány estate complex, see also Appendix 1.

For Product Safety Concerns and Information please contact our EU representative GPSR@taylorandfrancis.com
Taylor & Francis Verlag GmbH, Kaufingerstraße 24, 80331 München, Germany

www.ingramcontent.com/pod-product-compliance
Ingram Content Group UK Ltd.
Pitfield, Milton Keynes, MK11 3LW, UK
UKHW022027061126
477531UK00002B/70
* 9 7 8 9 4 6 3 7 2 7 9 3 8 *